Time Series Decomposition and Seasonal Adjustment

This book provides an in-depth examination of time series decomposition and seasonal adjustment, focusing on the X-13ARIMA-SEATS and TRAMO-SEATS methods. Seasonal adjustment removes distortions such as seasonal fluctuations and holiday effects from economic indicators (eg, GDP, CPI), enabling clearer insights into underlying trends, cycles, and shocks. These tools are vital for sound policymaking, accurate forecasting, and reliable international comparisons.

X-13ARIMA-SEATS, developed by the U.S. Census Bureau, combines empirical moving average filters with ARIMA/regARIMA modelling to handle outliers, calendar effects, and endpoint issues. TRAMO-SEATS, created by the Bank of Spain, uses a model-based strategy: TRAMO pre-adjusts data with ARIMA models, while SEATS applies signal extraction to decompose components. X-13ARIMA-SEATS excels with stable seasonal patterns, while TRAMO-SEATS provides rigorous solutions for complex holiday structures.

The book also examines modern challenges, including structural breaks from COVID-19, high-frequency data with multiple seasonalities, and the demand for real-time adjustments. It reviews innovations such as hybrid models that combine machine learning with traditional filters, Bayesian State Space approaches, and adaptive methods such as Kalman filters.

Intended for students, researchers, staff at national statistical agencies, central banks, and financial institutions, the book equips readers with methodological and practical tools to navigate evolving economic data landscapes.

Dr Ping Zong holds a bachelor's and master's degree in economics from Fudan University, China and a Ph.D in economics from Queen's University Belfast, UK. He has served as a research fellow at Queen's University Belfast, a British Academy K.C. Wong research fellow at the University of Newcastle Upon Tyne, and a senior research fellow at the University of Essex. From 2007 until his retirement in 2020, he worked as a senior research officer and methodologist at the UK Office for National Statistics, specialising in statistical methodology.

His primary research interests include parameter estimation, econometric modelling for economic forecasting, and time series analysis, with a particular focus on statistical and econometric methodology.

Dr Zong has authored 25 academic articles published in scholarly journals, as well as 15 books, including single-author works and collaborative volumes. His recent publications include *Economics of Marketable Surplus Supply* (reprinted by Routledge in 2018) and *The Art and Science of Econometrics* (published by Routledge in 2022). He has also been invited to serve as a peer reviewer for several academic journals and publishing houses.

Time Series Decomposition and Seasonal Adjustment

Ping Zong

CRC Press
Taylor & Francis Group
Boca Raton London New York

CRC Press is an imprint of the
Taylor & Francis Group, an **informa** business
A CHAPMAN & HALL BOOK

Designed Cover Image: UK Car Registration Series (1995–2002), Census.gov

First edition published 2026
by CRC Press
2385 NW Executive Centre Drive, Suite 320, Boca Raton FL 33431

and by CRC Press
4 Park Square, Milton Park, Abingdon, Oxon, OX14 4RN

CRC Press is an imprint of Taylor & Francis Group, LLC

ISBN: 978-1-041-20461-9 (hbk)
ISBN: 978-1-041-20464-0 (pbk)
ISBN: 978-1-003-71665-5 (ebk)

DOI: 10.1201/9781003716655

Typeset in Palatino
by Apex CoVantage, LLC

To my family: My wife, Xuedi Gong; daughter, Rong Zong;
My brother, He Zong; and two sisters, Qiqing Zong and Qisheng Zong

Contents

Figures

Tables

Preface

Time series decomposition is a vital field in modern statistical and econometric analysis. Like a meticulous watchmaker, it decomposes complex economic time series into trend, cycle, seasonal, and irregular components, enabling us to observe the various dimensions of economic activity. Seasonal adjustment, as the core process within this field, has become an indispensable fundamental technique in statistical and econometric analysis. This technique isolates and removes the influence of predictable seasonal patterns from time series data, enabling analysts and policymakers to look beyond seasonal fluctuations and more accurately discern the underlying trends and cyclical behaviour embedded in economic indicators.

From early simple moving average methods to today's sophisticated model-based seasonal adjustment techniques, the evolution of time series decomposition and seasonal adjustment methods spans over a century. In the early 20th century, economists began to recognise the influence of seasonal factors on the interpretation of economic data, prompting the development of seasonal adjustment techniques. In the 1920s, scientists at the U.S. Census Bureau developed the first seasonal adjustment methods, laying the foundation for this field. The 1950s witnessed a major leap forward with the advent of computer technology; the introduction of the X-11 method marked the beginning of a systematic and standardised era in seasonal adjustment. Subsequent developments, such as X-11-ARIMA, X-12-ARIMA, and now X-13ARIMA-SEATS, along with TRAMO-SEATS, which originated in the 1980s, have continuously advanced seasonal adjustment technology, significantly enhancing its data processing capabilities.

In modern socio-economic activities, seasonal adjustment technology plays an increasingly important role. Central banks formulate monetary policy using seasonally adjusted data; government departments evaluate the effectiveness of economic policies based on seasonally adjusted indicators; businesses analyse market trends and formulate commercial strategies using seasonal adjustment. Currently, key financial indicators such as GDP, CPI, and PMI released by national statistical offices worldwide undergo rigorous seasonal adjustment. These data serve as crucial references for various sectors of society to understand the economic situation and make decisions. Seasonal adjustment technology acts like a key, helping us unlock the door to understanding economic cycles and identifying turning points.

The core purpose of this book, which provides an in-depth analysis of the X-13ARIMA-SEATS and TRAMO-SEATS methods, is to help readers systematically master the core concepts, modelling principles, and operational procedures of modern seasonal adjustment methods. Against the backdrop of deepening globalisation and the transformation and upgrading of the economy, accurately understanding and utilising these internationally advanced seasonal adjustment tools is of great significance for enhancing the level of macroeconomic analysis and strengthening economic forecasting capabilities.

X-13ARIMA-SEATS, developed by the U.S. Census Bureau, represents one of the mainstream methods in the current seasonal adjustment field. It integrates the traditional X-11 filtering method with the signal extraction technique (SEATS) based on ARIMA models. It possesses powerful capabilities for handling complex seasonal patterns and automatically detecting outliers and calendar effects. This method is particularly suitable for processing monthly or quarterly economic data with stable seasonal patterns, such as retail sales and industrial production indices.

TRAMO-SEATS, widely adopted by institutions like the European Central Bank, was developed and maintained by the Bank of Spain. Compared to X-13ARIMA-SEATS, TRAMO-SEATS emphasises a more model-driven approach, preprocessing data through precise time series models (TRAMO part) before applying the SEATS method for seasonal adjustment. This method is theoretically more rigorous, with clearer assumptions about the data-generating process, making it especially suitable for handling the complex and diverse holiday patterns and calendar effects found in European countries.

Understanding the core concepts and operational details of these methods is paramount. Seasonal adjustment is not a simple 'black box' operation; improper seasonal adjustment can lead to misjudgments of trends and misinterpretations of cycles, potentially resulting in policy errors. This book will comparatively analyse the theoretical foundations, implementation pathways, and applicable scenarios of these two methods, focusing on discussions of their core concepts and operational details. Chinese readers may be particularly interested in adjustment methods for specific Chinese holidays, such as the Lunar New Year, Dragon Boat Festival, and Mid-Autumn Festival. This book will assist readers in selecting the most appropriate tool for specific problems, providing detailed introductions to user-defined regressors, calculation methods, modelling philosophy, parameter setting principles, and result interpretation methods, helping readers tackle these pressing concerns.

While this book focuses on the two most widely used seasonal adjustment tools, X-13ARIMA-SEATS and TRAMO-SEATS, it also explores emerging challenges and future directions in the field. In the digital economy era, the growing demand to analyse high-frequency data, such as daily or even real-time data, places new demands on traditional seasonal adjustment methods. The structural breaks caused by the COVID-19 pandemic present another major challenge for seasonal adjustment technology. During the pandemic, many economic time series exhibited anomalous fluctuations and structural changes that were difficult for traditional seasonal adjustment models to handle. This book explores approaches for addressing structural breaks, presenting the latest robust seasonal adjustment methods and technical solutions, such as the inclusion of pandemic dummy variables in models.

Methodological innovation in seasonal adjustment remains a key focus of this book. Whether in traditional methods (X-13ARIMA-SEATS and TRAMO-SEATS) or in the latest research trends, we are consistently committed to methodological innovation. The final chapter will introduce frontier explorations in the latest seasonal adjustment methods – trends such as the integration of traditional seasonal adjustment methods with machine learning techniques; using reinforcement learning to optimise seasonal adjustment parameter selection; combining state space models with deep learning; developing adaptive seasonal adjustment algorithms – and explore how Bayesian methods can be introduced into the seasonal adjustment process to quantify better uncertainty, or how to integrate mixed-frequency models to handle seasonal adjustment for data of different frequencies. These innovations not only expand the scope of application of seasonal adjustment methods but also enrich our understanding of time series decomposition theory.

It should be noted that many of the ideas discussed in this book are not my innovations. Most of the content is based on existing theories and methods developed by others. What I have done is add my commentary and attempt to connect and present these methods in a way that is more understandable and accessible for potential readers. There is a term currently popular online: 'distillation'. To me, this term captures something fundamental:

true innovation first requires mastery of existing knowledge. Only after we have thoroughly digested, absorbed, and 'distilled' the ideas and achievements of others can we hope to innovate upon them. This is why I have placed much emphasis on methodological learning in this book.

This book is compiled from work notes I made over ten years of work and research in the Time Series Analysis Branch (TSAB) at the UK Office for National Statistics (ONS). From my personal experience, gaining a deep understanding of complex theories and methods often requires significant time, and many of the ideas in this book are ones I only fully grasped after years of effort. By the time I finally 'get it', I am no longer young.

Unlike some young authors who might write books for career advancement or profit, I'm not in that position: I do not need professional advancement, and I certainly don't expect any financial income from this book. This is a highly professional and technical work, with its core objective being the dissemination of academic value and knowledge. My only hope is that it may be of some help to younger people interested in this field and that they won't have to spend quite as long retracing the same paths I did. That is one of my main motivations for writing, even if not everyone understands it.

This book focuses on X-13ARIMA-SEATS and TRAMO-SEATS, combining theoretical explanations with empirical case studies. It aims to help readers systematically master the core concepts, modelling principles, and operational procedures of modern seasonal adjustment methods, providing them with a rigorous and operational theoretical and methodological system for time series seasonal adjustment. By studying this book, readers will not only gain a thorough understanding of the technical details of seasonal adjustment but also develop the ability to evaluate, interpret, and apply adjustment results. This will enable them to handle economic data with greater rigour and contribute to more accurate economic analysis and forecasting. Promoting the development of both the theory and the practice of time series decomposition and seasonal adjustment has been my longstanding aspiration, original research intent, and strong academic mission in recent years.

This book is not intended to be a textbook on time series analysis. Instead, it serves as a reference for undergraduate and postgraduate students in relevant majors at colleges and universities, staff at national statistical agencies, banks, and related financial institutions, as well as researchers interested in this field.

The completion and publication of this book would not have been possible without the extensive help, support, and encouragement of numerous individuals. Firstly, I would like to express my gratitude to CRC's anonymous reviewers for their insightful comments and constructive suggestions. I am deeply grateful to Kristina Abbotts for her constructive comments and suggestions on the early draft of the book introduction. My sincere thanks also go to Lara Spieker, editor at Taylor & Francis, for her substantive and editorial insights, as well as for her patience and support. I also would like to thank Jayachandran Rajendiran, the project manager, and his team for their professional and hard work. Secondly, I would like to thank all my former colleagues at the ONS TSAB for many valuable discussions on seasonal adjustment methodology, with special thanks to Duncan Elliott, Neil Parkin, Danil Ayoubkhani, Kevin More, and Gary Brawn for their support and assistance. I also thank the researchers from the following institutions who participated in the biennial, two-day ONS-hosted workshops on seasonal adjustment theory and practice in London: Eurostat, Statistics Canada, and the U.S. Census Bureau (USCB). Special thanks to Brian C. Monsell (USCB) for his participation in these discussions, our valuable private exchanges, and the insightful information he generously shared. I am also grateful to Agustín Maravall from the Bank of Spain for visiting the ONS twice to provide detailed introductions and explanations of the principles and programs behind TRAMO-SEATS, as well as for our fruitful face-to-face discussions. Lastly, I extend my deepest gratitude to my family, my wife, XueDi Gong; my daughter, Rong Zong; my brother, He Zong; and my sisters, Qiqing Zong and Qishen Zong, for their unwavering support and understanding throughout the years. Without their understanding, support, and encouragement, and without the help of all these individuals, I would not have been able to complete this work. Naturally, I alone bear full responsibility for any errors or omissions that may remain in the book.

Ping Zong
June 2025

Book Highlights

1. Introduction

Statistical data are essential tools for modern economic policymaking, business cycle analysis, and economic forecasting. Socio-economic time series are used to examine and monitor trend developments, analyse the direction of socio-economic activity, and forecast potential turning points in financial conditions. However, this detection task becomes considerably difficult when the raw data contain variations attributable to seasonality, trading days, movable holidays, and irregular influences. Most monthly and quarterly macroeconomic data are affected by seasonal factors. Short-term statistical data are often characterised by seasonal fluctuations and other calendar/trading day effects, particularly high-frequency monthly and quarterly data. This can hinder a clear understanding of economic phenomena, such as trends, turning points, and consistency across indicators, and obscure the relevant short- and long-term fluctuations in the series.

To effectively monitor the national economy and estimate trends within time series, it is essential to process raw time series data using scientifically sound methods. Using methods of time series decomposition and seasonal adjustment, we remove seasonal variations, movable holidays, and trading day changes from the raw time series data, eliminating the effects of natural or socio-historical factors. Seasonally adjusted data serve as the basis for assessing the current actual economic situation, economic analysis, and forecasting. This makes the data comparable and correctly reflects the fundamental trends of the economic cycle and potential turning points in macroeconomic forecasting.

In modern socio-economic activities, seasonal adjustment technology plays an increasingly important role. Central banks formulate monetary policy using seasonally adjusted data; government departments evaluate the effectiveness of economic policies based on seasonally adjusted indicators; businesses utilise seasonal adjustment to analyse market trends and formulate commercial strategies. Currently, key financial indicators such as GDP, CPI, and PMI, released by national statistical offices worldwide, undergo rigorous seasonal adjustment. These data have become vital references for all sectors of society to understand the economic situation and make informed decisions. Seasonal adjustment technology acts like a key, helping us unlock the door to understanding economic cycles and identifying economic turning points.

The broader impact of seasonal adjustment extends beyond technical analysis, influencing the formulation and evaluation of economic policies and strategies. Accurate and timely seasonally adjusted data are essential for monitoring financial performance and responding to emerging trends and challenges. For example, central banks rely on seasonally adjusted indicators to assess the effectiveness of monetary policy interventions and identify potential risks to financial stability. Similarly, governments and international organisations use seasonally adjusted data to track progress towards economic goals, allocate resources, and design targeted interventions. Businesses and investors utilise seasonally adjusted data for strategic decisions, such as optimising production schedules, managing inventory, and identifying market opportunities. The widespread adoption and application of seasonal adjustment underscore its critical role in shaping economic outcomes and driving progress across different domains.

The short-term evolution of the economy is monitored accurately and in real-time, with a focus on international and inter-industry comparisons. In this context, annual economic indicators represent a significant portion of the output of national statistical agencies today. Seasonally adjusted data, advocated by international organisations, have become a natural and important input for policymaking activities, especially in monitoring cyclical movements, preventing or mitigating the impact of crises, supporting monetary policy, and ensuring stable and balanced growth. National statistical agencies worldwide, such as those in the United States, the UK, Canada, Australia, Japan, the European Central Bank (ECB), and Eurostat, require their member states to apply seasonal adjustment to raw statistical data. Seasonally adjusted data are published as official national statistics. The United Nations Industrial Development Organisation (UNIDO) requires the use of seasonally adjusted data as the actual statistical data for member states, serving as a unified method to (1) reveal the true movements and turning points of industrial activity; (2) calculate growth rates compared to the previous period and year-on-year growth rates compared to the same period of the previous year; (3) provide more reliable short-term forecasts; and (4) compare data from different countries.

Although an increasing number of people recognised the importance of seasonal adjustment in macroeconomic analysis, two prevalent issues persisted for a long time across both academic institutions and applied sectors: (1) Excessive focus was placed on the application of internationally recognised statistical software (such as X-12-ARIMA, X-13ARIMA-SEATS, or STL tools), with little in-depth exploration into the methodology of time series decomposition and seasonal adjustment techniques themselves; (2) there was a lack of specialised monographs systematically expounding the methodology of time series decomposition. This phenomenon has, to some extent, impeded research and progress in time series analysis and seasonal adjustment.

Seasonal adjustment methods constitute a complex analytical technique. Although modern seasonal adjustment software can automatically generate results, effective statistical practice still requires a deep understanding of core methods, such as model selection, parameter tuning, and error analysis, to accurately interpret and evaluate the outputs. Understanding the mathematical and statistical principles underlying seasonal adjustment is essential for fully grasping the process. In the practical process of seasonal adjustment, data may contain outliers, irregular fluctuations, or structural changes (such as abrupt changes, policy interventions, etc.). These issues often require identification based on theory and customised handling based on experience, rather than solely relying on the software's default output. Different data characteristics (such as stationarity, trend strength, and noise level) are suitable for different seasonal adjustment methods. If one relies solely on the software's default parameters or fixed methods, it may fail to capture the true patterns in the data. Seasonally adjusted data should reflect genuine changes in economic activity, rather than being artefacts of the adjustment process. Only when the adjustment results do not meet expectations (eg, residual seasonality or an overly smoothed trend) can one know to adjust parameters (such as filter width or seasonal frequency) or choose a more suitable model. Choosing an appropriate model, whether additive, multiplicative, ARIMA, or regARIMA, requires an understanding of the underlying assumptions and conditions of applicability. This knowledge is essential for selecting the right tool and for accurately interpreting and evaluating generated adjustment results. Today, there is a growing demand for knowledge of time series decomposition and seasonal adjustment, particularly in advancing methodological research and providing accessible explanations for broader audiences. This has been the primary motivation for writing this book: to contribute to the advancement of the academic field.

2. Time Series Decomposition Methodology and Seasonal Adjustment Techniques

Time series decomposition is a vital field in modern statistical and econometric analysis. Like a skilled watchmaker, it meticulously decomposes complex economic time series into their unobservable components – trend, cycle, seasonal, and irregular – enabling a clear examination of the various dimensions of economic activity. As the core process in this domain, seasonal adjustment has become an indispensable fundamental technique in statistical and econometric analysis. This technology allows analysts and policymakers to penetrate the fog of seasonal fluctuations and more accurately grasp the inherent underlying trends and cyclical behaviour in economic indicators by isolating and eliminating the effects of predictable seasonal patterns from time series data.

The primary purpose of seasonal adjustment is to reveal the true underlying movements and trends in time series data by removing distortions caused by seasonal effects. While seasonal variations are predictable, they may obscure the identification of important non-seasonal components, such as long-term growth, short-term fluctuations, and irregular events. By addressing this challenge, seasonal adjustment ensures that the remaining data more accurately reflects the true changes in economic activity, enabling policymakers, researchers, and businesses to make more reliable assessments and forecasts.

X-13ARIMA-SEATS and TRAMO-SEATS are currently the two most widely used seasonal adjustment tools. X-13ARIMA-SEATS utilises the X-11 moving average procedure and applies a set of fixed filters (Moving Averages – MA) to decompose a time series into trend, seasonal, and irregular components through an iterative process. In addition to continuing to use the X-11 moving average method for decomposition, X-13ARIMA-SEATS also incorporates SEATS as an alternative method.

Unlike the empirically based filter decomposition approach of X-13ARIMA-SEATS, TRAMO-SEATS is a model-based time series decomposition method. It employs model-based decomposition techniques to achieve high-precision estimation of seasonal and trend components, thereby providing a valuable decomposition of the time series data's structure. TRAMO-SEATS represents a major application of econometric methods in time series decomposition, integrating the theoretical foundations of statistics and econometrics to precisely separate time series into trend, seasonal, and irregular components.

The origins of time series decomposition and seasonal adjustment can be traced back to the 19th century. The work of Dutch meteorologist Buys-Ballot (1847)[1] on periodic signals (such as seasonal variations) can be considered the earliest exploration of time series decomposition. In 1847, Buys-Ballot modelled trends using polynomials and seasonality with dummy variables, implicitly employing linear regression techniques to estimate parameters while studying periodic temperature variations. Some 19th-century economists, such as Jevons (1862),[2] recognised that this study and analysis of 'cycles' might help explain and predict economic crises. However, the lack of precise definitions for these 'cyclical fluctuations' and computational resources hindered the development of relevant statistical methods for estimating underlying trend-cycle components for some time.

In the early 20th century, a significant part of time series research focused on estimating seasonal variation. Persons (1919)[3] was undoubtedly one of the earliest statisticians to define the various components of a time series and an algorithm for estimating them (the 'Link Relative Method'). Systematic research on seasonal adjustment primarily began with the improved decomposition process by Frederic R. Macaulay (1931)[4] at the National Bureau of Economic Research, who introduced the 'Ratio-to-Moving-Average Procedure' in the 1920s, variants of which are still widely used today. During the same period, King (1924)[5] and other authors still advocated using 'freehand curve fitting' to eliminate unwanted components.

In the 1920s, scientists at the U.S. Census Bureau developed the initial seasonal adjustment method, laying the foundation for this field. In the 1950s, with the development of computer technology, seasonal adjustment methods experienced their first major leap forward. The advent of the X-11 method marked the entry of seasonal adjustment into a systematic and standardised phase. In 1954, the manual 'Ratio-to-Moving-Average Procedure' was replaced by a computer program (Census I) developed by the U.S. Census Bureau. This program was modified and expanded in 1955, becoming known as the Census II method.[6] Since 1955, numerous variants of Census II emerged, evolving from X-1 to the famous X-11, followed by X-11-ARIMA, X-12-ARIMA, and up to the current X-13ARIMA-SEATS. Seasonal adjustment techniques have been continually refined, with an ever-greater capacity to handle data.

X-13ARIMA-SEATS, developed by the U.S. Census Bureau, represents one of the mainstream methods in the current field of seasonal adjustment. It integrates the traditional X-11 filtering approach with signal extraction techniques based on ARIMA models (SEATS), possessing powerful capabilities to handle complex seasonal patterns, automatically detect outliers, and account for calendar effects. This method is particularly well-suited for processing monthly or quarterly economic data with stable seasonal patterns, such as retail sales and industrial production indices.

TRAMO-SEATS, on the other hand, is an ARIMA-based method,[7] developed and maintained by the Bank of Spain. The prototype of the TRAMO part was developed in the early 1980s, followed by the SEATS part in the mid-to-late 1980s. The Bank of Spain officially released version 0.1 of TRAMO-SEATS in 1994, marking a significant milestone as the program's first public appearance as an official tool. The program has been continuously updated and iterated, with subsequent versions (TRAMO-SEATS+) adopted by institutions like Eurostat as one of the official seasonal adjustment methods.

Compared to X-13ARIMA-SEATS, TRAMO-SEATS emphasises a more model-driven approach. It uses the TRAMO part for data preprocessing through precise time series modelling and then applies the SEATS method for time series component decomposition. This method is theoretically more rigorous, with clearer assumptions about the data generation process, making it particularly suitable for handling the complex and diverse holiday patterns and calendar effects found in European countries.

Both X-13ARIMA-SEATS and TRAMO-SEATS have had a significant impact in both theoretical and applied fields, completing the toolbox of seasonal adjustment methods for statisticians.

2.1 X-13ARIMA-SEATS Method

X-13ARIMA-SEATS is an empirically based method for filter decomposition. Its foundation lies in the X-11 method, developed by the U.S. Census Bureau in the 1950s, which uses moving averages to decompose a time series into trend, seasonal, and irregular components. The X-11 approach applies moving averages and iterative procedures to the original data, gradually isolating each component.

However, moving average methods, due to the limitation of window length when calculating the trend or seasonal component of a time series, cannot fully cover the values at the beginning and end of the series, leading to data loss (the endpoint problem). To address the endpoint data loss inherent in moving averages, Statistics Canada introduced ARIMA models and developed the X-11-ARIMA method.

ARIMA models employ statistical modelling techniques, using extrapolation and interpolation to fit the entire time series and leverage their forecasting capabilities to fill in missing values at the endpoints.

The autoregressive (AR) part of ARIMA uses linear regression, predicting the current value based on past values (lags) of the time series. The Moving Average (MA) part of ARIMA adjusts the predictions using the error terms (noise terms) of the series, reducing the impact of noise. The Integration (I) part of ARIMA performs differencing operations on nonstationary series, transforming them into stationary series for better modelling. Unlike moving averages, ARIMA models do not rely on a fixed-length sliding window, thereby avoiding direct data loss and addressing the issue of moving averages failing to cover data at the window edges. By adjusting parameters (p, d, q), ARIMA can adapt to different types of time series, more flexibly capturing data characteristics.

The AR part of ARIMA predicts missing values by extrapolating from known sequence data. For instance, based on the lagged relationship of the initial values of the series, a regression model can be used for backcasting, extrapolating the earliest values of the series using the AR and MA parts of ARIMA to fill the first missing value. The ARIMA model uses existing endpoint values and their lags to forecast, generating future data points and thereby estimating the missing terminal values. This overcomes the limitation of moving averages, which rely solely on data within a fixed window before and after each point. By comparing the filled complete series with actual observations and checking if the forecasted data is reasonable, ARIMA parameters can be adjusted to improve fitting and forecasting accuracy.

The ARIMA model is essentially a univariate time series model, predicting future values based solely on the historical data of the time series itself. In reality, many time series (eg, consumption, employment, traffic flow) are affected by external factors such as (1) holiday effects (eg, the impact of Christmas or Spring Festival on retail data); (2) policy adjustments (eg, the effect of interest rate changes on economic data); or (3) unexpected events (eg, the impact of natural disasters or pandemics on data). ARIMA models cannot directly handle these external influences.

regARIMA is a model that integrates a regression model with an ARIMA model. While ARIMA's predictive power relies on the autocorrelation of the series and past trends, regARIMA (Regression with ARIMA errors) can explicitly model these external factors by introducing regressors (exogenous variables), thereby improving forecast accuracy.

Introducing the regARIMA model addresses the shortcomings of the ARIMA model. By combining regression analysis with ARIMA, regARIMA can explicitly model exogenous variables while retaining ARIMA's ability to model the time series's intrinsic dynamic characteristics. This enhancement makes regARIMA perform exceptionally well in seasonal adjustment, outlier handling, and complex time series modelling.

Methodologically, this specific model design of regARIMA, which uses an ARIMA model to model the residuals of the regression model, has special significance. The main reason is that after introducing a regression model into the seasonal adjustment model, the residuals may exhibit temporal dependence, which needs further modelling to eliminate.

In regression models, residuals are typically assumed to be independent and identically distributed (i.i.d.) white noise, meaning no temporal correlation. However, in reality, residuals may exhibit time-dependent patterns (such as trends, seasonality, or autocorrelation structures). Suppose the residuals display temporal dependence, such as autocorrelation or seasonal patterns, rather than behaving as white noise. In such cases, a standard regression model may prove insufficient to fully capture the underlying dynamics of the data. By applying an ARIMA model to the residuals, these temporal dependencies can be captured, improving the overall modelling effect.

Modelling the regression model's residuals with ARIMA enhances the randomness of the residuals. ARIMA can explain serial dependency structures, capturing the autoregressive (AR), Integrated (I) trend, and Moving Average (MA) characteristics within the series. By modelling the residuals, these correlations can be effectively removed, resulting in a residual sequence closer to white noise. Therefore, when there is no longer significant temporal dependence in the residuals, the model's assumptions (like independent normal distribution) are more easily satisfied, making model estimation and forecasting more reliable.

In the regARIMA model, the regression part describes the structural features in the data related to explanatory variables (ie, exogenous variables). It focuses on how exogenous factors affect the dependent variable, handling the explanatory variables. For the portion unexplained by regression, the ARIMA model can further explore the dynamic patterns in the residuals (such as trends, seasonality, or random fluctuations), thereby separating explanatory variables (structure) and randomness.

In economics and social sciences, many time series contain variations driven by explanatory variables (eg, policy factors for economic indicators) as well as time-dependent random fluctuations (eg, the inertia of market behaviour). In this context, the regARIMA model mitigates the risk of omitting key dynamic information by modelling residuals, using the regression component to account for policy effects or external variables, and the ARIMA component to capture the remaining time-dependent structure, thereby enhancing forecasting accuracy. This makes regARIMA a powerful tool widely applied in economics, meteorology, and social sciences for time series modelling and forecasting problems.

X-13ARIMA-SEATS uses a moving average procedure and applies a set of fixed filters (Moving Averages – MA) to decompose a time series into trend, seasonal, and irregular components through an iterative process. Typically, symmetric linear filters are applied to the middle of the series, while asymmetric linear filters are applied to the beginning and end. All filters used in these methods can be classified as local regressions, involving fitting a polynomial over a sliding interval (moving one point at a time) via least squares (weighted or unweighted), and smoothing the data at the centre of the interval (the smoothed value for the next date is obtained by fitting the polynomial at the next interval). Local regression is equivalent to applying specific moving averages when observations are regularly spaced and decomposing it into its components. This approach allows for robust and flexible handling of complex seasonal patterns, making it particularly suitable for various economic indicators. In addition to continuing to use the X-11 moving average method for time series decomposition, X-13ARIMA-SEATS also incorporates SEATS as an alternative method.

2.2 TRAMO-SEATS Method

Unlike the empirically based filter decomposition approach of X-13ARIMA-SEATS, TRAMO-SEATS is a model-based time series decomposition method. It employs model-based decomposition techniques to achieve high-precision estimation of seasonal and trend components, thereby providing a valuable decomposition of the time series data's structure. TRAMO-SEATS is a notable application of econometric methods in time series decomposition, integrating the theoretical foundations of statistics and econometrics to precisely separate a series into trend, seasonal, and random components. TRAMO-SEATS is based on the Box–Jenkins time series modelling theory. It decomposes a time series by fitting an ARIMA model into the trend, the seasonal, and the random components. TRAMO-SEATS consists of two parts: TRAMO and SEATS.

TRAMO (Time Series Regression with ARIMA Noise, Missing Observations, and Outliers) is the preprocessing stage for time series seasonal adjustment. Its main functions are as follows:

(1) Outlier Handling: Detecting and adjusting for outliers (eg, unexpected events, holiday effects) and using regression models to fit the impact of exogenous variables (eg, policy changes, weather factors);
(2) Missing Value Imputation: Estimating missing values based on the ARIMA model to ensure time series completeness;
(3) Model Selection: Automatically selecting the optimal ARIMA model parameters (p, d, q). This is similar to the ARIMA and regARIMA (pre-adjustment) functions in X-12-ARIMA and X-13ARIMA-SEATS.

SEATS (Signal Extraction in ARIMA Time Series) is the core of the TRAMO-SEATS method. Based on the ARIMA model provided by TRAMO, it decomposes the time series into trend, seasonal, and random signals. The SEATS part uses signal extraction theory from econometrics to achieve decomposition. SEATS employs the Best Linear Unbiased Estimator (BLUE) framework to extract trend and seasonal components from the ARIMA model; it uses filters to separate different components of the model, extracting stable trend and seasonal patterns while isolating random noise.

Unlike the traditional moving average decomposition used in X-12_ARIMA and X-13ARIMA-SEATS, SEATS's model-based decomposition method relies on the precise fitting of the ARIMA model and signal extraction. It employs algorithms such as Kalman filtering to optimise the decomposition process, producing smoother and more stable estimates of trend and seasonal components, thereby enhancing the stability and scientific rigour of the results.

TRAMO-SEATS combines regression analysis from econometrics with time series modelling, emphasising the modelling of exogenous variables (eg, holiday effects) and the optimal handling of random errors. In the analysis of macroeconomic data (eg, GDP, unemployment rate, industrial output), seasonal fluctuations may obscure true economic trends. TRAMO-SEATS provides a theoretically rigorous and widely applicable tool to address these issues. Furthermore, TRAMO-SEATS employs automated modelling techniques from econometrics, using information criteria (eg, AIC, BIC) to select the optimal model, thereby reducing manual intervention. It is thus a high-precision decomposition method based on econometrics and time series theory. Its theoretical foundations (eg, ARIMA modelling, signal extraction, filtering techniques) and scientific methodology make it an important tool in modern time series analysis and seasonal adjustment, particularly suitable for analysing complex economic and financial data and policymaking. It is also the primary seasonal adjustment software designated by Eurostat.

Understanding the core concepts and operational details of these methods is crucial. Seasonal adjustment is not a simple 'black box' operation; improper seasonal adjustment can lead to misjudging trends, misreading cycles, and policy errors. This book will comparatively analyse the theoretical foundations, implementation paths, and applicable scenarios of these two methods, focusing on a discussion of their core concepts and operational details. Chinese readers may be particularly concerned about adjustment methods for Chinese-specific holidays, such as the Lunar New Year, Dragon Boat Festival, and Mid-Autumn Festival. This book will help readers select the most appropriate tool for specific problems, detailing the calculation methods for user-defined regressors and regression variables, modelling philosophy, parameter setting principles, and result interpretation methods, assisting Chinese readers in solving these most pressing challenges.

3. Challenges Facing Traditional Standard Seasonal Adjustment Methods

While seasonal adjustment theory and methodology have evolved considerably over the past century, yielding more sophisticated tools, the established methods like X-13ARIMA-SEATS and TRAMO-SEATS, though robust for conventional monthly or quarterly economic data such as GDP, unemployment rates, and industrial production indices, now confront unprecedented challenges. These stem primarily from the profound disruptions caused by the COVID-19 pandemic, the increasing complexity of seasonal patterns inherent in high-frequency data, and the difficulties associated with real-time processing and the volatility of subsequent data revisions, particularly for real-time adjustments.

The COVID-19 pandemic fundamentally undermined a core assumption of traditional seasonal adjustment: the relative stability of seasonal patterns. The crisis triggered significant structural breaks, as seen during the global lockdowns of March–April 2020, which caused dramatic shifts in data. For example, the U.S. unemployment rate surged from 3.5% to 14.8%. Traditional seasonal peaks linked to holidays like Christmas disappeared due to restrictions, while new patterns emerged, such as pronounced peaks associated with e-commerce promotional events like Singles' Day. Furthermore, the scale and frequency of outliers generated by the pandemic overwhelmed the detection and adjustment capabilities of conventional methods, such as STL's IQR-based adjustments. Lockdown policies, supply chain breakdowns, and abrupt changes in consumer behaviour severely distorted the underlying seasonal rhythms of key time series (eg, retail sales, unemployment, energy consumption), rendering standard X-13ARIMA-SEATS or TRAMO-SEATS procedures ineffective without significant manual intervention.

Simultaneously, the rise of high-frequency data (hourly, daily, weekly) presents distinct complexities. While offering advantages such as near-real-time monitoring of market fluctuations (eg, immediate pandemic impacts on consumption) and the ability to detect very short-term seasonal patterns (eg, single-day sales spikes), this data also introduces new challenges. A major challenge is multiple seasonality, where nested cycles coexist. For example, daily data may exhibit both a weekly pattern (7-day cycle) and an annual pattern (365-day cycle), while hourly data, such as electricity demand, can display intraday (24-hour), weekly (7-day), and annual (365-day) seasonal patterns simultaneously. Traditional methods such as X-11 or SEATS, designed primarily for single-frequency seasonality (eg, 12 months, 4 quarters), struggle to model this complexity directly. Additionally, non-integer cycles pose problems, as seen in financial market data where the number of trading days per year varies (eg, 252 days in the U.S., 240 in China), disrupting the fixed-period assumption. The inherently higher noise levels and prevalence of transient outliers (eg, news shocks, market flash crashes) in high-frequency data also tax traditional filters: X-13ARIMA-SEATS's moving averages (like the Henderson filter) may over-smooth genuine short-term fluctuations, while TRAMO-SEATS's model-based approach can falter with the strong short-term autocorrelation often found in such data (eg, GARCH effects in finance). Finally, the computational burden of applying traditional methods like X-13's filters or SEATS's spectral decomposition to long runs of high-frequency data (eg, ten+ years of daily points) becomes prohibitive, making real-time adjustment for streaming data (eg, social media sentiment) impractical.

The need for real-time processing also exacerbates the inherent issue of revision volatility. Seasonal adjustments are typically revised as new data becomes available. For high-frequency series, these revisions can occur frequently, potentially undermining the consistency needed for timely decision-making. Early adjustments made during rapidly evolving situations, such as the onset of the pandemic, were often subject to substantial subsequent revisions as additional data clarified the picture (eg, repeated revisions to U.S. 2020 quarterly GDP and retail figures), creating a significant 'revision risk'. While statistical agencies implemented temporary solutions during COVID-19, such as marking pandemic periods as exceptional intervals, introducing specific pandemic dummy variables (eg, U.S. BLS for unemployment),[8] or manually overriding standard filters (eg, ECB using interpolation for 2020 Q2 data),[9] key legacy issues remain unresolved. These include the potential permanence of new seasonal patterns (eg, altered commuting/energy use due to remote work) and the fundamental challenge of achieving an appropriate balance between flexibility (to handle shocks and new patterns) and stability (to ensure reliable, comparable time series) in real-time adjustment contexts.

3.1 Methodological Innovations

Recent research has actively explored several promising frontiers, with a growing focus on methodological innovations that employ robust seasonal adjustment techniques and enhance model resilience to extreme values. (1) Statistics Canada uses 'adjusted LOESS smoothing parameters' in STL to reduce outlier weights,[10] while the Australian Bureau of Statistics (ABS) implemented 'partial seasonal adjustment', estimating seasonal factors only from pre-pandemic period data;[11] (2) the U.S. Census Bureau combines hybrid models with machine learning assistance, integrating X-13ARIMA-SEATS with Bayesian Structural Time Series (BSTS) for automatic detection of pandemic-related breaks;[12] (3) the OECD has investigated using Long Short-Term Memory (LSTM) networks to predict a 'normal' seasonal pattern and then compare it with actual pandemic data to identify distortions;[13] (4) the U.S. Bureau of Labour Statistics (BLS) adopted a two-step approach for Current Employment Statistics: firstly, estimating benchmark seasonality from 2015–2019 data, secondly, applying 'dynamic weights' to 2020–2022 data to reduce the filtering intensity during lockdown months;[14] (5) Eurostat's 'ESS Guidelines on Seasonal Adjustment' mandate manual adjustments for exceptional months, explicitly requiring 'suspension of automatic filtering' for pandemic data and instead adopting a 'rolling window approach' (estimating seasonal factors from 2017–2019 data), and (6) the International Monetary Fund (IMF) implemented its 'Real-Time Monitoring Framework', developed 'seasonal adjustment uncertainty bands' to quantify estimation errors caused by the pandemic.[15]

Despite progress in interim approaches, significant unresolved challenges persist in seasonal adjustment for the COVID-19 pandemic era. The most prominent issue is revision volatility. Data initially adjusted during the early pandemic phase subsequently required substantial revisions (eg, quarterly U.S. GDP revisions for 2020), while emerging seasonal patterns, such as permanent shifts in transportation/energy data cycles due to widespread remote work, complicate modelling. The core concern remains processing extreme fluctuations without overfitting seasonal patterns. Striking an appropriate balance between flexibility and stability proves particularly intricate for real-time adjustments.

3.2 Frontier Research Directions in New Methodologies

The study of time series decomposition and seasonal adjustment methodologies has emerged in many cutting-edge research directions: for example (1) the use of change point models to capture structural disruptions more effectively, which is critical during economic shocks such as the COVID-19 pandemic or major recessions; (2) responding to inflation and price volatility and being able to carefully distinguish between persistent inflationary trends and typical seasonal price changes, ie, enhanced models that incorporate seasonally adjusted inflation expectations (eg, CPI); (3) weekly and daily seasonal adjustments are becoming increasingly relevant with high-frequency data availability, with techniques such as the Fourier transform and wavelet being used for finer time adjustments; (4) the adaptive seasonal model, including embedded breakpoint detection within the state-space framework;[16] and (5) the 'hybrid AI approach' Transformer architecture captures long-term dependencies plus reliable statistical filtering;[17] (6) machine learning techniques are being experimented with to detect complex seasonality, but interpretability and reliability remain areas of concern; (7) adopting AI-driven models could revolutionise seasonal adjustments, but ensuring that they are explainable and meet statistical guidelines is critical.

Recent research on seasonal adjustment methodologies emphasises the adoption of advanced statistical techniques and robust tools, including the following.

(1) Hybrid approach combines machine learning with traditional seasonal adjustment, such as using LSTM or Transformer models to capture complex seasonality, and then combining with X-13ARIMA-SEATS for post-processing; the seasonal adjustment of high-frequency time series (such as daily, weekly, and intraday data) faces challenges such as multiple seasonality, noise interference, and nonlinear patterns, and it is difficult for traditional methods (X-13ARIMA-SEATS, TRAMO-SEATS) to be directly applicable. Deep learning models (such as LSTM and Transformer) can automatically learn complex time series patterns but often lack interpretability. Therefore, it is a feasible solution to combine the feature extraction ability of deep learning with the interpretability of traditional seasonal adjustment. Facebook's Prophet model[18] allows for custom multi-seasonality, partially replacing traditional methods.

The advantage of the hybrid approach of Deep Learning + X-13ARIMA-SEATS is that deep learning captures 'complex seasonality' (eg, multiple cycles, moving holidays) and X-13ARIMA-SEATS provides 'explainable decomposition' (trend, seasonal, irregular components). This hybrid LSTM/Transformer + X-13ARIMA-SEATS method combines 'nonlinear modelling capabilities' and 'explainable decomposition', which is suitable for seasonal adjustment of high-frequency data. It is the future direction of automatic machine learning (AutoML) optimisation model selection, and causal inference (eg, dealing with seasonal breaks under policy interventions).

(2) State Space Models (SSMs) employ structured time series frameworks to flexibly capture multiple seasonalities in high-frequency data. The Bayesian method uses Markov Chain Monte Carlo (MCMC) sampling to quantify seasonally adjusted uncertainty. The SSM provides a versatile and flexible framework for time series analysis, enabling effective decomposition and forecasting of high-frequency data by capturing trend, seasonal, and irregular components. At its core, it is essential to characterise the dynamic evolution of the system through state and observation equations, while allowing the model structure to be extended as needed to accommodate specific analytical requirements. The SSM model can also be easily extended as needed to include calendars, sampling errors, and outlier components. Outliers and holiday components can be specified in the model as required to address various effects, including missing values. Users can choose from six different maximum likelihood estimation optimisers that automatically detect level changes and additive outliers, with many options for working with time series data. The trend components are modelled using various approaches.

The SSM model can handle high-frequency data. It is a regression model with a coefficient of change over time and smoothing ability.[19] Additionally, SSM performs well in both Linux and PC environments. Seasonality components can be used to construct triangular seasonality for weekly data, using dummy variables or periodic splines that vary over time.

(3) A popular phenomenon is the growing preference for exploring the use of Bayesian methods. Bayesian methods offer a probabilistic framework for time series decomposition, integrating prior information with observed data to achieve full posterior inference of trend, seasonal, and noise components. The core of this problem is to transform the traditional decomposition problem into a joint probability distribution estimation of hidden states and parameters, and use MCMC sampling to solve it numerically, to provide a flexible framework for the application of seasonal adjustment and time series decomposition. The posterior distribution provides confidence intervals for trend and seasonal components, enabling quantification of uncertainty, and fuses prior information to introduce domain knowledge, such as seasonal amplitude limits, and to deal with complex trends through random walks or GPs.

(4) Online learning and incremental calculation, such as recursive filtering (such as Kalman filtering), achieve real-time seasonal adjustment and avoid batch processing delays. The fundamental differences between the traditional seasonal adjustment method and the online method are as follows: the batch processing method (X-11-ARIMA) requires a complete sample interval, and needs to use a symmetric moving average filter, the endpoint processing relies on backward prediction, and the seasonal component is fixed or extrapolated through the ARIMA model; the online learning method (state space model) can be recursively processed point by point, asymmetric filters can be used, state estimates can be updated in real-time, and more importantly, the seasonal components can obey random processes. This paradigm shift has enabled real-time monitoring of economic indicators. For example, the EUROLIB system adopted by the European Central Bank leverages this technology to assess the state of the economy daily. Subsequent developments, such as the 'Small Area Estimation' method proposed by Pfeffermann in 2013,[20] have further expanded the use of e-learning in official statistics. Bayesian methods and uncertainty quantification.

Addressing the demand for timely insights, online learning and incremental computation techniques represent a paradigm shift. AI-driven hybrid models hold potential, such as combining Transformer architectures (capable of capturing complex long-range dependencies) with robust statistical filters. However, a key challenge for these advanced techniques remains ensuring explainability and adherence to established statistical principles, moving beyond 'black box' solutions.

We believe that, despite the many challenges associated with traditional seasonal adjustment methods and, regardless of how the seasonal adjustment tools evolve, the foundations laid by X-13ARIMA-SEATS and TRAMO-SEATS, along with the profound understanding of the data generation process and the pursuit of interpretability in the adjusted results, will remain the core values in this field. Looking ahead, amid rapid technological advancements in the era of big data, X-13ARIMA-SEATS and TRAMO-SEATS are expected to remain widely used. At the same time, integrating traditional and modern methodologies is emerging as a key direction for the future of seasonal adjustment.

4. Book Purpose, Structure, and Scope

This book provides a systematic exposition of the theory and methodology of time series decomposition and seasonal adjustment, along with practical applications of related statistical software. Its primary aim is to equip users and data producers with the essential background knowledge to generate high-quality seasonally adjusted data and address the diverse challenges that may arise during the adjustment process. Amid deepening globalisation and ongoing economic transformation, mastering advanced international seasonal adjustment theories and tools is essential for strengthening the capacity of statistical and economic researchers in macroeconomic analysis and forecasting.

4.1 Structure and Organisation

The book is organised into nine chapters across four thematic sections.

Following the introductory Chapter 1, Part I (Chapters 2–4) establishes the conceptual framework and theoretical foundations of time series decomposition and seasonal adjustment. Chapter 2 reviews the evolution of the X-11 series of seasonal adjustment methods (X-12-ARIMA, X-13ARIMA-SEATS), emphasising their continuous pursuit of two objectives: improving decomposition accuracy and enhancing the stability of results. This progression reflects both advances in statistical theory and the demands of practical application. Chapter 3 examines the general theoretical framework of seasonal adjustment, defining core components of time series (trend-cycle, seasonality, outliers, trading day effects) and their impact on short-term economic assessment. It also explores 'pre-adjustment' techniques and various models for long-term trends, seasonal components, and irregular fluctuations. Chapter 4 focuses on the moving average methodology central to X-11 seasonal adjustment, detailing the time/frequency domain properties of moving averages and the iterative X-11 procedure. This groundwork prepares readers for later chapters on X-11 and SEATS methodologies while offering practical guidance on model selection and diagnostics.

Part II (Chapters 5–6) delves into the procedures and methods underpinning X-13ARIMA-SEATS.

Chapter 5 comprehensively explains the application of X-13ARIMA-SEATS – from basic operations to advanced configurations – empowering users to master this robust seasonal adjustment tool. It specifically addresses ARIMA and regARIMA specifications, as well as the identification and handling of various outliers, level shifts, and holiday effects (including Easter). Special attention is given to constructing user-defined regressors, such as variables for the Chinese Lunar New Year and other movable festivals. By emphasising appropriate runtime environments and rigorous result diagnostics, this chapter enables efficient time series analysis. Chapter 6 explores critical ancillary topics in seasonal adjustment, including optimal time series length requirements; reconciling seasonally adjusted data with annual aggregates ('seasonal adjustment and aggregation'); the 'direct versus indirect adjustment' dilemma; quality assessment of adjusted data; inherent revisions in adjustment processes; and non-statistical criteria influencing strategy selection.

Part III (Chapters 7–8) centres on the TRAMO-SEATS methodology.

Chapter 7 details TRAMO's pre-adjustment stage – encompassing outlier detection, missing value imputation, and ARIMA model fitting – and SEATS' subsequent decomposition of time series components via ARIMA modelling and Wiener–Kolmogorov filtering. Practical implementation workflows are thoroughly demonstrated.

Chapter 8 introduces JDemetra+, the primary software platform implementing TRAMO-SEATS. It explores JDemetra+'s core functionalities, operational procedures, and real-world applications. While JDemetra+ integrates both TRAMO-SEATS and X-13ARIMA-SEATS (covered in Chapter 5), this chapter focuses exclusively on its TRAMO-SEATS implementation. Readers learn to perform single-series and multi-series seasonal adjustments, utilise quality diagnostics, and apply optimisation strategies to ensure result reliability. The chapter also addresses best practices for publishing seasonally adjusted data and communicating effectively with end-users.

Part IV (Chapter 9) concludes the book by evaluating the contributions and limitations of traditional seasonal adjustment methods. It then examines contemporary challenges, such as structural breaks during economic crises or pandemic disruptions, and explores emerging hybrid approaches that integrate classical techniques with innovations like machine learning and state space modelling. The chapter concludes by outlining future trajectories for the field.

4.2 Distinctive Approach

This book transcends conventional software manuals by prioritising methodological understanding. Foundational theories, including moving averages, spectral decomposition, and signal extraction, are clearly explained to provide a solid conceptual basis for practical work. While software proficiency is essential, we maintain that methodological insight remains the core of rigorous analysis. Conversely, the book also moves beyond purely theoretical textbooks. It bridges theory and practice by providing concrete operational guidance, recommendations for model selection, and systematic solutions to complex scenarios such as seasonal breaks, indirect adjustments, and the construction of user-defined regressors (eg, for the Chinese Lunar New Year, Mid-Autumn Festival, and Dragon Boat Festival). This balanced structure accommodates both beginners seeking step-by-step guidance and advanced users requiring technical depth. Practical code examples further enhance intuitive understanding. Topics are presented pedagogically to ensure accessibility for non-specialists without sacrificing the rigour demanded by expert users. Formal mathematical treatments are introduced judiciously to avoid overwhelming readers.

4.3 Broader Implications

Our exploration of time series decomposition and seasonal adjustment reveals a field in dynamic evolution. Traditional paradigms are continuously refined to meet modern demands. Future progress will hinge on harmonising three critical dimensions: theoretical rigour, computational efficiency, and robustness against structural shocks. Achieving this balance requires sustained interdisciplinary collaboration, spanning statistics, econometrics, mathematics, and computer science, alongside continuous methodological innovation and empirical validation. Only through such concerted efforts can seasonal adjustment retain its vital role in providing reliable analytical foundations for economic monitoring and policymaking, ensuring it adapts to technological shifts and evolving economic realities. This ongoing evolution extends beyond methodological advancement; it fundamentally shapes our capacity to interpret economic phenomena, forecast future trends, and propel the discipline forward.

Notes

1 Box, G. E. P., Jenkins, G. M., & Reinsel, G. C. (2008). "Time Series Analysis: Forecasting and Control", 4th ed. Wiley; & Hassani, H. (2007). "Time Series Decomposition and Forecasting: A Review". *Journal of Computational Methods in Sciences and Engineering*, 7(2), 123–143.
2 Jevons, W. S. (1862). "On the Study of Periodic Commercial Fluctuations". Report of the British Association for the Advancement of Science, Cambridge.
3 Persons, W. M. (1919). "Indices of Business Conditions". *Review of Economics and Statistics*, 1(1), 5–107.
4 Macaulay, F. R. (1931). "The Smoothing of Time Series". National Bureau of Economic Research (NBER), New York.
5 King, W. I. (1924). "Elements of Statistical Method". The Macmillan Company, New York.
6 Shiskin, J., & Eisenpress, H. (1957). "Seasonal Adjustments by Electronic Computer Methods". *Journal of the American Statistical Association (JASA)*, 52(280), 415–449. https://doi.org/10.1080/01621459.1957.10501402.
7 Burman, J. P. (1980). "Seasonal Adjustment by Signal Extraction". *Journal of the Royal Statistical Society. Series A (General)*, 143(3), 321–337. https://doi.org/10.2307/2982132; Gómez, V., & Maravall, A. (1997). "Programs TRAMO and SEATS Instructions for the User (Beta Version)". Working Paper No. 97001. Banco de España, Madrid, Spain.
8 Hipple, S. F., & Sok, P. (2020). "Impact of the COVID-19 Pandemic on the Employment Situation". *BLS Monthly Labor Review*.
9 Eurostat. (2021). "ESS Guidelines on Seasonal Adjustment During the COVID-19 Crisis". Eurostat Methodologies and Working Papers. Eurostat.
10 Dagum, E. B., & Fortier, I. (2021). "Robust Seasonal Adjustment of Pandemic-Affected Data". *Journal of Official Statistics*, 37(2).
11 Australian Bureau of Statistics. (2020). "Partial Seasonal Adjustment for COVID-19 Affected Series". ABS Cat. No. 5204.0. ABS.
12 McDonald, J., & Thorsrud, A. (2021). "Nowcasting with Bayesian Structural Time Series During Crises". NBER Working Paper 28408. NBER.
13 OECD. (2022). "AI-Assisted Seasonal Adjustment: A COVID-19 Case Study". OECD Statistics Working Papers. OECD.
14 Hipple, S. F. (2021). "Adjusting Seasonal Factors During the Pandemic". BLS Working Paper 542. BLS.
15 IMF. (2022). "Real-Time Data Uncertainty in Post-Pandemic Macroeconomics". IMF Working Paper WP/22/89. IMF.

16 Durbin, J., & Koopman, S. J. (2012). "Time Series Analysis by State Space Methods", 2nd ed. Oxford University Press. https://doi.org/10.1093/acprof:oso/9780199641178.001.0001; Chib, S. (1998). "Estimation and Comparison of Multiple Change-Point Models". *Journal of Econometrics*, 86(2), 221–241; Barber, D., et al. (2011). "Bayesian Online Changepoint Detection". arXiv:1101.1438.
17 Zeng, A., et al. (2023). "Autoformer: Decomposition Transformers for Seasonal Adjustment". NeurIPS.
18 Taylor, S. J., & Letham, B. (2018). "Forecasting at Scale". *The American Statistician*, 72(1), 37–45. https://doi.org/10.1080/00031305.2017.1380080.
19 Harvey, A. C., Koopman, S. J., & Riani, M. (1997). "The Modelling and Seasonal Adjustment of Weekly Observations". *Journal of Business & Economic Statistics*, 15(3), 354–368. https://doi.org/10.1080/07350015.1997.10524712; 以及Harvey, A. C., & Koopman, S. J. (1993). "Forecasting Hourly Electricity Demand Using Time-Varying Splines". *Journal of the American Statistical Association*, 88(424), 1228–1236. https://doi.org/10.1080/01621459.1993.10476402.
20 Pfeffermann, D. (2013). "New Important Developments in Small Area Estimation". *Statistical Science*, 28(1), 40–68. https://doi.org/10.1214/12-STS395.

1

Introduction

1.1 Introduction

Statistical data serve as a fundamental tool in contemporary economic policymaking, business cycle analysis, and economic forecasting. Socio-economic time series are employed to examine and monitor trend developments, assess the direction of economic activity, and identify potential turning points in economic dynamics. However, variations in the raw data, caused by seasonal patterns, trading day effects, moving holidays, and irregular disturbances, complicate the analytical process. Most monthly and quarterly macroeconomic indicators are affected by seasonal effects. In particular, short-term statistics, especially those at monthly and quarterly frequencies, often exhibit seasonal patterns and calendar-related effects. These distortions can obscure the underlying economic signals, impeding the accurate identification of trends, turning points, and cross-indicator consistency, while also concealing important short- and long-term fluctuations in the data series.

To more effectively monitor national economic performance and estimate underlying trends in time series data, it is essential to subject raw data to rigorous statistical treatment. Through time series decomposition and seasonal adjustment, components attributable to seasonality, holidays, trading day effects, and other natural or socio-historical influences are removed. The seasonally adjusted data then serve as the basis for analysing the current state of the economy and for conducting economic forecasting. This process enhances data comparability and ensures that the underlying cyclical trends and turning points of the macroeconomy are accurately identified.

This chapter is structured as follows. Section 1.1 introduces the necessity of seasonal adjustment in economic statistical analysis. Section 1.2 reviews the historical development of seasonal adjustment methods and outlines the motivation for writing this book. Section 1.3 presents the book's objectives and core focus, offering an in-depth analysis of the X-13ARIMA-SEATS and TRAMO-SEATS methodologies. Section 1.4 provides a brief overview of the book's structure and scope. Finally, Section 1.5 provides a general summary of the chapter.

1.2 Motivation for Writing This Book

In contemporary socio-economic activities, seasonal adjustment techniques play an increasingly vital role. Central banks rely on seasonally adjusted data to formulate monetary policy, government agencies use seasonally adjusted indicators to evaluate the effectiveness of economic policies, and enterprises apply seasonal adjustment to analyse market trends

DOI: 10.1201/9781003716655-1

and develop business strategies. Today, national statistical offices worldwide publish key financial indicators, such as GDP, CPI, and PMI, only after rigorous seasonal adjustment. These adjusted datasets have become essential tools for various sectors of society to understand the economic situation and make informed decisions. In this sense, seasonal adjustment functions as a key that unlocks our understanding of the economic cycle and helps identify turning points.

The impact of seasonal adjustment extends far beyond the realm of technical analysis; it directly influences the design and evaluation of economic policy and strategy. Accurate and timely seasonally adjusted data are critical for monitoring economic performance and responding to emerging trends and challenges. For instance, central banks in many countries rely on seasonally adjusted indicators to assess the effectiveness of monetary policy interventions and to identify potential risks to financial stability. Likewise, governments and international organisations use such data to track progress toward economic objectives, allocate resources, and design targeted policy interventions. Businesses and investors employ seasonally adjusted figures to support strategic decisions, such as optimising production schedules, managing inventories, and identifying market opportunities. The widespread adoption and application of seasonal adjustment underscore its essential role in shaping economic outcomes and advancing progress across multiple sectors.

In today's context, short-term economic developments are monitored with increasing precision and timeliness, with growing emphasis on international and sectoral comparisons. Against this backdrop, annualised economic indicators still represent a substantial portion of national statistical outputs. Seasonally adjusted data, as advocated by international organisations, have become a natural and indispensable input in the policymaking process, particularly in monitoring cyclical movements, preventing or mitigating the impact of economic crises, supporting monetary policy, and promoting stable and balanced growth.

National statistical agencies in countries such as the United States, the United Kingdom, Canada, Australia, and Japan, as well as institutions such as the European Central Bank and Eurostat, now require seasonally adjusted data to be published as official statistics. The UNIDO mandates the use of seasonally adjusted data as the actual statistical basis for member states. Seasonally adjusted series are employed as a standardised method to (1) reveal the true dynamics and turning points of industrial activity; (2) calculate quarter-on-quarter or year-over-year growth rates; (3) provide more reliable short-term forecasts; and (4) enable meaningful international comparisons.

Despite growing recognition of the importance of seasonal adjustment in macroeconomic analysis, two long-standing issues persist in both academic and applied settings: (1) an excessive focus on the application of internationally recognised software tools (such as X-12-ARIMA, X-13ARIMA-SEATS), with insufficient attention paid to the underlying methodology of time series decomposition and seasonal adjustment; and (2) a lack of comprehensive and systematic monographs on the core techniques of time series decomposition. These shortcomings have, to some extent, hindered research progress and limited the wider application of seasonal adjustment techniques.

Moreover, seasonal adjustment is inherently a complex analytical task. A proper understanding of the mathematical and statistical principles underlying the adjustment process is essential. In practical applications, data may exhibit outliers, irregular fluctuations, or structural breaks (such as abrupt regime changes or policy interventions), which must be diagnosed through theoretical insight and addressed through expert customisation, rather than relying solely on default software outputs. Different data characteristics, such as stationarity, trend strength, and noise levels, necessitate the application of alternative adjustment approaches. Relying exclusively on fixed procedures or default parameter settings may fail to capture the actual patterns in the data.

Seasonally adjusted data should reflect the genuine movements of economic activity, not merely the artefacts of an automated adjustment process. When results deviate from expectations, such as residual seasonality persisting or the trend component appearing overly smoothed, it becomes necessary to fine-tune parameters (eg, filter length or seasonal frequency) or consider alternative model specifications. The choice among additive or multiplicative models, ARIMA or regARIMA models, and other specification details depends on a clear understanding of the assumptions and conditions under which each method is appropriate. Only through such understanding can practitioners correctly interpret automatically generated outputs and exercise sound judgment.

Thus, the main motivation for this book is to provide a systematic exposition of the theoretical foundations of time series decomposition and seasonal adjustment, to equip readers with accurate and practical knowledge of statistical methods, and to conduct a critical evaluation of widely used tools and to share the author's experience with their application in real-world settings.

1.3 The Focus of This Book

Time series decomposition represents a critical area in modern statistical and econometric analysis. Much like a master watchmaker meticulously disassembling a complex mechanism, time series decomposition separates economic data into its unobservable components – trend, cycle, seasonal, and irregular fluctuations – providing a clearer view of the multiple dimensions of economic dynamics. At the heart of this analytical framework lies seasonal adjustment, a foundational technique that has become indispensable in both statistical and econometric practice. By isolating and removing predictable seasonal patterns from time series data, seasonal adjustment enables analysts and policymakers to look beyond the noise of recurring fluctuations, allowing for a more accurate identification of the underlying trends and cyclical patterns in economic indicators.

The principal objective of seasonal adjustment is to eliminate distortions caused by seasonal effects, thereby revealing the genuine underlying movements and trends within a time series. Although seasonal variation is typically predictable, it can obscure the detection of important non-seasonal components, such as long-term growth, short-term volatility, and irregular shocks. By addressing this issue, seasonal adjustment ensures that the remaining data more accurately reflects the true evolution of economic activity. This, in turn, enables policymakers, researchers, and business leaders to conduct more reliable evaluations and make sound, forward-looking decisions.

1.3.1 Methodologies for Time Series Decomposition and Seasonal Adjustment

Among the most widely used tools for seasonal adjustment today are X-13ARIMA-SEATS and TRAMO-SEATS. X-13ARIMA-SEATS builds upon the traditional X-11 moving average algorithm, employing a set of fixed filters, specifically, moving averages (MAs), to iteratively decompose a time series into its trend, seasonal, and irregular components. In addition to the classical X-11 procedure, X-13ARIMA-SEATS incorporates the SEATS method as an alternative decomposition approach, thereby integrating both filter-based and model-based perspectives within a unified framework.

In contrast to the empirical filter-based decomposition employed by X-13ARIMA-SEATS, TRAMO-SEATS adopts a model-based methodology. It utilises statistical modelling to estimate trend and seasonal components with high precision, providing a structured and theoretically grounded decomposition of time series data. TRAMO-SEATS represents a significant application of econometric methods in the field of time series decomposition, combining the theoretical foundations of statistics and econometrics to accurately identify and extract trend, seasonal, and stochastic components.

The origins of time series decomposition and seasonal adjustment can be traced back to the 19th century. Dutch meteorologist Buys-Ballot (1847) is regarded as one of the earliest contributors to the field due to his study of periodic signals, including seasonal variations.[1] In his 1847 work, Buys-Ballot modelled trends using polynomials, incorporated seasonal patterns using dummy variables, and implicitly applied linear regression techniques for parameter estimation, thereby analysing the cyclical behaviour of temperature data.

In the latter half of the 19th century, economists such as Jevons (1862)[2] began to recognise the potential of cyclical analysis for understanding and predicting economic crises. However, the lack of precise definitions for 'cyclical fluctuations' and the limited computational tools available at the time hindered the development of effective statistical methods for estimating latent trend-cycle components. These limitations delayed progress in the formal application of decomposition techniques for economic analysis.

In the early 20th century, a significant portion of time series research focused on estimating seasonal variation. Persons (1919)[3] was undoubtedly one of the first statisticians to establish a formal definition of time series components and develop an algorithm for their decomposition, the 'method of link relatives', for estimating those components. The systematic study of seasonal adjustment began primarily with the work of Frederic R. Macaulay, whose pioneering efforts in the early 20th century laid the foundation for modern approaches to time series decomposition. Macaulay (1931)[4] of the National Bureau of Economic Research (NBER), who refined the decomposition process and introduced the *ratio-to-moving-average* procedure in the 1920s, a method whose variants remain widely used today. During the same period, researchers such as King (1924)[5] continued to advocate for *freehand curve fitting* as a means to eliminate unwanted components from data.

In the 1920s, scientists at the U.S. Census Bureau developed some of the earliest formal methods for seasonal adjustment, laying the groundwork for future advancements. The 1950s marked a breakthrough, coinciding with the advent of digital computing. The emergence of the X-11 method signalled a shift toward the systematisation and standardisation of seasonal adjustment techniques. In 1954, the previously manual ratio-to-moving-average procedures were replaced by a computer-based program developed by the U.S. Census Bureau, Census I, that was subsequently revised and expanded in 1955 into Census II.[6]

Since 1955, a series of improvements has been made to these procedures, evolving through X-1 to the well-known X-11, and further extended to X-11-ARIMA, X-12-ARIMA, and ultimately the current X-13ARIMA-SEATS. With each iteration, seasonal adjustment methods have become increasingly refined, offering greater precision and capacity to handle complex data structures.

X-13ARIMA-SEATS, developed by the U.S. Census Bureau, is one of the leading methodologies in contemporary seasonal adjustment. It integrates the classical X-11 filtering technique with SEATS, a model-based approach for signal decomposition. The method features powerful capabilities for handling intricate seasonal patterns, automatically detecting outliers, and adjusting for calendar effects. It is particularly well-suited for high-frequency economic data with stable seasonal structures, such as monthly retail sales or industrial production indices.

By contrast, TRAMO-SEATS is an ARIMA-based seasonal adjustment method[7] developed and maintained by the Bank of Spain. The initial development of the TRAMO (Time Series Regression with ARIMA noise, Missing values, and Outliers) module began in the early 1980s, and it was followed by the development of SEATS in the late 1980s. In 1994, the Bank of Spain officially released TRAMO-SEATS version 0.1, marking its debut as a publicly available official tool. The software has continued to evolve, with updated versions, such as TRAMO-SEATS+, being adopted by institutions including Eurostat as one of the official methodologies for seasonal adjustment.

Compared to X-13ARIMA-SEATS, TRAMO-SEATS emphasises a more model-driven approach. The process begins with rigorous time series modelling via the TRAMO component, which carries out pre-adjustment tasks such as outlier detection, missing value imputation, and calendar effect modelling. The adjusted series is then decomposed into components using the SEATS methodology. This approach is theoretically more robust, with clearer assumptions about the data-generating process. It is particularly well-suited for handling the complex and diverse holiday patterns and calendar effects found in European countries.

Both X-13ARIMA-SEATS and TRAMO-SEATS have had a significant influence in both theoretical and applied domains, forming the core of the modern seasonal adjustment toolkit available to statisticians.

1.3.2 X-13ARIMA-SEATS Methods

X-13ARIMA-SEATS is an empirical, filter-based decomposition method. Its foundation lies in the X-11 seasonal adjustment procedure developed by the U.S. Census Bureau in the 1950s. The X-11 method utilises moving averages and iterative algorithms to decompose a time series into its three fundamental components: trend, seasonality, and irregularity.

However, the use of moving averages introduces a limitation at the endpoints of the series. Because moving averages rely on fixed-length windows, the start and end values of a series may not be fully covered, leading to what is known as the endpoint problem or data loss at the edges. To address this issue, Statistics Canada introduced ARIMA modelling, leading to the development of X-11-ARIMA.

The ARIMA (AutoRegressive Integrated Moving Average) model addresses the endpoint issue by statistically modelling the time series and using extrapolation and interpolation to estimate missing values at the boundaries. Specifically, the autoregressive (AR) component predicts current values based on past observations (lags). In contrast, the moving average (MA) component corrects these predictions using past forecast errors, thus reducing the influence of noise. The integrated (I) component applies differencing to transform nonstationary series into stationary ones, allowing for more accurate modelling.

Unlike fixed-window moving average methods, ARIMA does not suffer from direct data loss at the endpoints and provides more flexibility. By adjusting the model parameters (p, d, q) (P, D, Q), ARIMA can accommodate a wide variety of time series characteristics. This enables more effective modelling of trends and seasonality, especially at the edges of the series, where conventional moving average filters may fail.

The ARIMA model predicts missing values at the endpoints of a time series through its autoregressive (AR) component, which extrapolates from known data points. For example, by utilising lag relationships derived from initial observations, the model can perform backcasting, that is, backwards prediction, using both the AR and MA components to estimate the earliest missing values in the series. Similarly, ARIMA generates forecasts of future values based on the latest available data and their lags, a process known as forecasting, to recover values at the end of the series. This approach overcomes the limitations of traditional moving average filters, which rely solely on data within a fixed window.

To evaluate model performance, the imputed (completed) series can be compared with actual observations. If discrepancies arise, ARIMA model parameters can be adjusted to improve both the fit and predictive accuracy.

Fundamentally, ARIMA is a univariate time series model; it relies solely on the historical values of the series itself to predict future behaviour. However, in real-world applications, many time series (eg, consumption, employment, traffic volume) are influenced by external factors, such as the following:

1. Calendar effects (eg, the impact of Christmas or the Lunar New Year on retail sales),
2. Policy interventions (eg, interest rate changes affecting economic indicators), or
3. Exogenous shocks (eg, natural disasters or pandemics).

The standard ARIMA model is not inherently structured to directly account for external influences or exogenous variables. This limitation is addressed by the regARIMA model, which integrates regression analysis with ARIMA errors. While ARIMA leverages internal autocorrelation and past trends for forecasting, regARIMA enhances predictive capability by incorporating exogenous variables into the model. These variables allow for the explicit modelling of external effects, such as holidays or policy changes, while still capturing the underlying autocorrelated structure of the time series.

By combining the strengths of regression analysis and ARIMA modelling, regARIMA significantly enhances the handling of complex time series, particularly in seasonal adjustment, outlier detection, and modelling structural breaks. Its flexibility makes it a powerful tool in both official statistics and applied economic analysis.

Methodologically, the design of the regARIMA model, which applies an ARIMA model to the residuals of a regression model, holds particular significance. The key reason is that after introducing a regression component in seasonal adjustment models, the residuals may exhibit temporal dependence that requires further modelling to eliminate.

In classical regression models, residuals are typically assumed to be independently and identically distributed (i.i.d) normal random variables, ie, white noise with no temporal correlation. However, in practice, residuals often exhibit temporal patterns including trends, seasonality, or autocorrelation structures. If residuals are not white noise but instead exhibit time-dependent structures (eg, autocorrelation or seasonal patterns), a standard regression model alone may fail to adequately capture the underlying behaviour and dynamic properties of the data.

By modelling the residuals with an ARIMA process, these temporal dependencies can be explicitly accounted for, improving the overall modelling performance. The ARIMA component captures autoregressive (AR), differencing (I), and moving average (MA) characteristics of the residuals, effectively removing their serial correlation and yielding residuals closer to white noise.

When residuals no longer contain significant temporal dependence, model assumptions, such as the independence and normality of errors, are more likely to hold, thereby making parameter estimation and forecasting more reliable.

Within the regARIMA framework, the regression part models the structural effects related to explanatory (exogenous) variables, focusing on how these external factors influence the dependent variable. For the portion of variation unexplained by the regression, the ARIMA model further explores the residual dynamic patterns (such as trends, seasonality, or stochastic fluctuations), thus separating the deterministic structural components from the stochastic noise.

In the fields of economics and social sciences, many time series contain both structural variations driven by explanatory variables, such as policy factors and macroeconomic indicators, and time-dependent random fluctuations, such as market inertia effects. In this complex context, the regARIMA model achieves deep exploration of the dynamic characteristics of time series by modelling the temporal dependence structure of residuals. Specifically, the regression component captures the systematic effects of exogenous explanatory variables on the series, while the ARIMA component models the regression residuals using autoregression, differencing, and moving average operations to identify and remove temporal dependencies. This approach effectively avoids the risk of omitting important dynamic information, significantly enhancing the model's predictive accuracy and explanatory power. For this reason, regARIMA has become a powerful and widely applied tool for time series modelling and forecasting across economics, meteorology, social sciences, and other disciplines that deal with complex dynamic data.

On the other hand, the X-13ARIMA-SEATS method combines traditional moving average decomposition techniques with modern statistical modelling to systematically and efficiently decompose time series data. At its core, it employs a set of fixed moving average (MA) filters in an iterative process to decompose the series into three fundamental components: trend, seasonal, and irregular. Symmetric linear filters are applied to the central part of the series to ensure smoothness and stability, while asymmetric filters are employed at the series' endpoints to address boundary data limitations and prevent information loss. These filters can essentially be viewed as a special form of local regression: Within a sliding window, weighted or unweighted least squares polynomial fitting is applied to smooth the data point at the centre of the window. This approach enables X-13ARIMA-SEATS to flexibly handle complex and evolving seasonal patterns while effectively separating trend and irregular components, thereby improving the accuracy and stability of seasonal adjustment.

Notably, X-13ARIMA-SEATS not only inherits the classical strengths of the X-11 moving average method but also incorporates the ARIMA model-based signal extraction technique known as SEATS. This combination preserves the intuitive and practical advantages of filter decomposition while adding the theoretical rigour of model-driven approaches, enabling superior handling of economic data with complex seasonal and calendar effects. Due to its robust functionality and flexible applicability, X-13ARIMA-SEATS has become a standard seasonal adjustment tool widely adopted by statistical agencies, central banks, and various research institutions.

1.3.3 TRAMO-SEATS Methods

Unlike the empirical, filter-based decomposition approach of X-13ARIMA-SEATS, TRAMO-SEATS is a model-based time series decomposition method that achieves high-precision estimation of seasonal and trend components by leveraging a theoretical modelling framework. It offers a valuable structural decomposition of time series data by integrating statistical and econometric foundations, thus representing an important application of econometric methods in time series decomposition. TRAMO-SEATS precisely separates a time series into trend, seasonal, and irregular components through rigorous modelling.

The method is grounded in the Box–Jenkins ARIMA modelling theory, fitting an ARIMA model to the observed time series and then decomposing the series into its underlying components: trend, seasonal, and random fluctuations. TRAMO-SEATS consists of two main parts: TRAMO and SEATS.

TRAMO (Time Series Regression with ARIMA Noise, Missing Observations, and Outliers) serves as the preprocessing and adjustment phase for seasonal adjustment. Its core functions include the following:

1. Outlier detection and adjustment: Identifying and correcting abnormal observations caused by extraordinary events such as shocks or holiday effects. It also employs regression models to capture the effects of exogenous variables such as policy changes or weather conditions.
2. Missing data imputation: Employing ARIMA-based estimation methods to fill in missing values, ensuring a complete and consistent time series.
3. Model selection: Automatically selecting the optimal ARIMA model parameters (p, d, q), similar to the ARIMA and regARIMA components in X-12-ARIMA and X-13ARIMA-SEATS preprocessing.

The SEATS component forms the core of the TRAMO-SEATS methodology. Based on the ARIMA model identified by TRAMO, SEATS performs the actual decomposition of the time series into its constituent signals, trend, seasonal, and irregular using econometric signal extraction theory. SEATS operates within the framework of the BLUE, extracting the trend and seasonal components directly from the ARIMA model parameters. It applies specially designed filters to isolate these components, enabling the identification of stable seasonal and trend patterns while segregating random noise.

In contrast to X-12-ARIMA and X-13ARIMA-SEATS, which rely on traditional moving average filters for decomposition, SEATS employs a fully model-based approach. This method depends heavily on the precise ARIMA model fit and advanced signal extraction techniques, including algorithms such as Kalman filtering, which optimise the decomposition process. As a result, SEATS produces smoother and more stable estimates of trend and seasonal components, yielding decomposition outputs that are not only statistically rigorous but also scientifically robust.

TRAMO-SEATS integrates econometric regression analysis with time series modelling, placing strong emphasis on modelling exogenous variables, such as holiday effects, and optimally handling stochastic errors. In macroeconomic data analysis, such as GDP, employment rates, and industrial production, seasonal fluctuations can obscure the underlying economic trends. TRAMO-SEATS provides a theoretically rigorous and widely applicable toolset to address these challenges.

Moreover, TRAMO-SEATS employs automated modelling techniques from econometrics, using information criteria such as the Akaike Information Criterion (AIC) and the Bayesian Information Criterion (BIC) to select the optimal model, thereby minimising the need for manual intervention. This makes it a high-precision decomposition method grounded firmly in econometric and time series theory. Its scientific foundation, including ARIMA modelling, signal extraction, and advanced filtering techniques, establishes TRAMO-SEATS as an essential instrument in modern time series analysis and seasonal adjustment.

TRAMO-SEATS is particularly well-suited for analysing complex economic and financial datasets and plays a critical role in policy formulation. Notably, it has been designated as the primary seasonal adjustment software by the European Statistical Office (Eurostat).

Understanding the core concepts and operational details of these methods is a key focus of this book. Seasonal adjustment is not a simple 'black box' process; improper adjustment can lead to misinterpretation of trends and cycles, potentially causing policy errors. This book will comparatively analyse the theoretical underpinnings, implementation pathways, and appropriate use cases of TRAMO-SEATS and other methods, with a focus on their conceptual frameworks and practical applications.

For Chinese readers, particular attention will be given to adjusting for culturally specific holidays, such as the Lunar New Year, Dragon Boat Festival, and Mid-Autumn Festival. The book will guide readers in selecting the most appropriate tools for their specific issues and provide detailed explanations of user-defined regressors, modelling philosophy, parameter setting principles, and result interpretation. This tailored approach aims to help Chinese analysts address the most pressing challenges they face in seasonal adjustment.

X-13ARIMA-SEATS and TRAMO-SEATS are currently the two most widely used software tools for seasonal adjustment. Advances in computational tools and algorithms have further expanded the scope and accessibility of seasonal adjustment techniques.

JDemetra+ (JD+ for short) is an open-source, platform-independent seasonal adjustment and time series analysis tool. JD+ was developed jointly by the National Bank of Belgium, the Deutsche Bundesbank, and Eurostat. Since its official release in 2012 (version 1.0), JD+ has undergone continuous updates and iterations, introducing new features such as support for the X-13ARIMA-SEATS method, an improved user interface, and enhanced processing capabilities. It has released multiple major versions (eg, v2.x, v3.x) and has rapidly become a core tool for official statistical agencies and central banks across Europe and globally.

JD+ implements the two principal seasonal adjustment methodologies – TRAMO-SEATS and X-13ARIMA-SEATS – allowing users to select either method for their seasonal adjustment needs.

This book provides an in-depth introduction and analysis of X-13ARIMA-SEATS and TRAMO-SEATS, while offering brief overviews of other seasonal adjustment methods in later chapters.[8]

1.4 Purpose, Structure, and Scope of This Book

This book aims to provide a systematic introduction and explanation of the theories and methods related to time series decomposition and seasonal adjustment, as well as the application of relevant statistical software. It intends to equip both users and developers with the necessary background to produce high-quality data and address various issues that may arise during the adjustment process. In the context of deepening globalisation, economic transformation, and upgrading, a precise understanding and effective use of these internationally advanced seasonal adjustment tools are essential to enhance macroeconomic analysis capabilities and improve economic forecasting accuracy.

The book is organised into nine chapters grouped into four parts. Following the introductory Chapter 1, Part I – comprising Chapters 2 to 4 – presents the general conceptual framework and the theoretical foundations of time series decomposition and seasonal adjustment.

Chapter 2 provides an overview of the time series decomposition methods based on the X-11 family, highlighting the development trajectory centred around two major objectives:

improving decomposition accuracy and enhancing result stability. This evolution not only reflects advances in statistical theory but also responds to the practical demands of application.

Chapter 3 discusses the general theoretical framework of seasonal adjustment, defining the various components of a time series, such as trend-cycle, seasonality, outliers, and trading day effects, and explaining their contributions and implications for short-term economic assessment. It also covers preprocessing techniques for seasonal adjustment, along with possible models for long-term trends, seasonal components, and irregular fluctuations.

Chapter 4 focuses on the moving average methods underlying the X-11 seasonal adjustment procedure, emphasising both time domain and frequency domain properties of moving averages, as well as the iterative process of X–11. These chapters provide the theoretical foundation for the subsequent discussions on the X-11 and SEATS methods, as well as practical guidance on model selection and diagnostic techniques.

Part II of the book, covering Chapters 5 and 6, is dedicated to the procedures and methods used in X-13ARIMA-SEATS. Chapter 5 provides a detailed introduction to the application of X-13ARIMA-SEATS, guiding users from basic operations to advanced configurations. It aims to help users master this powerful seasonal adjustment tool.

Chapter 5 specifically covers ARIMA and regARIMA specifications, identification of various outliers, level shifts, and Easter effects and their treatment. Special attention is given to the construction of user-defined regression variables and regressors, such as those for the Chinese Lunar New Year. By choosing an appropriate running environment and conducting rigorous result diagnostics, users can efficiently complete time series analysis tasks.

Chapter 6 focuses on several other important topics related to seasonal adjustment, including the required length of series for adjustment, the issue of 'seasonal adjustment and aggregation' and temporal consistency for annual data, the 'direct versus indirect' adjustment problem, quality assessment of adjusted data, inherent revisions during the adjustment process, their impacts, and some non-statistical criteria for strategy selection.

Part III, comprising Chapters 7 and 8, centres on the TRAMO-SEATS modelling approach.

Chapter 7 introduces the pre-adjustment methods of the TRAMO model and the time series decomposition approach used in SEATS. TRAMO focuses on outlier detection, missing value treatment, and ARIMA model fitting for pre-adjustment, while SEATS decomposes the series using ARIMA models and Wiener–Kolmogorov filtering. The chapter also details the practical implementation and functionalities of the TRAMO-SEATS program.

Chapter 8 offers a comprehensive introduction to the platform JD+, which implements TRAMO-SEATS and X-13ARIMA-SEATS methods. It covers the core functions, operational workflow, and practical applications of the software. JD+ provides a complete seasonal adjustment solution, including data preparation, pre-adjustment modelling, result analysis, and user communication. Emphasis is placed on how to perform seasonal adjustment for single and multiple series, along with quality diagnostics and optimisation recommendations to ensure the reliability of results. Additionally, the chapter addresses data release strategies and user communication methods to help statistical agencies efficiently produce and disseminate seasonal adjustment data. Although JD+ includes both leading seasonal adjustment tools, X-13ARIMA-SEATS and TRAMO-SEATS, Chapter 8 focuses exclusively on the use of TRAMO-SEATS.

Finally, Part IV of the book, Chapter 9, briefly evaluates the contributions and limitations of traditional seasonal adjustment methods, then focuses on the modern challenges in time series decomposition and seasonal adjustment. It explores future directions involving the integration of conventional and contemporary techniques.

The mission of this book is to provide an in-depth analysis of X-13ARIMA-SEATS and TRAMO-SEATS methods, helping readers systematically grasp the core concepts, modelling principles, and operational procedures of modern seasonal adjustment techniques, enabling the correct application of these two tools. Unlike typical statistical software manuals, this book emphasises methodological learning. It thoroughly introduces core theories such as moving averages and spectral decomposition, providing a solid theoretical foundation for practical operation. From our perspective, software is merely a tool, while methodology is the core of analysis.

Moreover, this book differs from traditional theoretical textbooks by not only discussing theory but also offering extensive practical operation methods and model selection advice. It proposes targeted and systematic solutions for complex data scenarios, such as seasonal breaks, indirect seasonal adjustment, user-defined regression variables, and regressors for Chinese festivals, including the Lunar New Year, Mid-Autumn Festival, and Dragon Boat Festival. The structure is designed to guide beginners through a gradual learning process while also meeting the technical details and flexibility requirements of advanced users.

Additionally, the book provides relevant program code examples to help readers, especially beginners, build a deeper and more intuitive understanding of the methods. Topics are presented in a pedagogical style to facilitate accessibility for both non-experts and experienced users. A complex formalism is introduced without compromising the technical understanding of beginners and non-specialists.

1.5 Summary and Conclusions

Seasonal adjustment serves not only as a statistical tool to refine time series data but also as a critical foundation for improving the accuracy of economic interpretations and the effectiveness of policy responses. With increasing demands for more precise and timely macroeconomic monitoring, the impact of seasonal fluctuations on short-term indicators has become more pronounced. This underscores the need for a deeper theoretical understanding and more proficient application of seasonal adjustment techniques.

Among these techniques, X-13ARIMA-SEATS and TRAMO-SEATS stand out as internationally recognised standards. Their proper use enhances the quality, stability, and interpretability of adjusted time series, bridging the gap between statistical analysis and policy formulation.

Despite progress, there is still a lag in certain aspects of seasonal adjustment, particularly in building a localised theoretical system, adapting international tools to national conditions, and developing expert capacity. Thus, establishing a modern seasonal adjustment framework that meets both domestic statistical requirements and global standards is of pressing importance.

This book is designed to meet that need. It offers a systematic introduction to the principles, models, and implementation of seasonal adjustment, with a special focus on X-13ARIMA-SEATS and TRAMO-SEATS.

Through theoretical discussion, hands-on modelling guidance, and practical case studies, readers will acquire the skills necessary to conduct seasonal adjustment, evaluate results, and apply them effectively in economic research and policymaking.

Notes

1 Box, G. E. P., Jenkins, G. M., & Reinsel, G. C. (2008). "Time Series Analysis: Forecasting and Control", 4th ed. Wiley; and Hassani, H. (2007). "Time Series Decomposition and Forecasting: A Review". *Journal of Computational Methods in Sciences and Engineering*, 7(2), 123–143.
2 Jevons, W. S. (1862). "On the Study of Periodic Commercial Fluctuations". Report of the British Association for the Advancement of Science, Cambridge.
3 Persons, W. M. (1919). "Indices of Business Conditions". *Review of Economics and Statistics*, 1(1), 5–107.
4 Macaulay, F. R. (1931). "The Smoothing of Time Series". National Bureau of Economic Research (NBER), New York.
5 King, W. I. (1924). "Elements of Statistical Method". The Macmillan Company, New York.
6 Shiskin, J., & Eisenpress, H. (1957). "Seasonal Adjustments by Electronic Computer Methods". *Journal of the American Statistical Association (JASA)*, 52(280), 415–449. https://doi.org/10.1080/01621459.1957.10501402.
7 Burman, J. P. (1980). "Seasonal Adjustment by Signal Extraction". *Journal of the Royal Statistical Society. Series A (General)*, 143(3), 321–337. https://doi.org/10.2307/2982132; Gómez, V., & Maravall, A. (1997). "Programs TRAMO and SEATS Instructions for the User (Beta Version)". Working Paper No. 97001. Banco de España, Madrid, Spain.
8 For other seasonal adjustment tools, we will briefly introduce them in Chapter 9.

2

The Evolution of the X-11 Family Seasonal Adjustment Methods

2.1 Introduction

The X-13ARIMA-SEATS method represents the latest development in the X-11 family of seasonal adjustment techniques, following in the lineage of X-11, X-11-ARIMA, and X-12-ARIMA. To understand and effectively apply X-13ARIMA-SEATS, it is essential to begin with a solid grasp of the X-11 family of methods, especially its decomposition based on moving averages. While X-13ARIMA-SEATS continues to rely on the X-11 approach for time series decomposition, it also integrates the SEATS methodology, which will be discussed in in detail in Chapter 7. This chapter focuses on the historical evolution and application of the X-11 family of seasonal adjustment techniques.

The structure of this chapter is as follows: Section 2.2 reviews the theoretical foundations of seasonal adjustment; Section 2.3 traces the development of moving average techniques within the original X-11 framework; Section 2.4 discusses the integration of statistical modelling in X-11-ARIMA; Section 2.5 explores the enhancements brought by X-12-ARIMA, particularly the inclusion of regression and diagnostic tools; Section 2.6 covers the unification of parametric and nonparametric approaches in X-13ARIMA-SEATS; finally, Section 2.7 presents a summary and concluding remarks. Through this structure, the chapter aims to provide a clear understanding of how seasonal adjustment techniques have evolved in response to increasing analytical demands and advances in statistical methodology.

2.2 Theoretical Foundations of Seasonal Adjustment Methods

Seasonal adjustment is a statistical technique that decomposes an observed time series, $\{y_t\}$, into several unobservable components: trend, cycle, seasonal, and irregular variations. The primary goal of seasonal adjustment is to isolate and remove seasonal effects from raw data, thereby revealing the underlying long-term trends and short-term dynamics more clearly. As such, time series decomposition has become an essential analytical tool in both economics and statistics, particularly in macroeconomic analysis and short-term forecasting.

2.2.1 Origins and Early Development of Time Series Decomposition

The conceptual roots of time series decomposition can be traced back to astronomy, where early models were developed to explain the long-term trends and cyclical variations in

DOI: 10.1201/9781003716655-2

planetary motion.[1] As early as the 17th century, astronomers observed that the measured positions of planets deviated slightly from their theoretical orbits.[2] These deviations were considered to be composed of a 'true' theoretical path and random fluctuations. This conceptual framework, distinguishing between systematic and irregular components, was gradually accepted in astronomical research.

By the late 18th and early 19th centuries, this analytical framework had attracted the attention of leading mathematicians, like Euler, Lagrange, and Laplace.[3] In 1801, the British astronomer William Herschel[4] published a pioneering study that identified a correlation between the periodicity of sunspots and fluctuations in wheat prices, marking one of the earliest attempts to link natural cycles with economic data. In 1847, the Dutch meteorologist Buys-Ballot applied polynomial smoothing techniques to extract trends and seasonal indices, implicitly employing linear regression methods to estimate parameters.[5] During this period, the focus was primarily on eliminating short-term fluctuations to reveal more persistent long-term trends.

At the same time, developments in mathematics provided tools for advancing time series analysis beyond mere visual inspection. Of particular importance was the work of Jean-Baptiste Joseph Fourier, who introduced the idea of decomposing functions into sums of trigonometric terms, now known as Fourier analysis.[6] His contributions laid the foundation for harmonic analysis and, later, spectral analysis,[7] which became central to the study of stochastic processes once the concept of randomness was formalised.

These ideas soon migrated into economics, a field that, like the natural sciences, sought deterministic laws governing social and economic phenomena. The identification and analysis of cycles, daily, weekly, quarterly, or annual, became a major focus. Analysts aimed not only to understand cyclical behaviour but also to remove such patterns from the data to uncover non-cyclical or irregular movements. This line of inquiry made a significant contribution to early efforts aimed at explaining and predicting economic crises.[8]

By the late 19th and early 20th centuries, increasing attention was given to the decomposition of economic time series, the estimation of individual components, and the formal definition of those components.[9] In 1905, Lucien March introduced an early classification of variations, distinguishing between annual variations, multi-year cycles (eg, decennial changes), long-term shifts, and sub-annual fluctuations.[10]

Among the pioneers of empirical techniques were the Mabille sisters,[11] whose innovative work attempted to isolate turning points in time series using correlograms, a precursor to modern autocorrelation analysis. However, perhaps the most influential and systematic early contribution was made by Wesley Mitchell Persons in 1919.[12] His decomposition model sought to *define*, *formalise*, and *estimate* the unobservable components of economic time series using structured graphical and statistical methods. His work remains a cornerstone in the evolution of seasonal adjustment techniques.

Persons proposed that any time series could be conceptually decomposed into four distinct unobservable components, an idea that continues to underpin modern statistical practice:

1) Trend component (T_t): representing the long-term direction of the series, such as overall economic growth;
2) Cyclical component (C_t): capturing medium-term fluctuations, often associated with business cycles;
3) Seasonal component (S_t): accounting for regular, predictable patterns that recur annually;
4) Irregular component (I_t): representing random or unpredictable influences, often due to extraordinary events like wars or natural disasters.

These components were typically combined using either an additive model, $Y_t = T_t + C_t + S_t + I_t$, or a multiplicative model,

$$Y_t = T_t \times C_t \times S_t \times I_t \text{ or.}$$
$$Y_t = T_t \times (1+C)_t \times (1+S_t) \times (1+I_t)$$

These formulations laid the theoretical foundation for all subsequent developments in seasonal adjustment methodology. The choice between additive and multiplicative forms remains an essential modelling decision in practical applications, depending on the characteristics of the observed series.

At the time, these definitions and decomposition models were largely accepted by the academic community with minimal theoretical debate. Meanwhile, several other key concepts also gained broad acceptance. These include (1) recognising that seasonality may evolve; (2) the principle that all components should be jointly considered when estimating the seasonal component; and (3) the recognition that simple or explicit mathematical formulas cannot always represent trend and cycle components. It was also acknowledged that special attention must be given to atypical or outlier observations. However, the focus during this period remained primarily on the practical techniques of seasonal adjustment and cycle extraction, rather than on formal theoretical development.

2.2.2 Early Estimation Methods for Seasonal Adjustment

In terms of practical techniques for seasonal adjustment and cycle extraction, two distinct estimation methods[13] were prevalent during the early development of the field:

(1) The 'Link Relative' Method: Developed by W.M. Persons in the 1910s, the *link relative* method gained widespread popularity among statistical economists. This method involved calculating the ratio of each month's value to the preceding month's value. These monthly ratios were then organised into 12 separate series, one for each month of the year. For each series, the median was computed. These medians were multiplicatively linked, starting with a base of 100 for January. The resulting seasonal factors were then adjusted using a correction factor to ensure internal consistency.[14]

(2) The Moving Average Method:[15] First employed by the U.S. Federal Reserve as early as 1922[16] and later popularised by Frederick R. Macaulay,[17] this approach was based on computing a 12-month centred moving average to estimate the trend component. The difference between the raw data and this estimated trend yielded an initial estimate of the seasonal component. To reduce irregular fluctuations, medians (or means) were computed for each month's component, and the resulting values were scaled so that their average equalled 1, producing the final seasonal factors.

Although both methods were widely used at the time, they were not without theoretical criticism. Notably, Slutsky and Yule pointed out that moving averages could introduce artificial periodicity into the data.[18] Irving Fisher lamented the reliance on empirical or ad hoc procedures due to the lack of rigorous mathematical tools.[19]

Perhaps the most significant development of this period came in the late 1920s with the emergence of *autoregressive* (AR)[20] and *moving average* (MA) models[21] for time series analysis. These models marked the first major shift away from deterministic frameworks, introducing stochastic processes into economic statistics. However, they were not immediately applied effectively to seasonal adjustment and only gained traction many years later.

In the 1930s, regression-based approaches to seasonal adjustment experienced mixed success. On one hand, these methods relied on additive decomposition of the original series, or simple transformations thereof, and employed parametric functions to model each component, with parameters estimated using Ordinary Least Squares (OLS). On the other hand, the practical application of these techniques was limited by the difficulty of specifying suitable models and the strong assumptions required to identify unobservable components. These challenges proved to be major obstacles to the adoption of regression-based seasonal adjustment at the time.[22]

2.2.3 The Emergence of Modern Seasonal Adjustment Methods

Following World War II, advancements in scientific information technology played a pivotal role in improving and popularising seasonal adjustment techniques. The early 1950s witnessed two breakthroughs that shaped the development of modern methods:

(1) The introduction of exponential smoothing techniques, and
(2) The adoption of computers in statistical analysis.

The implementation of exponential smoothing significantly simplified the estimation process, while the growing availability of computing power allowed researchers to experiment with and refine more complex procedures. Tasks that once required days of manual computation could now be completed in seconds. This computational leap enabled large-scale testing of alternative algorithms across many time series, paving the way for data-driven selection of optimal methods.

It was in this context that modern seasonal adjustment methods began to take form. The Census I method, introduced in 1954 by the U.S. Census Bureau, marked a departure from earlier techniques such as Macaulay's, which suffered from data loss at the endpoints due to centred 12-term moving averages. To address this limitation, Census Method I incorporated forecasting and backcasting steps into the decomposition framework, extending the series beyond its endpoints to fill in missing values at the beginning and end.

Census Method I was soon succeeded by Census Method II in 1955, developed by Julius Shiskin. This version effectively automated earlier manual procedures, adapting them into an electronic format suitable for computer implementation. While Census II improved on its predecessor, it was not without criticism. Analysts raised concerns about its handling of various components and irregularities in the data.

In response to these critiques, a more sophisticated iteration, X-11, was developed and released in 1965. X-11 represented a major evolution in seasonal adjustment methodology and remained the most widely used approach throughout the 1970s and 1980s. Its development reflected a critical turning point: the transition from heuristic, manually intensive methods to automated, model-based procedures capable of robust adjustment across diverse economic and statistical datasets.

2.3 The X-11 Method: A Milestone in Moving Average Techniques

The X-11 method, developed by the U.S. Census Bureau during the 1950s and officially released in 1965, stands as a milestone in the evolution of seasonal adjustment techniques. Throughout the 1960s, X-11 became the standard tool for decomposing and adjusting economic time series, particularly in identifying and removing seasonal components. As a refined successor to the earlier Census Methods I and II, the X-11 procedure was developed under the leadership of Julius Shiskin and marked a significant advancement in automating seasonal adjustment.

Originally, X-11 was referred to as an extension of 'Census Method II', itself a refinement of the early mechanical and heuristic approaches to seasonal decomposition. The X-11 methodology formalised the use of moving averages, iterative filtering, and diagnostic tools, laying the groundwork for future developments in seasonal adjustment software.

2.3.1 Census Method I: The Initial Application of Moving Averages

In 1954, Julius Shiskin introduced Census Method I, which represented the first systematic application of moving averages in seasonal adjustment. The method proposed estimating seasonal factors by calculating seasonal-to-irregular ratios – ratios of the original series values to their corresponding 12-month centred moving averages, which were assumed to represent the trend-cycle component.

The seasonal component was then derived by smoothing these ratios using a set of moving averages tailored to reduce irregular fluctuations while preserving recurring seasonal patterns. In addition to this approach, an alternative adjustment was using a 5-term moving average instead of the 12-term one, generating a preliminary version of the seasonally adjusted series.

A key innovation of Census Method I was the integration of forecasting and backcasting to address endpoint data problems. This allowed for the estimation of missing values at the beginning and end of the series, enabling more reliable decomposition by extending the available data range. This concept would later become a cornerstone of subsequent seasonal adjustment procedures.

2.3.2 Census Method II: Improvements and Extensions

In 1957, under the leadership of Julius Shiskin, *Census Method I was subsequently refined and expanded into Census Method II.* This new version introduced several significant enhancements that laid the groundwork for modern seasonal adjustment techniques:

1. Extension of moving averages to the endpoints using asymmetric weights: One of the most critical innovations was the development of asymmetrically weighted moving averages, which enabled the estimation of seasonal components at the beginning and end of the series, traditionally problematic due to a lack of symmetric neighbouring data.
3. Smoothing of the trend-cycle component using a 15-term moving average: Replacing the 5-term moving average used in the second stage of Method I, Census Method II applied a 15-term weighted moving average to derive a smoother and more flexible trend-cycle estimate. While the second-round seasonal factors in Method I were optional alternatives, they became the preferred estimates in Method II, providing a more stable and refined seasonal decomposition.
5. Outlier adjustment before seasonal factor estimation: To improve the robustness of the seasonal factors, Method II introduced a pre-adjustment step in which extreme values in the monthly seasonal-to-irregular ratios were identified and corrected before further smoothing.

Part of the success of Census Method II stemmed from its practical usability. The method provided a graphical output of the decomposed components, facilitating visual inspection and evaluation of the results by users. Its algorithm was relatively robust and did not require users to possess a deep background in statistics or mathematics.

Seasonal adjustment, as implemented in Method II, served two main purposes:

(1) To estimate seasonal variations for use in production planning and inventory control;
(2) To remove seasonal effects from the overall variation so that underlying trends and cyclical movements, particularly those related to the business cycle, could be more clearly identified.

Following the release of Method II, Duane Evans at the U.S. Bureau of Labour Statistics developed a 'moving average ratio' method, which later influenced outlier treatment procedures in X–11.[23] Around the same time, the U.S. Census Bureau, under the direction of Harry M. Rosenblatt, began exploring parameter-based methods for seasonal adjustment. However, neither of these alternative approaches gained traction, partly due to the volatility of seasonal factor revisions they produced.[24]

The separation of seasonal and cyclical movements remained an important goal, especially in contexts where policymakers require timely assessments of economic conditions and short-term forecasts. Census Method II supported this by offering seasonal factor forecasts for up to one year ahead, allowing real-time seasonal adjustment of current data. Nonetheless, these forecasts often required substantial revisions as new data became available.

The X-11 method originated as a variant of Census Method II,[25] emerging from a series of 11 experimental versions labelled X-1 through X–11. Through continual iteration and refinement, versions such as X-3 and X-9 were selected as official releases. The final release of X-11 in 1965, benefiting from the growing availability of computing power, incorporated features like automatic trading day adjustment and outlier detection, enabling more sophisticated and automated seasonal adjustment.

Another notable enhancement in X-11 was the incorporation of regression techniques to estimate calendar effects (eg, trading days, fixed holidays, and moving holidays). These effects can be automatically adjusted within the X-11 software framework, and users can also choose between additive or multiplicative seasonal decomposition and specify the type of moving averages to be applied.

The Census X-11 method rapidly became the international standard in seasonal adjustment, thanks to its computational efficiency, practical reliability, and adaptability to different time series structures.[26]

2.3.3 Structure and Process Optimisation of the X-11 Method

The release of the X-11 method in 1965 marked a milestone in seasonal adjustment methodology, representing a significant refinement of Census Method II. Until the 1980s, X-11 remained the most widely used seasonal adjustment tool globally. Its core structure was built upon recursive computations using moving averages, systematically organised into three key stages:

1. Trend-cycle extraction;
2. Outlier detection and adjustment; and
3. Seasonal factor estimation.

The method uses multiple iterative loops to progressively refine the estimates of each component. One of X-11's notable innovations was the option to choose between additive and multiplicative decomposition models, allowing it to accommodate different data structures with greater flexibility. For instance, the additive model is suitable when the trend exhibits linear growth, whereas the multiplicative model better fits series with exponential growth patterns.

Through successive moving average calculations, X-11 extracted the trend-cycle, seasonal, and irregular components, while incorporating outlier treatment and revisions constraints directly into the computational procedure. The method's output included graphical displays of the decomposition results, allowing users to evaluate the quality of the seasonal adjustment and effectively identify anomalies.

Due to its robustness, automation, and user-friendly diagnostics, X-11 rapidly became the global standard for seasonal adjustment in official statistics and macroeconomic time series.

Despite its operational strength, X-11 remained fundamentally an empirical filtering method, lacking any explicit statistical model of the underlying data-generating process. This limitation became increasingly apparent over time. One of its most significant drawbacks was the instability of its seasonal factor forecasts. As new observations were added to the series, recalculated seasonal factors often led to substantial revisions, sometimes even altering the seasonal adjustment of observations from several years earlier.

Thus, while X-11 simplified and standardised seasonal adjustment in practice, it struggled with forecast consistency and revision control. Its sensitivity to new data meant that seasonal factor estimates required frequent and often substantial revisions, particularly in real-time applications. These shortcomings eventually motivated the development of model-based approaches, such as X-11-ARIMA, which sought to improve endpoint stability and enhance the method's forecasting performance.

2.4 X-11-ARIMA: A Breakthrough in Statistical Modelling

In the 1970s, the ARIMA model developed by Box and Jenkins introduced a new theoretical framework for time series analysis. Building on this foundation, Estela Bee Dagum of Statistics Canada achieved a major theoretical breakthrough by integrating ARIMA modelling with the X-11 seasonal adjustment procedure, leading to the development of the X-11-ARIMA method.[27]

This innovative approach involves first fitting a Box–Jenkins ARIMA model to the original time series to generate forecasted values for the next one to two years. These extrapolated values are then appended to the original data, effectively extending the series. The extended time series is then passed through the traditional X-11 program, which performs seasonal adjustments using moving average techniques.

By combining model-based forecasting with nonparametric seasonal adjustment, X-11-ARIMA significantly reduces the magnitude of revisions to seasonal factors and adjusted values, particularly at the endpoints of the series. For most economic time series, seasonally adjusted values produced using X-11-ARIMA can be considered final within two years, rather than the three-year revision window required by the original X-11 method.

These two key advantages, improved endpoint stability and reduced revision, have been supported by both theoretical validation and extensive empirical evidence. By extending the time series through both forecasting and backcasting, X-11-ARIMA effectively overcomes the sensitivity of traditional X-11 methods to the lack of data at the beginning and end of the series.

2.4.1 Background and Principles of X-11-ARIMA

X-11-ARIMA was officially adopted by Statistics Canada in January 1975, primarily for the seasonal adjustment of labour force series and other key economic indicators. The method was implemented through the coordinated operation of two computational programs:

1. An ARIMA module, which estimates model parameters based on a user-specified Box–Jenkins model;
2. The X-11 program performs seasonal adjustments on the extended series, incorporating both forecasted and backcasted values.

This integration of time series modelling and nonparametric decomposition marked a turning point in seasonal adjustment methodology. It improved the handling of data at the endpoints, enhanced the consistency of seasonal factors, and enabled more reliable real-time analysis, all without sacrificing the empirical strengths of the X-11 filtering framework.

With continued experimentation in ARIMA modelling, it became clear that a limited set of ARIMA models was sufficient to effectively fit and forecast a wide range of economic time series. Responding to the increasing demand for this methodology, both within Canada and internationally, a new automated version of the X-11-ARIMA procedure was developed. This integrated system combined the previously separate ARIMA modelling and X-11 seasonal adjustment programs into a unified software package.

The automated version is capable of selecting the most appropriate ARIMA model for the extrapolation stage from a predefined set of three candidate models, based on standardised selection criteria and user guidelines. Furthermore, for users who prefer to specify their models, the software provides the flexibility to input custom ARIMA specifications.

In addition to automating model selection and integration, the program enhanced user experience by incorporating advanced graphical and tabular outputs. These outputs, together with a comprehensive suite of diagnostic statistics, facilitate rigorous assessment and validation of the quality of the seasonally adjusted data.

Consequently, the year 1975 marked a pivotal advancement in seasonal adjustment methodology with the formal transition from the traditional Census X-11 method to the more sophisticated and statistically grounded X-11-ARIMA framework.[28] This evolution represented a significant milestone, blending classical moving average techniques with modern time series statistical modelling to improve accuracy, reduce revisions, and enhance the interpretability of seasonal adjustments in economic data.

2.4.2 Advantages of ARIMA Modelling in Seasonal Adjustment

The popularisation of the ARIMA model by Box and Jenkins in the 1970s[29] opened up new directions for the development of seasonal adjustment tools. On one hand, ARIMA models were used to extend the initial time series before applying the X-11 decomposition method, thereby limiting the magnitude of revisions to estimates at the end of the series when additional observations became available. On the other hand, ARIMA modelling was also incorporated into seasonal adjustment methods based on signal extraction theory. Numerous studies[30] have demonstrated the application of ARIMA modelling combined with spectral analysis for seasonal adjustment. These methodological and theoretical advancements have contributed to the growing adoption and development of model-based decomposition techniques in contemporary practice.

The introduction of ARIMA models effectively addressed the problem of insufficient data at the ends of the time series. By forecasting future observations and backcasting missing values, the X-11-ARIMA method significantly reduced revision sizes and enhanced the stability of adjustment results. The initial technical features of the X-11-ARIMA approach can be summarised as follows:

(1) ARIMA Modelling: The time series is extended through ARIMA modelling by filling in missing data before adjustment, thereby reducing uncertainty at both ends of the series. The model employs both seasonal and non-seasonal differencing to capture data dynamics and generate forecasts and backcasts for series extension.
(2) Automation Tools: X-11-ARIMA integrates automatic model selection and parameter optimisation capabilities, minimising user intervention and workload;
(3) Enhanced Diagnostic Functions: Improvements include sliding span analysis and revision history analysis, providing users with comprehensive tools to assess the quality and reliability of the seasonally adjusted data.

Over time, X-11-ARIMA underwent several upgrades, notably the X-11-ARIMA/88 and X-11-ARIMA/2000 versions. These enhanced releases expanded diagnostic capabilities, introduced stronger outlier detection mechanisms, and incorporated advanced trading day effect adjustments, thereby supporting a broader range of data applications and improving overall seasonal adjustment accuracy and robustness.

2.5 X-12-ARIMA: Enhanced Regression and Diagnostic Capabilities

The X-11-ARIMA seasonal adjustment program developed by Statistics Canada (Dagum, 1980) incorporated all the functionalities of the original X-11 method while introducing important improvements. Most notably, X-11-ARIMA enabled the extension of the time series through forecasting and backcasting based on ARIMA models before seasonal adjustment. This extension, using forecasting and backcasting, generated an initial seasonal adjustment that, on average, resulted in smaller revisions[31] when recalculated after new data became available. By addressing a series of data deficiencies inherent in the X-11 trend estimation procedure, particularly in the first year and the last six months of the series, this extension markedly improved the stability and reliability of seasonal adjustments. Under additive decomposition, the optimal use of symmetrical half-season lengths for forecasting and backcasting minimised revisions in the mean squared error sense.

2.5.1 regARIMA Modelling and Outlier Detection

In 1997, the U.S. Census Bureau introduced the X-12-ARIMA method as an extension and refinement of the X-11-ARIMA method. This enhanced program executes a sequence of steps: first, it modifies the time series according to any user-specified prior adjustments; second, it fits a regARIMA model to the series to detect and adjust for outliers and other distortions, thereby improving forecasting accuracy and seasonal adjustment quality; finally, it decomposes the series into three components using a series of moving averages. The final step also yields a comprehensive set of diagnostic statistics that describe the final seasonal adjustment and offer guidance for potential improvements.

X-12-ARIMA's key innovations lie in four main areas: enhanced calendar effect adjustments, improved outlier detection, diagnostic functionality, expanded diagnostic output, and greater user-friendliness. The software provides a flexible user interface that enables users to customise parameters to accommodate diverse data characteristics, thereby facilitating more precise and robust seasonal adjustment results

2.5.2 Improvements in Calendar Effect Adjustments

X-12-ARIMA significantly enhanced the correction of calendar effects, substantially improving the handling of trading day and holiday effects, thereby achieving more precise calendar effect adjustments. A core component of X-12-ARIMA is its regARIMA modelling subsystem,[32] which integrates a linear regression model for ARIMA time series errors with various factors and events that influence seasonal adjustment before the seasonal decomposition process. This integration enables forecasting, backcasting, and prior adjustments, serving as effective diagnostics for the chosen modelling and seasonal adjustment options.

The regARIMA models for trading day and holiday effects provide more reliable diagnostics of the irregular component in seasonal decomposition,[33] such as those observed in X-11, compared to the F-statistics of the conventional regression model. The prior (pre-) adjustments generated by regARIMA represent a significant enhancement over the earlier pre-adjustment steps used in the X-11 seasonal adjustment method.

A key functionality of regARIMA within X-12-ARIMA is its advanced capability of outlier detection. The model can test for changes in seasonal patterns and trading day effects and can incorporate additive outliers (AO) and level shifts (LS) as regression variables within the regARIMA framework.[34] This allows for the identification and correction of AO and LS through regression-based detection methods.

By detecting outliers, regARIMA contributes to robust parameter estimation and improves both forecasting and backcasting accuracy by accounting for additive outliers and level shifts. Its approach to handling AO also provides an effective means for dealing with small amounts of missing data, as precisely handling missing observations is roughly equivalent to replacing them with estimated AO values.

Compared to the earlier methods used in X-11-ARIMA and its 1988 update (X-11-ARIMA/88), which applied trading day pre-adjustments and other regression adjustments mostly for point forecasts, regARIMA's modelling approach offers a more comprehensive and flexible framework. Additive outliers (AO) and level shifts (LS) notably impact parameter estimation in seasonal adjustment models. For example, sudden jumps or drops in sales revenue caused by policy changes can create difficulties for trend components to track abrupt level changes, a common challenge in seasonal adjustment. X-11-ARIMA could not model such AO and LS effects effectively or incorporate regression variables for extended forecasting.[35]

The novel approach in X-12-ARIMA enables more precise identification and correction of anomalies in data, including extreme values and sudden trend shifts, thus substantially improving the quality and reliability of seasonal adjustments.

2.5.3 Expansion and Application of Diagnostic Functions

X-12-ARIMA offers diagnostic capabilities that significantly surpass those of its predecessors, X-11 and X-11-ARIMA, by introducing new tools such as revision history analysis, sliding span diagnostics, and spectral analysis. These enhancements allow users to assess the quality and stability of seasonal adjustments more intuitively. Among the new diagnostics are spectral indicators designed to detect seasonal and trading day effects, sliding span analysis to evaluate the consistency of seasonal adjustments over time, and revision history diagnostics to track changes in adjusted data as new observations are added.

Although the diagnostic improvements in X-11-ARIMA represented a major advancement over X-11, these tools sometimes fell short in identifying problematic series or achieving satisfactory adjustments. Unlike X-11-ARIMA, X-12-ARIMA computes indirect seasonal adjustments and compares them with direct adjustments. This comparison enables users to evaluate the consistency and reliability of seasonal adjustment results across different aggregation levels. The inclusion of these systematic and comprehensive diagnostics enables a more thorough evaluation of adjustment quality, supporting users in achieving more reliable results.

Building upon the foundational seasonal adjustment methodology of X-11-ARIMA, X-12-ARIMA provides greater flexibility and additional options. New moving average options include longer seasonal moving averages, Henderson trend filters of any odd length, and subtle modifications to some asymmetric moving averages. These were originally used in X–11. These changes stem from a unified optimisation principle, thereby offering a broader range of tailored choices. Additionally, X-12-ARIMA introduces a 'pseudo-additive' decomposition, which has proven particularly useful for series with small or zero-valued cycles.

Further improvements include a refined estimation of trading day effects and other regression variables, incorporating user-defined effects as a novel feature. The regARIMA modelling capabilities in X-12-ARIMA enable direct estimation of such effects from observed time series, thereby improving model accuracy and adaptability. An essential aspect of X-12-ARIMA is not only the predictive extension offered by the regARIMA model but also the expanded scope of the regARIMA framework itself. This includes the ability to incorporate any seasonal effects and user-specified regression variables, broadening model selection and diagnostic options. The program offers extensive diagnostics based on out-of-sample forecast performance, thus making diagnostic analyses in seasonal adjustment clearer and more actionable. Compared to X-11-ARIMA, the new X-12-ARIMA essentially encompasses all the functionalities of its predecessor while addressing and improving upon its modelling and diagnostic limitations.[36]

Due to these enhancements, X-12-ARIMA has become the standard seasonal adjustment tool for many national statistical offices worldwide, especially excelling at processing quarterly and monthly data. The methodology of the X-12-ARIMA workflow has been well illustrated in process diagrams by experts like David Findley and institutions (the Deutsche Bundesbank).[37]

2.6 X-13ARIMA-SEATS: Integration of Parametric and Nonparametric Approaches

In 2006, the U.S. Census Bureau released the X-13ARIMA-SEATS method, an advanced successor to X-12-ARIMA that further integrates the SEATS signal extraction theory. X-13ARIMA-SEATS is distinguished by several key features, including (1) comprehensive time series modelling and model selection capabilities based on linear regression models with ARIMA errors (regARIMA models); (2) the ability to generate ARIMA model-based seasonal adjustments using the SEATS software originally developed by Víctor Gómez and Agustín Maravall at the Bank of Spain, alongside the nonparametric adjustments derived from the X-11 procedure; (3) extensive diagnostics to assess the quality and stability of adjustments under the chosen options; and (4) efficient batch processing capability allowing the simultaneous handling of multiple series.

2.6.1 Choice between X-11 and SEATS Signal Extraction

As an enhancement over its predecessor, X-12-ARIMA, X-13ARIMA-SEATS introduces new functionality that enables users to produce seasonal adjustments via either the X-11 or SEATS methods within a unified framework. X-13ARIMA-SEATS effectively integrates these two fundamentally different approaches: the X-11 method, which is a nonparametric decomposition based on moving averages, and the SEATS method, which relies on parametric signal extraction theory to decompose trend and seasonal components directly from the ARIMA model.

The advantage of X-13ARIMA-SEATS lies in its combination of the robustness of the X-11 approach with the theoretical rigour of SEATS. This dual capability empowers users to select seasonal adjustments based on their preference for either the proven stability of X-11 or the formal modelling advantages of SEATS. Furthermore, the software incorporates an automated model selection process that allows users to generate both X-11 and SEATS seasonal adjustments through the same interface, supported by a common set of diagnostic tools to facilitate direct comparison and improve the intelligence of parameter selection.

The diagnostic toolkit of X-13ARIMA-SEATS has been expanded to include advanced spectral analysis and diversified outlier detection methods, significantly enhancing its capacity to identify data irregularities and improve the quality of seasonal adjustment.

X-13ARIMA-SEATS has been widely adopted by major international organisations such as the IMF and Eurostat, becoming a standardised solution for economic data analysis and fostering greater consistency in global economic cooperation.

2.6.2 Automation and Model Selection Enhancements

X-13ARIMA-SEATS introduces several new functionalities that are aimed at improving automation and model selection precision. Among these enhancements are expanded options for generating spectral diagnostics, improved model-based seasonal adjustment diagnostics, and a suite of new regression tests. Notably, according to Findley (2009), the method includes an Akaike's Information Critique, Biased Correction (AICC)-based test for

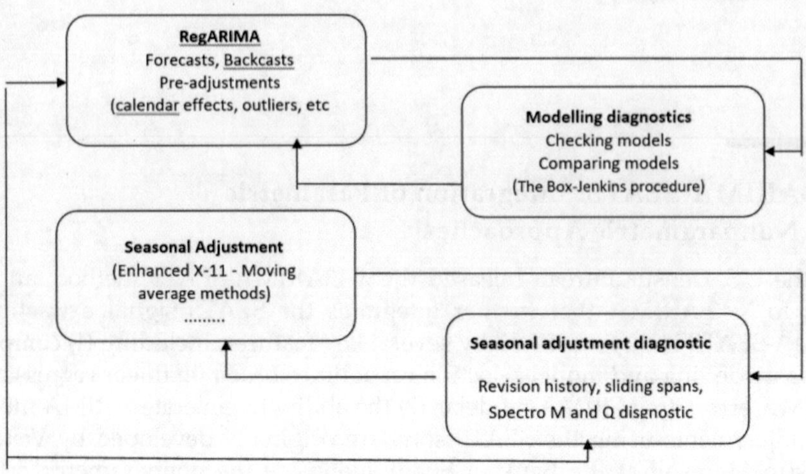

FIGURE 2.1
X-12-ARIMA Operational Procedures.

monthly-length regressors and introduces new regressors for end-of-month stock and Easter effects.[38] Additionally, it offers technical capabilities for defining the grouping of user-specified holiday regressors employing a novel Chi-square testing procedure to determine whether particular regressor groups should be incorporated into the regARIMA model.

Performance improvements in the X-13ARIMA-SEATS program have been achieved on two fronts. First, the restructuring of SEATS routines for producing sliding span and revision history diagnostics has significantly accelerated the generation of stability diagnostics for model-based seasonal adjustments. The speed of producing revision history diagnostics with SEATS is now roughly nine times faster than in previous versions. Second, corrections to the differencing identification algorithm within the 'automdl' specification have greatly enhanced the program's ability to accurately identify ARIMA models, substantially reducing convergence issues during automatic model selection. Extensive simulation testing has further refined error handling in this routine.

2.6.3 Handling of High-Frequency Data and Complex Structures

X-13ARIMA-SEATS has also expanded its diagnostic suite by incorporating an F-statistic tailored for models with stable seasonal components. This statistic is used to test for stable seasonality in the original time series, offering a model-based criterion for assessing seasonal stability. The F-test is derived from a Chi-square test on groups of regression variables, a feature carried over from earlier versions of X-12-ARIMA and X-13ARIMA-SEATS, which was designed to determine whether specific groups of regression parameters in the regARIMA model are collectively zero. These predefined regression groups typically represent fixed seasonal regressors, which may be specified either as monthly (or quarterly) indicator variables or as triangular representations of fixed monthly patterns.[39]

When such regression terms are included in the regARIMA model, the resulting Chi-square test on seasonal regressors serves as an indicator of stable seasonality within the series. A new lognormal adjustment functionality has been added (implemented via the lognormal parameter in the forecast spec), allowing users to generate forecasts from the regARIMA model that are consistent with the assumption that predictions originate from a lognormal distribution (lognormal = yes).

Additional refinements include new parameter variations within the spectral specification (spectrum spec), several new regressors, and options for testing and evaluating regressive effects. For instance, temporary level shift regressors can now be specified within the outlier specification (outlier spec). The computation of t-statistics for testing the null hypothesis of zero temporary level shifts, up to a user-specified maximum number (via the lsrun parameter in the outlier spec), has been introduced. This test is based on summing the estimated parameters for level shifts in each run, divided by their corresponding standard error.

Moreover, the regression spec for the regARIMA model now supports user-defined moving holiday regression groups, including the Chi-square test parameter for statistical testing. The AICC test rules for monthly or quarterly length regressors have been formalised, allowing for more rigorous model evaluation and selection.[40]

X-13ARIMA-SEATS is widely used for seasonal adjustment of economic and social data worldwide. It demonstrates excellent performance in handling high-frequency and complex structured data, making it one of the leading tools for seasonal adjustment today. With increasing data complexity, calendar effects and outlier treatment have become critical issues in seasonal adjustment. X-13ARIMA-SEATS significantly improves the accuracy of adjustment results by automatically identifying and correcting these effects. The latest version of X-13ARIMA-SEATS, Build 61 version 1.1, was released on July 10, 2024.

2.7 Summary and Conclusions

This chapter reviewed the evolution and development of the X-11 series of seasonal adjustment methods. From X-11 to X-13ARIMA-SEATS, seasonal adjustment techniques have evolved from empirical rule-based approaches to sophisticated statistical modelling. Core advancements include the introduction of ARIMA models, enhanced regression techniques, and the integration of parametric and nonparametric methods. Looking ahead, the increasing complexity of data will drive seasonal adjustment methods to incorporate artificial intelligence and modern statistical techniques, providing stronger support for economic and social data analysis.

1. Early astronomical decomposition theories and moving average techniques laid the theoretical foundation for time series analysis by decomposing series into trend, cycle, seasonal, and irregular components. Persons' four-component decomposition method and the additive/multiplicative models established a conceptual framework for subsequent research.
2. The X-11 method, based on moving averages, iteratively extracts trend, seasonal, and irregular components. Its key innovations include the choice between additive and multiplicative models, outlier handling, and graphical output. While X-11's robustness made it a global standard, its sensitivity to data endpoints and frequent revisions of seasonal factors remained major limitations.
3. The theoretical breakthrough of X-11-ARIMA addressed X-11's endpoint data deficiency by employing ARIMA models for forecasting and backcasting to extend the series. This approach significantly reduced the magnitude of revision and enhanced the stability of adjustment, marking a shift from empirical rules to statistical modelling and laying the groundwork for future methods.
4. Building on X-11-ARIMA, X-12-ARIMA introduced the regARIMA model, improving calendar effect correction and outlier detection. Key innovations included handling the trading day effect, sliding span analysis, and indirect adjustment comparison. These enhancements improved adjustment accuracy and diagnostic power, making X-12-ARIMA widely adopted by statistical agencies.
5. X-13ARIMA-SEATS integrates X-11's moving average approach with SEATS's signal extraction theory, offering dual parametric and nonparametric seasonal adjustment options. New features included automated model selection, spectral analysis, and diverse outlier detection methods. Excelling in handling complex and high-frequency data, X-13ARIMA-SEATS has become one of the mainstream seasonal adjustment tools today.

Notes

1 Armatte, M. (1992). "Conjonctions, conjoncture et conjecture. Les baromètres économiques (1885–1930)". *Histoire et Mesure*, 7, 99–149; Bell, W. R., & Hillmer, S. C. (1984). "Issues Involved with the Seasonal Adjustment of Economic Time Series". *Journal of Business and Economic Statistics*, 2, 291–394; Hylleberg, S. (1992). "The Historical Perspective". In *Modelling Seasonality*, pp. 15–25. Oxford University Press, London; Nerlove, M., Grether, D. M., & Carvalho, J. L. (1979). *Analysis of Economic Time Series: A Synthesis*. Academic Press, New York.

2 Nerlove, M., Grether, D. M., & Carvalho, J. L. (1979). "Analysis of Economic Time Series: A Synthesis". Academic Press, New York.
3 Cournot, A. (1838). "Recherches sur les principes mathématiques de la théorie des richesses". Librairie philosophique J. Vrin, 1980; 和Jevons, W. S. (1862). "On the Study of Periodic Commercial Fluctuations". In *Investigations in Currency and Finance*. Macmillan, London, 1884.
4 Herschel, W. (1801). "Observations Tending to Investigate the Nature of the Sun in Ordre to Find the Causes or Symptoms of its Variable Emission of Light and Heat with Remarks on the Use that May Be Possibly Be Drawn from Solar Observation". *Philosophical Transactions of the Royal Society of London*, 91, 265–318.
5 Buys-Ballot, C. (1847). "Les changements périodiques de température". Kemink et Fils, Utrecht; 参阅 Jevons, W. S. (1862). "On the Study of Periodic Commercial Fluctuations". In *Investigations in Currency and Finance*. Macmillan, London, 1884.
6 Fourier, J. B. (1822). "Théorie analytique de la chaleur". Jacques Gabay, Paris, 1988.
7 Fourier, J. B. (1822). "Théorie analytique de la chaleur". Jacques Gabay, Paris, 1988.
8 Armatte, M. (1992). "Conjonctions, conjoncture et conjecture. Les baromètres économiques (1885–1930)". *Histoire et Mesure*, 7, 99–149.
9 For example, Lucien March (1905) distinguishes between "annual variation, multi-annual variation (eg, decennial), M. Armatt (1992) refers to periods shorter than a year. Maballée (1906) attempts to use the correlogram to separate a series of turning points. Voir March, L. (1905). "Comparaison numérique de courbes statistiques". *Journal de la Société Statistique de Paris*, 255–277; Armatte, M. (1992). "Conjonctions, conjoncture et conjecture. Les baromètres économiques (1885–1930)". *Histoire et Mesure*, 7, 99–149; Yule, G. U. (1921). "On the Time Correlation Problem, with Especial Reference to the Variate-Difference Correlation Method". *Journal of The Royal Statistical Society*, 84, 497–526; Bell, W. R., & Hillmer, S. C. (1984). "Issues Involved With the Seasonal Adjustment of Economic Time Series". *Journal of Business and Economic Statistics*, 2, 291–394; Maballée, Colette et Berthe. (1906). "Peaks and Peaks and Correlograms". *Zeitschrift für Wahrscheinlichkeitstheorie und Verwandte Gebiete*, 3, 139–167.
10 March, L. (1905). "Comparaison numérique de courbes statistiques". *Journal de la Société Statistique de Paris*, 255–277; Yule, G. U. (1921). "On the Time Correlation Problem, with Especial Reference to the Variate-Difference Correlation Method". *Journal of the Royal Statistical Society*, 84, 497–526; Bell, W. R., & Hillmer, S. C. (1984). "Issues Involved with the Seasonal Adjustment of Economic Time Series". *Journal of Business and Economic Statistics*, 2, 291–394.
11 Maballée, Colette et Berthe. (1906). "Peaks and Peaks and Correlograms". *Zeitschrift für Wahrscheinlichkeitstheorie und Verwandte Gebiete*, 3, 139–167.
12 Persons, W. M. (1919). "Indices of Business Conditions". *Review of Economic Statistics*, 1, 5–107.
13 Menderhausen, H. (1937). "Annual Survey of Statistical Technique: Methods of Computing and Eliminating Changing Seasonal Fluctuations". *Econometrica*, 5, 234–262.
14 Persons, W. M. (1919). "Indices of Business Conditions". *Review of Economic Statistics*, 1, 5–107.
15 A full introduction to moving averages will be given in Chapter 3.
16 Poynting, J. H. (1884). "A Comparison of the Fluctuations in the Price of Wheat and in the Cotton and Silk Imports into Great Britain". *Journal of the Royal Statistical Society*, 47, 34–64.
17 Macaulay, F. R. (1931). *The Smoothing of Time Series*. National Bureau of Economic Research.
18 Slutsky, E. (1927). "The Summation of Random Causes as the Source of Cyclical Processes". *Econometrica*, 84, 105–146 (1937), translation of a Russian paper (Conjuncture Institute, Moscow); Yule, G. U. (1927). "On a Method of Investigating Periodicities in Disturbed Series with Special Reference to Wolfer's Sunspot Numbers". *Philosophical Transactions of the Royal Society, Series A*, 226, 267–298.
19 Fisher, A. (1937). "A Brief Note on Seasonal Variation". *Journal of Accountancy*, 64, 54–59.
20 Yule, G. U. (1927). "On a Method of Investigating Periodicities in Disturbed Series with Special Reference to Wolfer's Sunspot Numbers". *Philosophical Transactions of the Royal Society, Series A*, 226, 267–298.
21 Slutsky, E. (1927). "The Summation of Random Causes as the Source of Cyclical Processes". *Econometrica*, 84, 105–146 (1937), translation of a Russian paper (Conjuncture Institute, Moscow).
22 Bell, W. R., & Hillmer, S. C. (1984). "Issues Involved with the Seasonal Adjustment of Economic Time Series". *Journal of Business and Economic Statistics*, 2, 291–394.
23 Ladiray, D., & Quenneville, B. (2001). "Seasonal Adjustment with the X-11 Method". Springer.
24 Ladiray, D., & Quenneville, B. (2001). "Seasonal Adjustment with the X-11 Method". Springer.

25 Shiskin, J., Young, A., & Musgrave, J. C. (1967). "The X-11 Variant of the Census Method II Seasonal Adjustment Program". Technical Paper No. 15. Bureau of the Census, US Department of Commerce, Washington, DC.
26 Young, A. H. (1965). "Estimating Trading-Day Variations in Monthly Economic Series". Technical Paper No. 12. Bureau of Census, US Department of Commerce, Washington DC.
27 Dagum, E. B. (1975). "Seasonal Factor Forecasts from ARIMA Models". Proceedings of the International Institute of Statistics, 40th Session, Contributed Papers, 3, pp. 206–219, Warsaw; Dagum, E. B. (1980). "The X-11-ARIMASeasonal Adjustment Method". Catalogue 12-564E. Statistics Canada.
28 At the same time, significant progress has been made in parametric modeling and spectral analysis of time series, mainly due to the development of stochastic process theory, from which seasonally adjusted tools are gradually benefiting. Harmonic analysis has long been used to solve series factorisation problems, it is placed in a framework of absolute determinism and is used to identify exact periodicity, although it is well known that cycles may not be strictly cyclical, or seasonality may change. Similarly, it was not until the 1960s that computers made theoretical advances and spectral analysis was used better: The estimation of spectral density, and the study of nonstationary processes was improved. See Bell, W. R., & Hillmer, S. C. (1984). "Issues Involved with the Seasonal Adjustment of Economic Time Series". *Journal of Business and Economic Statistics*, 2, 291–394.
29 Box, G. E. P., & Jenkins, G. M. (1970). "Time Series Analysis: Forecasting and Control". Holden-Day, San Francisco.
30 Bell, W. R., & Hillmer, S. C. (1984). "Issues Involved with the Seasonal Adjustment of Economic Time Series". *Journal of Business and Economic Statistics*, 2, 291–394.
31 Huot, G., Chiu, K., Higginson, J., & Dagum, E. B. (1986). "The X-11-ARIMA Seasonal Adjustment Method". Statistics Canada, Time Series Research and Analysis Centre, (Technical Report, Often cited as a Foundational Methodological Document); Bobbitt, L. M., & Otto, M. C. (1990). "Effects of Filter Selection on Revision Histories of Seasonally Adjusted Data Using X-11-ARIMA", *Proceedings of the Business and Economic Statistics Section, American Statistical Association*, 197–202.
32 We will introduce and discuss the regARIMA method in more detail in a later section.
33 Bell, W. R., & Hillmer, S. C. (1984). "Issues Involved with the Seasonal Adjustment of Economic Time Series". *Journal of Business and Economic Statistics*, 2, 291–394.
34 Chang, I., & Tiao, G. C. (1983). "Estimation of Time Series Parameters in the Presence of Outliers". Technical Report #8. Statistics Research Center, University of Chicago; Bell, W. R. (1983). "Seasonal Adjustment Using Regression Models with ARIMA Errors". Census Bureau Technical Report. U.S. Bureau of the Census; Chang, I., Tiao, G. C., & Chen, C. (1988). "Estimation of Time Series Parameters in the Presence of Outliers". *Technometrics*, 30(2), 193–204. https://doi.org/10.1080/00401706.1988.10488367.
35 Findley, D. F., & Monsell, B. C. (1986). "New Capabilities and Methods of the X-11-ARIMA Seasonal Adjustment Program". *Journal of Business & Economic Statistics*, 4(2), 127–152. https://doi.org/10.1080/07350015.1986.10509510; Findley, D. F., Monsell, B. C., Bell, W. R., Otto, M. C., & Chen, B.-C. (1990). "New Capabilities and Methods of the X-12-ARIMA Seasonal Adjustment Program". *Journal of Business & Economic Statistics*, 16(2), 127–177. https://doi.org/10.1080/07350015.1998.10524752.
36 Dagum, Estela Bee. (1988). "The X-11-ARIMA Seasonal Adjustment Method". Catalogue No. 12-564. Statistics Canada.
37 PC and Unix Versions Are Available on the U.S. Census Bureau Website. Retrieved from http://www.census.gov/srd/www/x12a/. In addition, the SAS/Graph program (X12-Graph) can be used, allowing users to generate useful diagnostics from the X12-ARIMA output.
38 Findley, D. F. (2009). "Some Recent Developments and Directions in Seasonal Adjustment". *Journal of Official Statistics*, 25(1), 1–32.
39 For more details on these regression variables, see U.S. Census Bureau. (2007). *X-12-ARIMA Reference Manual*, version 0.3. U.S. Department of Commerce, Census Bureau, Statistical Research Division, Washington, DC.
40 We will not discuss this information here but will provide more detailed information in the section on the related technical discussion at a later date.

3

A Conceptual Framework for the General Theory of Seasonal Adjustment

3.1 Introduction

The decomposition of time series constitutes the general theoretical framework for seasonal adjustment. Understanding time series decomposition is essential for analysing economic data. Mastery of the theories and methods of seasonal adjustment not only enhances data interpretability but also improves forecasting capabilities and model fit. It helps isolate cyclical effects within the data, identify long-term trends, and avoid being misled by short-term fluctuations. A clear grasp of these theoretical foundations directly affects the accuracy and effectiveness of seasonal adjustment.

Seasonal adjustment procedures typically follow a two-stage estimation framework:

(1) Pre-adjustment: The goal is to select a regression model with ARIMA errors that best captures the characteristics of the original series. The chosen model is used to adjust the series for deterministic effects and to extend it through backcasting and forecasting in preparation for the decomposition process.

(2) Seasonal adjustment: The two most commonly used decomposition methods are as follows:

   (a) The X-11 method, a nonparametric approach based on moving averages; and

   (b) The SEATS method, a model-based parametric estimation approach.

These methods decompose the pre-adjusted series into three unobserved components: trend, seasonal, and irregular. The X-11 and SEATS decomposition methods will be discussed in detail in Chapter 4 (X-13ARIMA-SEATS) and Chapter 7 (TRAMO-SEATS), respectively.

Although the pre-adjustment and seasonal adjustment steps are implemented separately, they are considered to be fully integrated parts of any seasonal adjustment procedure. As shown by Findley et al. (1998),[1] the sensitivity of model specification in the pre-adjustment phase systematically affects the decomposition results. Moreover, the results of time series decomposition may, in turn, suggest modifications to the pre-adjustment stage. A careful diagnostic analysis of both stages is thus crucial for assessing the quality of seasonal adjustment.

This chapter presents the general conceptual framework for seasonal adjustment, with a particular focus on pre-adjustment procedures. The seasonal adjustment part will be discussed in Chapter 4. Following this introductory Section 3.1, Section 3.2 presents the theoretical foundations of seasonal adjustment, including time series decomposition models, seasonal and non-seasonal components, and irregular elements. Section 3.3 discusses

DOI: 10.1201/9781003716655-3

pre-adjustment in X-13ARIMA-SEATS, covering both ARIMA and regARIMA approaches. Section 3.4 provides a summary and conclusion.

3.2 Theoretical Framework of Seasonal Adjustment: Decomposition and Models

A time series is a collection of data recorded sequentially over time. Economic data, in particular, are typically obtained at regular intervals (eg, monthly or quarterly) and represent continuous measurements of economic activity (ie, variables) over time, a series of phenomena evolving temporally. Examples include employment figures, the Retail Price Index (RPI), balance of payments, and Gross Domestic Product (GDP). Time series are primarily used for economic and social analysis, and for such data to be meaningful, they must remain comparable over time. Thus, the concepts being measured and the methods of measurement must be consistent over time. Time series analysis aims to reveal behavioural patterns that evolve, to analyse and estimate these dynamic characteristics, and to use the resulting estimates for forecasting purposes.

Raw data represent a collection of observed values of certain characteristics of a time series at fixed time intervals. Time series can be classified into two major types: (1) Stock series, and (2) Flow series. Stock series represent the state of an economic variable measured at a specific point in time, such as the unemployment rate on a particular date. In contrast, flow series reflect activities measured over a given period, such as retail sales over a quarter, current account deficits, or balance of payments. Most flow series are subject to calendar effects. The key distinction between stock and flow series lies in the fact that flow series are generally more affected by calendar-related factors. Nevertheless, both types of series can be adjusted using the same seasonal adjustment methods.

3.2.1 Decomposition of Time Series and Its Components

Observed raw time series often display a high degree of variability. This variability may stem from underlying cyclical, seasonal, and irregular influences. These embedded components are typically unobservable and must be inferred from the data. To accurately interpret the fluctuations in a raw time series, it is essential to decompose the series into its constituent components. A standard decomposition breaks a time series into four unobserved components:

$$Y_t = T_t + C_t + S_t + I_t \tag{3.1}$$

where Y_t is the observed time series; T_t is the trend component, which captures long-term, low-frequency structural changes in the series; C_t is the cyclical component, representing medium-term fluctuations; S_t the seasonal component; and I_t is the irregular component.

The seasonal component S_t reflects periodic variations within the year that recur in a more or less stable pattern in terms of timing, direction, and magnitude; this is referred to as *seasonality* in a time series. It accounts for regular intra-year effects such as seasonal weather changes, variations in month length, and holidays like Christmas. Seasonality may also include calendar-related systematic effects that do not occur at regular intervals throughout the year and are induced by calendar variations from year to year.

The irregular component (I_t) is assumed to follow a white noise process, $I_t \sim WN(0,\sigma^2)$, and includes unpredictable effects that are considered stochastic. These effects are assumed to have an expected value of zero (in additive models) or one (in multiplicative models). The irregular component is the residual after removing the trend, cycle, and seasonal effects (including calendar effects). In some cases, specific calendar-related elements such as trading day (TD) effects, Easter effects, or outliers (AO) are treated as separate, explicitly modelled components.

In deterministic time series analysis, the trend-cycle component typically consists of two related but distinct elements: (1) the trend refers to the long-term movement in the series, indicating the underlying level of the data; (2) the cycle denotes recurring movements longer than one year, reflecting periods of economic expansion or contraction.

Cyclical fluctuations are usually smaller in magnitude than seasonal fluctuations and require long time series for reliable estimation. For this reason, seasonal adjustment procedures typically combine the trend and cycle into a single component, commonly referred to as the trend-cycle or simply the trend. In contrast to irregular and seasonal fluctuations, the trend is relatively smooth and rarely changes direction within short periods. Turning points in the trend generally mark shifts in the business cycle, although they are not strictly periodic. Conversely, seasonal peaks and troughs occur at regular intervals and exhibit consistent directional patterns.

Assuming the trend component includes the cyclical component, a typical decomposition expresses the original time series as three major, mutually independent, unobserved components:

$$Y_t = T_t + S_t + I_t \tag{3.2}$$

where the trend T_t captures both long-term and medium-term changes in the data, the seasonal component S_t captures short-term cyclical fluctuations lasting no more than one year, and the irregular component I_t captures erratic, unpredictable movements.

Seasonality, in a broader sense, encompasses all recurring fluctuations in socio-economic data that arise from factors such as typical weather patterns, fixed-date holidays, industry conventions, the distribution of working days across months or quarters, and social customs. Seasonal effects are assumed to be *constant*, meaning that the seasonal pattern's amplitude is consistent over the entire length of the time series. Among the various sources of variation in a time series, the seasonal component is typically the most regular and stable.

However, under stochastic analysis, the assumption of constant seasonality is often violated. In such cases, so-called moving seasonality approaches are employed. The advantage of these approaches lies in avoiding the risks of under- or over-correction that may arise from imposing a fixed seasonal pattern.

For a given time series, identifying and estimating the trend, seasonal, and irregular components form the core of seasonal adjustment. Their distinct statistical characteristics enable them to be separately identified and quantified within a given dataset.

3.2.2 Time Series Decomposition Models: Additive, Multiplicative, and Pseudo-Additive

Seasonal adjustment decomposes a time series into trend (T_t), seasonal (S_t), and irregular (I_t)components through a process known as time series decomposition. Seasonal adjustment depends on two models of time series decomposition: (1) additive models, and (2) multiplicative models.[2]

The choice of model depends on the relationship between the trend and seasonal components. If the time series components behave independently of one another, the pattern of seasonal variation is generally stable throughout the year. For example, an increase in the trend does not lead to an increase in the magnitude of the seasonal decline, the difference between the trend and the raw data remains approximately unchanged over a similar period (month, quarter), in which case an additive model is generally used; Multiplicative models are especially useful when seasonal and irregular fluctuations vary in response to trend behaviour, with the amplitude of seasonality increasing (or decreasing) as the trend rises (or falls). In contrast to the additive case, these components are not independent of each other. In many cases, charting economic data can provide a clearer picture of the decomposition model of the time series.

To calculate each component of a time series, the model must be fitted to the time series. In the additive model, the original series can be Y_t thought of as the sum of the unobserved components:

$$Y_t = T_t + S_t + C_t + I_t \tag{3.3}$$

Where Y_t is the raw data; T_t is the trend-cycle component; S_t is a seasonal portion; C_t is a calendar component; I_t is an irregular component.

In an additive model, trends in the time series do not affect the seasonal component; if the time series increases or decreases over time, the size of the peaks and troughs remains (roughly) constant. The sum of seasonal fluctuations over a year is relatively small, as seasonal effects tend to offset each other over the annual cycle.

Assuming that the unobserved components are independent of each other, the seasonality and calendar adjustment series of the additive model are derived by subtracting the seasonal and calendar components from the original series, so that each component of the time series has the same dimension or unit as the original time series. The seasonally adjusted data are as follows:

$$SA_t = Y_t - S_t + C_t = T_t + I_t \tag{3.4}$$

In a multiplicative model, the time series Y_t is broken down into the product of the unobserved components

$$Y_t = T_t \times S_t \times C_t \times I_t \tag{3.5}$$

The multiplication model assumes that the magnitude of the unobserved components is proportional to the level of the series. If the time series increases or decreases over time, the size of the peaks and troughs also increases. Trends in the time series influence the seasonal component, causing the amplitude between peaks and troughs to change over time. For example, for seasonal components, the multiplicative model means that the seasonal peaks increase as the series level increases. Since the trend-cycle component determines the entire level of the series, the other unobserved components are expressed as T_t percentages, usually referring to the seasonal factor, which is used to represent the proportional relationship between the seasonal change and the trend-cycle component.[3]

Because the original time series is the sum of its constituent parts, each component has the same dimensions or units as the original time series. The seasonality and calendar adjustment series of the multiplicative model are the ratios between the original series and the seasonality and calendar factors. Assuming multiplicative factorisation, the seasonally adjusted data is

$$SA_t = \frac{Y_t}{S_t \times C_t} = T_t \times I_t \tag{3.6}$$

In a multiplicative model, the sum of seasonal changes in a year is approximately equal to the number of periods in a calendar year, ie

$$\sum_{j=1}^{n} S_{t+j} \approx n \tag{3.7}$$

where $n = 12$ is a monthly frequency time series, and $n = 4$ is a quarterly frequency time series. Thus, the average magnitude of the seasonality factor is 1, and there is no total contribution to the series level over the year. This constraint ensures that seasonality factors do not capture longer (more than a year) item periods in the time series.

In practice, most economic time series exhibit a multiplicative relationship, so multiplicative decomposition usually provides the best fit. However, if there are any zeros or negatives in the time series, multiplicative decomposition cannot be performed. The multiplicative model is linearised by log-additive transformation

$$logY_t = \log(T_t \times S_t \times C_t \times I_t) = \log(T_t) + \log(S_t) + \log(C_t) + \log(I_t) \tag{3.8}$$

After completing the additive and decomposition of the logarithmic series, the seasonally adjusted series is obtained, and the exponents of the logarithmic trend period and irregular components are taken:

$$SA_t = \exp\left[\log(T_t) + \log(I_t)\right] = T_t \times I_t \tag{3.9}$$

In addition to the most commonly used additive decomposition and multiplicative decomposition models, the most popular one is the pseudo-additive decomposition model. The ONS has combined some features of the additive and multiplicative models to develop pseudo-additive decomposition, expressed by the following equation:

$$Y_t = T_t(S_t + I_t - 1) \tag{3.10}$$

The pseudo-additive model assumes that both seasonal and irregular components depend on trend behaviour while remaining independent of each other. Pseudo-additive models are designed for time series that primarily exhibit a multiplicative relationship between components, where the series values may be zero or near zero. In this case, multiplicative decomposition cannot be applied. Conversely, pseudo-additive models permit seasonal or irregular components to take on zero values, while leaving trend periods unaffected. In the pseudo-additive model, both seasonal and irregular components are centred around 1. The initial equation can be rewritten as follows:

$$Y_t = T_t + T_t(S_t - 1) + T_t(I_t - 1) \tag{3.11}$$

Among them, $T_t(S_t - 1)$ and $T_t(I_t - 1)$ can be interpreted as an additive seasonal effect and an additive irregular effect, respectively.

The pseudo-additive model is only supported by the X-12-ARIMA method. TRAMO/SEATS does not support this model. This type of model is primarily used for time series that exhibit strong and fluctuating seasonal effects.

3.2.3 Model Identification and Selection Methods

The best way to identify the best decomposition model for a time series is to perform a graphical inspection of the series. If the graph indicates that seasonal fluctuations increase with the level of the series, a multiplicative adjustment is recommended, or a logarithmic transformation followed by additive decomposition. Such approaches help stabilise seasonal variation and mitigate potential heteroskedasticity in the irregular component and regression residuals. If the seasonal pattern remains stable over time and does not vary with changes in the trend, no transformation is necessary, and an additive decomposition is appropriate. If neither knowledge of the series nor examination of the diagram aids in making the decision, the X-13ARIMA-SEATS automatic program can be used. As long as the processed series has all positive values, X-13ARIMA-SEATS performs an AIC-based selection to decide whether to convert logarithmically or not. To determine the appropriate transformation to apply, and thus which decomposition model to use, X-13ARIMA-SEATS fits a regARIMA model to both untransformed and transformed series, then selects logarithmic transformations, unless the following conditions are met.

$$AICC_{nolog} - AICC_{log} < \Delta_{AICC}$$

where $AICC_{nolog}$ is the AICC value obtained by fitting the regARIMA model to the unconverted series, and $AICC_{log}$ is the AICC value obtained by fitting the regARIMA model to the transformed series.[4]

Seasonally adjusted time series (SA) are formed by estimating and removing seasonal components. For additive models, SA is the seasonal effect value associated with a given period (month, quarter) subtracted from the original time series, ie, SA = $Y - S$; for a multiplicative model, divide the initial time series value by the seasonal component, ie, SA = Y/S. In multiplicative decomposition, seasonal effects vary proportionally to the trend. If the trend is up, the seasonal effect will also increase, and if the trend is down, the seasonal effect will also decrease. In additive decomposition, seasonal effects remain roughly constant regardless of which direction the trend moves. If there is neither a seasonal effect nor a calendar effect in the raw data Y, the seasonal adjustment series is given by the original data. For a series without a recognisable seasonal pattern but with identifiable calendar effects, the seasonally adjusted series is represented by the calendar adjustment series.

3.2.4 Seasonal and Calendar Effects

Seasonal effects include seasonal fluctuations that repeat at a similar annual time, direction, and amplitude. Seasonal variations can be influenced by climatic factors, administrative or legal regulations, social and cultural traditions, as well as annual calendar effects such as public holidays or national celebrations. Changes in any of these effects can alter the properties of seasonal patterns.

- The calendar effect depends on the length of the period (month, quarter) and the distribution of working days and holidays in a particular period, and the calendar effect causes fluctuations in daily activities, affecting the records of each month or

TABLE 3.1

Distribution of Weekdays in January (2008–2011)

Year-Month	2008–01	2009–01	2010–01	2011–01
Monday	4	4	4	5
Tuesday	5	4	4	4
Wednesday	5	4	4	4
Thursday	5	5	4	4
Friday	4	5	5	4
Saturday	4	5	5	5
Sunday	4	4	5	5

quarter. Calendar seasonality refers to effects that recur during the same season each year. For example, except for February, the length of the months is seasonal: ie, March always has 31 days, April always has 30 days, and so on. The changes caused by the seasonal calendar effect are not real changes in a given time series, but are due to an uneven number of working days in different periods. Only the seasonal portion of the calendar effect is part of the seasonality of the series and is removed by the seasonal adjustment. Calendar effects include trading day or business day effects.

- Trading days or working days (TD) represent the difference between the number of business days (eg, Monday through Friday) and the number of weekend days (eg, Saturday and Sunday) within a given month. This effect is caused by differences in the length of the months and the distribution of weekends and holidays between the months. It relates to differences in the number of days in each week across the quarters and months of the year. Generally, there are four full weeks in each month, plus one, two, or three days. So, for each month, there will be at least 4 occurrences on a given day of the week, but there will be 5 occurrences on some days. The trading day effect occurs because the number of occurrences of each day of the month in a week varies from year to year. This is shown in Table 3.1.

The workday effect distinguishes between two types of days: weekdays and weekends. This effect can occur if the data depends on the number of working days and the level of economic activity is similar for each weekday, but significantly different from that of the weekend. Thus, the working day effect can be seen as a simplified version of the trading day effect. An example of a weekday change is shown below:

TABLE 3.2

Distribution of Working and Non-Working Days in January (2008–2011)

Year-Month	2008–01	2009–01	2010–01	2011–01
The working day	23	22	21	21
Non-working day	8	9	10	10

Similar to seasonal effects, it is necessary to estimate and remove trading day effects to help explain the time series. This is because the changes caused by the seasonal calendar effect are not real changes in a given time series, but are due to an uneven number of working days in different periods.

- The calendar effect is both seasonal and non-seasonal. Non-seasonal calendar effects include leap year (LP). The extra days in February of a leap year need to be calculated, with one day added every four years, which may result in a four-year cycle that peaks in the first quarter of a leap year;
- Moving holidays (h) refer to holidays that occur on different days and months each year: eg, Easter, which may occur in March or April, and Chinese New Year, which may occur in January or February, and are associated with events of religious or cultural significance in a country;
- The non-seasonal calendar effect also includes certain data that are not strictly aligned by calendar month, such as the 4–4–5 week data collection pattern of the UK Retail Tip Index;
- The non-seasonal part of the calendar effect is usually included in the calendar component and treated separately, and the time series is corrected by regression with the appropriate regression variables, or pre-adjustment, before the seasonal adjustment, such as using the regARIMA model to correct the time series or 'data cleansing'.[5]

3.2.5 Types and Handling of Outliers

Additive outliers are data points that are beyond trends and general patterns of seasonal components. Outliers can be caused by random effects (ie, extreme irregularities) or other identifiable causes (eg, natural disasters, strikes, or economic and financial crises). This chapter discusses only the second type of outliers, which are cases where there is a potential economic reason to explain the anomalous behaviour of a data point. The seasonally adjusted method is based on a moving average. Moving averages can be affected by extreme values or outliers. For example, the presence of anomalous data points can make the average not representative of a series of patterns. If outliers are not adjusted or handled, they will result in distorted estimates of all components in the time series.

Four main types of outliers are commonly used in economic time series:

(a) Additive outlier (AO), or pulse outliers: associated with only one period. Outliers may result from random fluctuations or identifiable events such as strikes, severe weather, or war. For example, in the pharmaceutical and medical products sector, the number of drugs sold often increases sharply before price hikes or during seasonal outliers.

(b) Level shift (LS), or trend break, refers to a sudden but sustained change in the underlying level of a time series, while the annual seasonal pattern remains unchanged. There are numerous potential causes of level shifts in time series, such as changes in the concept or definition of the surveyed population, modifications in data collection methods, shifts in economic behaviour, legislative changes, or evolving social traditions. If there is a sudden change in the level of the series, the estimates will be distorted when applying the moving average to calculate the preliminary trend period. Trend period estimates before abrupt changes will be underestimated, while estimates after abrupt changes will be overestimated. Since the irregular and seasonal components are calculated based on this initial trend period, they may be distorted. The final seasonally adjusted estimate will become more volatile. Additive outliers and temporary changes are added to the irregular components, and the level shift is transferred to the trend.[6] or temporary shift.

(c) Temporary change (TC), ie, an observation that is extremely high or very low, and then gradually decreases in size (exponentially) throughout subsequent observations until the time series returns to its initial level. TC can also be seen as a series of additive outliers. For example, in the construction industry, if the weather is better than usual in winter (ie, warmer temperatures and no snow), yields will be higher. When weather conditions return to normal, production also reverts to its usual level, and the associated effects on the series gradually dissipate.

(d) Slope outliers (Ramp, RP) are linear increases or decreases at the level of the series within a specified time interval t0 to t1, and RP can be considered LS, the effects of which are distributed over several periods. Linear increases or decreases, as well as temporary level changes, are allowed only for certain quarters or months of the year. For example, wage increases in different economic sectors occurring at different times can lead to a gradual rise in average wages across the economy within a given year.

These effects are usually pre-adjusted by regARIMA using a specific regression variable model.[7] These outliers can be identified more intuitively from Figure 3.1.

In the absence of outliers, the irregular components are treated as white noise and assumed to have a normal distribution of random variables that are independent of the constant variance at all times. In statistical terms, such a variable is called a white noise process. Unlike outlier effects, the white noise process is assumed to be part of the underlying linear data generation process for the series.

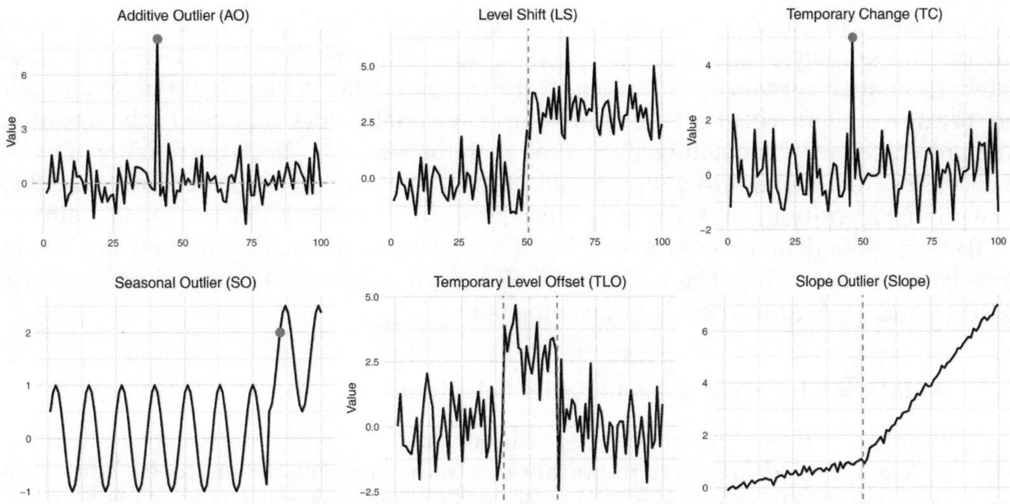

FIGURE 3.1
Different Types of Outliers.

Data Source: The simulated data

A basic premise for applying the seasonal adjustment method is that the time series being processed should show a clear and sufficiently stable seasonal effect. Seasonal adjustments should not be made for series that do not have seasonal effects, or for series where seasonal effects are difficult to identify from the original series. Thus, the first step in seasonal adjustment is to test the original series to determine if there is a recognisable seasonality. At the same time, the series should also be tested for calendar effects. Calendar effects are generally less pronounced than seasonal effects, so their identification relies on statistical tests to determine whether their contribution to the series is statistically significantly different from zero. Seasonal adjustments are only necessary for a seasonal series.

It is important to note that: (1) The purpose of seasonal adjustments is not intended to smooth out time series. The seasonally adjusted series is the sum of the trend-cyclical component and the irregular component. Thus, when the irregularity component is strong, the seasonally adjusted series may not show a smooth pattern over time. To extract the trend-cycle component, the irregular component needs to be further removed from the seasonally adjusted series. However, trend-cycle extraction is a complex task with greater uncertainty than seasonal adjustments, especially in the final stages of the time series. (2) The seasonally adjusted data represents the analysis and processing of the original data. Designed to detect potential changes that could otherwise be obscured by seasonal fluctuations. (3) There is no single solution for performing seasonal adjustments. X-13ARIMA-SEATS adopts two methods: the moving average-based nonparametric method (X-11 method) and the model-based parametric method (SEATS). (4) Seasonally adjusted data may also be revised as future data becomes available. Keeping unadjusted data may make it easier to identify the source of errors and correct them.

3.3 Pre-Adjustment: ARIMA versus regARIMA Model

In the analysis of actual economic and financial data, seasonal factors may have a greater impact on the analysis and prediction of data. To make seasonal adjustments more accurately, it is usually necessary to pre-adjust the original data, including (1) removing outliers: because outliers will affect the estimation of seasonal factors, they need to be identified and processed before modelling; (2) dealing with the workday effect: the number of workdays varies from month-to-month or quarter-to-quarter, which can lead to systemic fluctuations; (3) identifying and removing trends: trend changes can mask seasonal patterns; (4) flattening the data involves smoothing it, using methods such as differencing, to better adjust for seasonality. The X-13ARIMA-SEATS pre-adjustment relies primarily on the ARIMA and regARIMA models to clean the data.

3.3.1 ARIMA Model: Structure and Order Selection

As a widely used time series modelling method, the ARIMA (AutoRegressive Integrated Moving Average) model plays an important role in the pre-adjustment process before seasonal adjustment. The ARIMA model includes (1) AutoRegression (AR): AR is a linear regression that relates the past value of the time series with the future value; (2) Differential (Integrated-I): most of the economic time series are nonstationary series, and I represents how many times the data series must be differential to obtain the stationary series; and (3) the Moving average (MA): MA is the error from the future values of the past forecast data series.

Following the Box–Jenkins (1976) framework, the ARIMA model is expressed as (p, d, q) (P, D, Q). (p,d,q) stands for non-seasonality, and (P, D, Q) stands for seasonality. where p represents the autoregressive (AR) order, which indicates how the value of the past p period affects the current value; d represents the difference order, which is used to remove the trend to smooth the data; q stands for Moving Average (MA) order and represents how the error in the past q period affects the current value. These are the three main parameters that the ARIMA model must estimate.

3.3.1.1 AR Model

The autoregressive process of the p-order of the economic time series $\{Y_t\}$can be written as

$$Y_t = \varphi_1 Y_{t-1} + .. + \varphi_p Y_{t-p} + \varepsilon_t \tag{3.12}$$

or

$$\phi(B) Y_t = \theta(B) \varepsilon_t \tag{3.12a}$$

Y_t *regresses according to* Y_t *past value,* Y_{t-1}. If we take the AR(1) process as an example, it can be written as follows:

$$Y_t = \varphi_1 Y_{t-1} + \varepsilon_t \tag{3.13}$$
$$Y_{t-1} = \varphi_1 Y_{t-2} + \varepsilon_{t-1} \tag{3.14}$$

(3.14) successive substitution (3.13), it can be written as

$$\begin{aligned} Y_t &= \varphi(\varphi Y_{t-2} + \varepsilon_{t-1}) + \varepsilon_t \\ &= \varphi^2(\varphi Y_{t-3} + \varepsilon_{t-2}) + \varphi\varepsilon_{t-1} + \varepsilon_t \end{aligned} \tag{3.15}$$

Ultimately, Y_t *can be expressed in the form of an infinite-order MA process, provided by the* assumption.

$$\begin{gathered} -1 < \varphi < +1, \\ Y_t = \varepsilon_t + \varphi\varepsilon_{t-1} + \varphi^2\varepsilon_{t-2} + ... \end{gathered} \tag{3.16}$$

For simplicity, the backwards shift operator B can be used for equation (3.16), which can then be written out

$$(1 - \varphi B) Y_t = \varepsilon_t \tag{3.17}$$

thus

$$\begin{aligned} Y_t &= \varepsilon_t / (1 - \varphi B) \\ &= (1 + \varphi B + \varphi^2 B^2 + ...)\varepsilon_t \\ &= \varepsilon_t + \varphi\varepsilon_{t-1} + \varphi^2\varepsilon_{t-2} + ... \end{aligned} \tag{3.18}$$

In this form, $E(Y_t) = 0$ and $Var(Y_t) = \sigma_\varepsilon^2(1 + \varphi^2 + \varphi^4 + ...)$, if$|\varphi| < 1$, the variance is finite, where

$$Var(Y_t) = \sigma_x^2 = \sigma_\varphi^2(1 + \varphi^2) \tag{3.19}$$

The AR process is reversible, but not smooth.[8]

3.3.1.2 MA Model

The moving average MA process of the q-order of the economic time series $\{Y_t\}$can be expressed as follows:

$$Y_t = \theta_0 \varepsilon_t + \theta_1 \varepsilon_{t-1} + \ldots + \theta_q \varepsilon_{t-q} \tag{3.20}$$

where $\{\theta_t\}$ are constants because ε_s is usually scaled proportionally, the $\{\theta_t\} = 1$, $E(Y_t) = 0$ variance Y_t is

$$Var(Y_t) = \sigma_\varepsilon^2 \sum_{i=0}^{q} \theta_i^2 \tag{3.21}$$

The MA process is smooth, but it can be irreversible. For example, if the first-order MA process is in the case of A and B, it is as follows:

$$\text{A: } Y_t = \varepsilon_t + \theta \varepsilon_{t-1}$$

$$\text{B: } Y_t = \varepsilon_t + \frac{1}{\theta} \varepsilon_{t-1}$$

If these two different processes have the same autocorrelation function, the MA process cannot be uniquely identified from the given autocorrelation function. If the above A and B models ε_t are denoted as Y_t, Y_{t-1}, then A and B models can be represented as successive substitutions, ie

$$\text{A: } \varepsilon_t = Y_t - \theta Y_{t-1} + \theta^2 Y_{t-2} - \ldots$$

$$\text{B: } \varepsilon_t = Y_t - \frac{1}{\theta} Y_{t-1} + \frac{1}{\theta^2} Y_{t-2} - \ldots$$

If $|\theta| < 1$, the series of A converges, and the series of B does not, ie, model A is reversible, and model B is irreversible. The estimation procedure involves estimating the residuals, which will naturally lead to Model A. Imposing a reversibility condition on the MA process is a fundamental requirement that ensures the uniqueness of the MA representation for a given autocorrelation function.

3.3.1.3 ARMA Model

The combination of the MA process and AR process forms a useful class of time series ARMA models, including both the p-order AR term and the q-order MA term. ARMA is a process of order (p,q), which can be expressed as follows:

$$Y_t = \varphi_1 Y_{t-1} + \ldots + \varphi_p Y_{t-p} + \varepsilon_t + \theta_1 \varepsilon_{t-1} + \ldots + \theta_q \varepsilon_{t-q} \tag{3.22}$$

Using the backwards operator B, Equation (3.21) can be written in the following form:

$$\varphi(B) Y_t = \theta(B) \varepsilon_t \tag{3.23}$$

where $\varphi(B)$ and $\theta(B)$are polynomials of order p and q, respectively, such as

$$\varphi(B) = 1 - \varphi_1 B - .. - \varphi_p B^p \tag{3.24}$$

and

$$\theta(B) = 1 + \theta_1 B + .. + \theta_q B^q \tag{3.25}$$

$$Y_t = \psi(B)\varepsilon_i \tag{3.26}$$

3.3.1.4 ARIMA Model

Since most economic time series are nonstationary, to remove the source of nonstationary variation, the series needs to be differentiated, so that the original time series can be $\{Y_1, Y_2, ...Y_N\}$ transformed into a new time series $\{\nabla Y_{1,}, \nabla Y_2 .....\nabla Y_{N-1}\}$until it becomes stationary, ie, $(\nabla^d Y_t)$. Usually, a first- or second-order difference is sufficient to achieve significant stability. For example, if the original data model is $Y_t = \beta_0 + \beta_1 X_t + \varepsilon_t$, and both Y and X are nonstationary, and the error term is assumed ε_t to be nonstationary as well, then take the first difference of the equation to produce three new variables, ie,

$$\begin{aligned} Y_{t-1} &= \beta_0 + \beta_1 X_{t-1} + \varepsilon_{t-1} \\ \nabla Y_t &= Y_t - Y_{t-1} \\ \nabla Y_t &= \beta_0 - \beta_0 + \beta_1 X_t - \beta_1 X_{t-1} + \varepsilon_t - \varepsilon_{t-1} \\ &= \beta_1 (X_t - X_{t-1}) + (\varepsilon_t - \varepsilon_{t-1}) \\ \nabla X_t &= X_t - X_{t-1}, \ \nabla \varepsilon_t = \varepsilon_t - \varepsilon_{t-1} \end{aligned} \tag{3.27}$$

In this way, after the first-order difference, the regression to the new time series of the difference becomes

$$\nabla Y_t = \beta_1 \nabla X_t + \nabla \varepsilon_t \tag{3.28}$$

If all variables and their sums are stationary after first-order differencing, ie, I(1), then we can regress the β difference series, which is reliable with OLS estimation.

If you replace it withY_t^* a substitute $\nabla^d Y_t$to smooth out the original data, the comprehensive model can be written as

$$Y_t^* = \varphi_1 Y_{t-1}^* + ... + \varphi_p Y_{t-p}^* + \varepsilon_t + .. + \theta_q \varepsilon_{t-q} \tag{3.29}$$

By analogy with Equation (3.22), we can write Equation (3.29) in the following form

$$\varphi(B)(1-B)^d Y_i = \theta(B)\varepsilon_t \tag{3.30}$$

Thus, we have Y_t^* an ARMA (p, q) model for the Y_i sub-difference model, ie, the ARIMA process of order (p, d, q). If the value of d is taken as 1 and Y_i stationarity is achieved, the process can be represented as an ARIMA (0,1,0) model. After converting the nonstationary time series into stationary data, the most important stage in the ARIMA process is the identification and selection of the ARIMA model, that is, finding the appropriate order of AR(p) or MA(q) that is most suitable for the new differential series data.

3.3.1.5 Determination of the AR(p) Order

The order of an autoregressive (AR) model is usually denoted by p, ie, the AR(p) model. Determining the order of an AR model is generally combined with several methods:

1. Information Criteria: AIC, BIC, HQC of Information Criteria can be used to balance the goodness-of-fit and complexity of a model, and common methods include the following:
   a) AIC (Akaike Information Criterion):

$$AIC = -2\ln(L) + 2_P \tag{3.31}$$

   where L is the maximum likelihood function value, and p is the model order.
   b) BIC (Bayesian Information Criterion):

$$BIC = -2\ln(L) + P\ln(N) \tag{3.32}$$

   where N is the sample size, and the BIC penalises the model order more strictly than the AIC, favouring a simpler model.
   c) HQC (Hannan-Quinn Criterion):

$$HQC = -2\ln(L) + 2P\ln(\ln(N)) \tag{3.33}$$

   HQC sits between AIC and BIC and is suitable for order selection for larger samples. Typically, the order that minimises AIC, BIC, or HQC is chosen as the final AR model order.
2. Autocorrelation Functions (ACF) and Partial Autocorrelation Functions (PACF).
   a) Autocorrelation Function (ACF): A measure of correlation at different lag orders, the ACF of an AR process is usually exponentially decayed or oscillationally decayed. Autocorrelation (ACF) is a useful tool in the Box–Jenkins method for estimating ARIMA models. The strength and direction of the linear relationship between two variables are typically assessed using the correlation coefficient. For time series, correlation can refer to the relationship between their observations, for example, between the current observation and the observation of one or more units of time lag. In this case, all observations come from a single variable, so the similarity between a given time series and its k-lagged version over a continuous time interval is called autocorrelation. The autocorrelation coefficient at the lag k is defined as follows:

$$\rho(k) = \frac{\sum_{t=k+1}^{n}(x_t - \overline{x})}{\sum_{t=1}^{n}(x_t - \overline{x})^2} \tag{3.34}$$

   Among them, x_t is the time series; n is the total number of observations; $\overline{x}$ is the average of the time series. ρ(k) is the autocorrelation coefficient with a lag k.

ρ(1) x_t shows the x_{t-1} correlation of continuous values and ρ(2) shows how values x_t and values separated by two x_{t-2} time points are related to each other, and finally, ρ(k) is shown x_t to be related to the x_k point in time, ie, the values that are k points apart.

b) Partial Autocorrelation Function (PACF): A measure of the pure correlation of the lagged k-order to the current value after excluding the influence of intermediate orders.
    - In the AR(p) process, the PACF is truncated after the p-order, ie, the PACF is close to zero after the p-order.
    - Observing the order in which the PACF drops to zero can serve as a preliminary p-candidate.

3. Combine stepwise regression or cross-validation for final confirmation:
    - Stepwise regression: Use statistical testing methods, such as t-test or F-test, to perform stepwise regression on AR models of different orders, selecting significant lag orders.
    - Maximum Likelihood Estimation (MLE): The maximum likelihood estimation method is used to calculate the model of different orders, and the order with the highest likelihood value is selected.
    - Cross-validation: Evaluate the prediction error of AR models of different orders using cross-validation, such as selecting the optimal p through rolling window cross-validation.

3.3.1.6 Determination of the MA Order

The autocorrelation function of the MA(q) process is easy to identify because its 'truncation' is at the lag q point, while the autocorrelation function of the AR(p) process is a mixture of damped exponentials and sinusoids, and slowly disappears or decays. If the MA process is deemed appropriate for a given dataset, the order of the process is usually evident from the sample autocorrelation function. The theoretical autocorrelation function of the MA(q) process has a very simple form in that its 'truncation' is right at the lag q, so the analyst should look for a lag beyond which r_k the value is close to zero. The partial autocorrelation function is usually of little help in identifying MA models due to its attenuated form.

3.3.1.7 Determination of the Order of the ARMA Model

The autocorrelation function of hybrid ARMA models is also often attenuated rather than 'truncated'. For example, suppose we find that we are ρ_k significantly different from zero, but the ρ_k subsequent values are all close to zero, then the MA (1) model is indicated because its theoretical autocorrelation function is of this form. Alternatively, if ρ_1, ρ_2. . . . . . is decreasing exponentially, the AR (1) model may be appropriate. An experienced analyst should be able to look at the estimated correlation plots and justify a particular AR, MA, ARMA, or ARIMA model based on their perception and experience. While some statistical software can provide automated ARIMA modelling, this method is primarily used for non-automated methods, in which the analyst uses his or her subjective judgment to select the appropriate model from the large ARIMA model family, based on the characteristics of the individual series analysed. Thus, although the program is more versatile than many of its competitors, it is also more complex and requires considerable experience to determine the appropriate ARIMA model.

In the X-13ARIMA-SEATS, an automatic selection procedure is implemented to identify the ARIMA order. The default model is estimated. The default model for quarterly series is (0,1,1) (0,1,1), also known as the 'airline' model. This model is concise (only two parameters are estimated) and is generally well-suited to economic time series. Regression effects are also identified and removed using the default model. The difference orders (d and D) are estimated by performing a series of unit root tests. The ARMA[9] Order (p, q) (P, Q) is selected by comparing the statistical standard values of multiple models,[10] up to the maximum order of the user-specified regular and seasonal ARMA polynomials. Compares the residual diagnostics of the selected ARIMA model with the residual diagnostics in the default model. Based on these tests, the final model is selected and validated.

Choosing the right ARIMA model has a significant impact on the seasonal adjustment process:

a) The order in which ARIMA models are selected can affect the automatic selection process for calendar effects and outliers.
b) ARIMA models are used to generate predictions and back-predictions that are essential for solving a series. The asymmetry of the filter at the endpoint is necessary.
c) The ARIMA order chosen for the original series is central to the SEATS decomposition because of seasonality and trend.

The potential filter is derived from the coefficients of the estimated ARIMA model.

Generally, the user should consider the ARIMA order of the X-13ARIMA-SEATS automatic selection program as a benchmark model. When model statistics are not satisfactory, changes should be made to the benchmark model. For example, the model should be changed when the diagnostic signal for the residuals is incorrectly specified in the automatically recognised model. Simple models should generally perform better than complex ones. In most cases, the differential orders d and D are either 0 (if the component is stationary) or 1 (if the component is nonstationary). For some series of continuous fluctuations (eg, price data or deflators), multiple differences may occur. In general, you should avoid hybrid ARMA models, where both AR and MA operators exist in both seasonal and nonseasonal portions. For most series, the default model (0,1,1) (0,1,1) usually works well with seasonal economic time series and should be considered when no other model provides satisfactory results.

ARIMA can also use dummy variables to detect outliers such as additive outliers (AO), level shift (LS), trend change (TC), etc. For example,

1. Additive Outliers (AO):

$$Y_t = \mu + \omega I_t + e_t \tag{3.35}$$

The I_t value is 1 at time t, and the rest is 0

2. Level Shift (LS):

$$Y_t = \mu + \lambda S_t + e_t \tag{3.36}$$

Among them, S_t has been taken as 1 after the change.

3. Trend Change (TC):

$$Y_t = \mu + \gamma T_t + e_t \tag{3.37}$$

where T_t represents the trend change term.

Outliers are identified by residual analysis, statistical tests, etc., and corrected by interpolation or regression methods. However, since the introduction of regARIMA in X-12-ARIMA, both X-12-ARIMA and X-3ARIMA-SEATS have relied heavily on regARIMA to identify and adjust for calendar effects (trading day effects) and other corrections.

3.3.2 regARIMA Model Structure: Regression and ARIMA

regARIMA refers to the statistical modelling tools in X-12-ARIMA and X-13ARIMA-SEATS, which are mainly used for pre-adjustment of seasonal adjustments. regARIMA is essentially a generalised linear model with an ARIMA error structure:

$$y_t = x_t' \beta + z_t, \quad \Phi(B) z_t = \Theta(B) e_t \tag{3.38}$$

The ARIMA part of regARIMA, an autoregressive integrated moving average model, models the trend and seasonal features of the series. In X-13ARIMA-SEATS, the program can select the largest number of time series, allowing the appropriate ARIMA model form to be chosen for a single series using the program's built-in model fitting criteria, or the user can specify the form of the ARIMA model to apply. The regression portion of regrARIMA refers to the option to augment the basic ARIMA model with other regression variables. These variables may take the form of dummy variables used to model the effects of outliers, level shifts, and similar phenomena, or they may be direct explanatory variables representing special events or other external influences. Technically, regARIMA can be described as a linear regression, where $y_t = x_t' \beta + z_t, x_t$ is a vector of regression variables, but the error term is not white noise z_t process, but rather an ARIMA process that follows.

The transformation of the series is the first step in fitting the regARIMA model to the series, depending on whether the series is additive or multiplicative. When the series is additive, the regARIMA model fits directly to the original series; when the series is multiplicative, the model fits to the original series after logarithmic transformations. The effect of the logarithmic transformation is to change the scale of the fitting model series, then convert any predictions or previous adjustments estimated by the model back to the original scale, and convert the multiplicative effect into an additive effect. Once you specify a model, you can use it to generate predictions and backtrack predictions. The variables in the regression section of the model give estimates of calendar and other effects.

The regARIMA model is a regression model with ARIMA time series errors, which uses a regression framework to model the observed time series and assumes that the modelled residuals follow the ARIMA process. The regression part of the model is used for seasonality-related effects, such as additive outliers (AO), level shifts (ls), seasonal breaks (sb), temporary changes (TC), and trading day effects (trading day, TD), leap year (ly), Christmas holiday, Easter holiday, and other user-defined variables as regression inputs to the model (eg, Chinese New Year, Dragon Boat Festival, Mid-Autumn Festival). They are variables that explain the behaviour of the time series in some way, and in the time series seasonally adjusted model, these influencing factors can be used as regression variables for the regARIMA model ($x_1, x_2 \ldots, x_k$) included in the model. The regARIMA model can be seen as a generalised regression model that allows the error to follow an ARIMA process, or as an ARIMA model generalised to allow a regression mean function. regARIMA is applied in X-13ARIMA-SEATS.[11]

3.3.2.1 regARIMA Model Structure

For a general 'multiplicative' seasonal ARIMA model of a time series Y_t, it can be written as follows:

$$s\varphi_p(B)\Phi_P(B^s)(1-B)^d(1-B^s)^D Y_t = \theta_q(B)\Theta_Q(B^s)\alpha_t \tag{3.39}$$

Thereinto:
Y_t is the original time series;
B is the backwards shift operator, $By_t = y_{t-1}$
p and P are non-seasonal and seasonal AR orders, respectively;
q and Q are non-seasonal and seasonal MA orders;
d and D are non-seasonal and seasonal I orders, respectively;
s is the seasonal period (s = 4 is quarterly, s = 12 is monthly);
t is the period.

$\varphi_p(B)$ is a non-seasonal AR operator, and $\varphi(B) = 1 - \varphi_1 B - ... - \varphi_p B^p$;
$\Phi_p(B^s)$ is a seasonal AR operator, and $\varphi(B^s) = 1 - \varphi_1 B^s - ... - \varphi_p B^{ps}$;
$\theta_p(B)$ is a non-seasonal MA operator, and $\theta(B) = 1 - \theta_1 B - ... - \theta_p B^q$;
$\Theta_p(B^s)$ is a seasonal MA operator, and $\Theta(B^s) = 1 - \Theta_1 B^s - ... - \Theta_p B^{Qs}$;
α_t is an error term with *the property i.i.d.* $(0, \sigma^2)$

If the factors influencing the seasonal adjustment of the time series Y_t [12] include additive outliers, level shifts, seasonal disruptions, temporary variations, 'trading day effects', 'leap year effects', Easter effects, and other holiday effects such as moving bank holidays, etc. These factors can be included in modelling the series in a regression framework.

Assuming that the observed time series has been transformed as needed (eg, taking a logarithmic pattern (logs) in the case of 'multiplicative' seasonal decomposition, the series can be modelled using a regression framework, and the regression model can be

$$y_t = \sum \beta_i x_{it} + z_t \tag{3.40}$$

Where: y_t is the observed time series; x_{it} is a regression variable; β_i is a regression parameter; z_t is the model residual, ie, the regression error term. The regression parameter can be thought of as the magnitude of each regression effect. Thus, each model residual can be considered as the total value of the observed time series minus the regression effect. Model residuals can be expressed as

$$z_z = y_t - \sum \beta_i x_{it} \tag{3.41}$$

Model residuals can be swapped z_t Y_t and merged into the ARIMA model as shown below (3.39):

$$\varphi_p(B)\Phi_p(B^s)(1-B)^d(1-B^s)^D Y_t = \theta_q(B)\Theta_Q(B^s)\alpha_t$$

regARIMA can be expressed as

$$\varphi_p(B)\Phi_P(B^s)(1-B)^d(1-B^s)^D\left(y_t-\sum_{i=1}^{r}\beta_i x_{it}\right)=\theta_q(B)\Theta_Q(B^s)\alpha_t \tag{3.42}$$

Where: s is the length of the seasonal period, $s=4$ is the quarterly data, and $s=12$ is the monthly data. p, q and P, Q are the order of the polynomial, respectively $(\varphi_p(z),\Phi_P(z),\theta_q(z),\Theta_Q(z))$, and the constant term is equal to 1.

The residual regression model is used because the residuals may be autocorrelated, which violates one of the standard assumptions of the regression framework, ie, the error terms must be independent. Therefore, the residuals of the regression model must first be differenced in order to obtain a stationary time series. For example, if $p\geq 1$, we have $\varphi_p(z)=1-\varphi_1 z-\ldots-\varphi_p z^p$. These polynomials are subject to constraints such that the moduli of the zeros of $\theta_q(z)$ and $\Theta_Q(z)$ are greater than or equal to 1, and (during the estimation process, where this is set by default) the moduli of the zeros of $\phi_p(z)$ and $f_P(z)$ are greater than 1. Since α_t is assumed to be a series of independent variables with mean 0 and constant variance σ_α^2, it follows from these constraints that

$$w_t=(1-B)^d(1-B^s)^D(y_t-\sum_{i=1}^{r}\beta_i x_{it}) \tag{3.43}$$

is to satisfy the difference equation

$$\phi_p(B)\Phi_P(B^s)w_t=\theta_q(B)\Theta_Q(B^s)\alpha_t \tag{3.44}$$

of the covariance-stationary time series. Thus, we can represent Y_t's model (3.42) as

$$(1-B)^d(1-B^s)^D y_t=\sum_{i=1}^{r}\beta_i\{(1-B)^d(1-B^s)^D x_{it}\}+w_t \tag{3.45}$$

where w_t is to follow the stationary ARMA model requirements.

Equation (3.45) highlights the fact that both the observed time series and the time series of the regression variables require the use of a new time series, ie, the time series after the 'difference' has been 'differential' to eliminate the trend, to smooth the new time series so that the regression model residuals follow a stationary process. w_t is stationary and can thus be incorporated into the ARMA model. We can obtain the adjusted time series y_t^* by recursively applying w_t to the observations in the observed time series y_t.

The regARIMA model implies that (1) the combined regression effect is subtracted from the observed time series; (2) the resulting series is 'differential' (non-seasonal and seasonal); (3) it is then assumed that the series follows a stationary ARMA model.[13]

3.3.3 The regARIMA Model Parameter Estimation

Particular attention should be paid to the regARIMA model parameter estimation method. regARIMA is a regression model with a steady-state ARMA *error* for y_t after appropriate differences w_t. Its regression is generated by x_{it} applying the same difference operation to the pair. The assumption of the error w_t in model (3.45) and error in model (3.44) are *i.i.d.* N $(\sigma_\alpha^2 0,)$. The estimated regression coefficients in the model are β_i, and σ_α^2; the estimated ARMA parameters in the model are $\varphi_p(B),\Phi_p(B^s)$ and coefficient $\theta_q(B)_Q,\Theta_Q(B^s)$. The estimate method uses the maximum likelihood function, and the default likelihood setting used in the X13-ARIMA-SREATS program is the perfectly accurate Gaussian likelihood to help circumvent the convergence problem in numerical maximisation.

The X13-ARIMA-SEATS program did not use the OLS method to estimate regression parameters outside of the ARIMA model. Instead, it employs maximum likelihood estimation using the iterative generalised least squares (IGLS) algorithm proposed by Harvey (1990) to estimate both the regression parameters and the ARMA parameters.[14] The process consists of two regular steps:

1. For ARMA parameters, an initial value is given, and then the regression parameter is estimated, or
2. For the regression parameters, an initial value is given, and then the ARMA parameters are estimated.

If the regARIMA model includes only the regression-AR (regAR) component and excludes the regression-MA (regMA) part, OLS can be used to maximise (1) the likelihood function of the AR parameter given the initial regression parameter, and (2) the likelihood function of the regression parameter given the initial AR parameter. A change in the estimated AR parameters will result in a change in the estimated regression parameters, which in turn will lead to a change in the estimated ARMA parameters, and so on. According to some convergence criteria, the iterative process continues until data convergence is achieved. In the case of the given observational data, the selected ARMA model, convergence criterion, and initial parameter values, the ARMA and regression parameters that ultimately achieve convergence are regarded as the most probable and optimal estimates.

If the regARIMA model involves only regMA parameters and does not include the regAR part, the estimation process is similar to that of the regAR model, although not identical, because the error term cannot be specified as an independent variable in the linear regression model.

For regARIMA models that involve both AR and MA parameters as well as regression parameters, the iterative routines are slightly more complex: estimated jointly using nonlinear least squares for ARMA parameters and generalised for regression parameters least squares), which is a linear regression method that estimates regAR and regMA separately. In this way, as with the regAR and regMA models, a relatively difficult problem is reduced to two relatively simple problems.

Another approach is to use nonlinear least squares to jointly estimate the AR or MA parameters along with the regression parameters; however, this method requires considerably more computation. Because the ARMA parameter estimation is nonlinear, it is difficult to obtain an accurate estimate of the ARMA model parameters. Theoretically, co-estimating AR or MA parameters and regression parameters using the nonlinear least squares method is also an estimation method for ARMA model parameters. However, they have the disadvantages of high computational complexity and no guarantee of convergence. Thus, a suboptimal approach commonly used in engineering involves estimating the AR and MA parameters separately rather than simultaneously performing optimal parameter estimation. In estimating ARMA parameters, the AR parameters are first estimated, followed by the estimation of MA parameters based on the AR-estimated parameters. Estimates of the ARMA model are then obtained. Thus, the parameter estimation of the MA model is typically carried out as part of the broader ARMA parameter estimation process. The model was then statistically tested and analysed. However, an iterative approach may be better than a non-iterative approach, which has the advantage of reducing a relatively difficult nonlinear problem to two relatively simple linear problems.

One implication of this estimation method is that the size of the adjustments made to the observed time series depends on (among other factors) the choice of the ARMA model that fits the data. Re-identifying the order of the ARIMA model will result in a re-estimation of the model parameters, including regression parameters, until a new combination of parameter values is deemed optimal, resulting in more accurate estimates.[15]

3.3.4 The regARIMA Parameter Estimation Process

1. Y is first log-transformed, differential, and then Y is regressed to X (log-transformed, differential, hysteresis), just like in a standard OLS regression model, and after determining the structure of the regression residuals, the corresponding regARIMA model is fitted. In this model, (log-transformed, differential) *Y* regresses on (log-transformed, *differential, hysteresis)* X, and the regression residuals are modelled as some ARMA process. This process cannot be estimated using OLS and requires processes such as maximum likelihood estimation or iterative GLS.
2. Once the final regARIMA model has been determined, check the significance and direction of the estimated X-regression coefficients. An absolute value of about 2 for t indicates significance at the 5% level. Alternatively, if the computer output does not show *a p*-value, a *p*-value can be calculated from a standard statistical table for the t-distribution; The appropriate calculation of degrees of freedom is $n - k - 1$, where n is the number of valid observations (sample size) and k is the number of parameters estimated in the regARIMA model (the number of regression parameters plus the number of ARMA parameters), and then the error variance is checked. If the coefficient on X is significant, then its orientation should also be checked accordingly. For example, if the coefficients of interest should be positive (by substantial reasoning about positive correlations), but the estimated results are negative, in this case, if there are multiple *X* lags in the model, perform a backwards deletion check on the regression variables, removing them from the model one at a time until all the regression variables are statistically significant.
3. The regression variables ($X_1, X_2 \ldots, X_n$) in the regression residuals were filtered, and only those with statistically significant test values were retained and incorporated into the regression component of the model. Estimation was then performed using methods such as maximum likelihood estimation or iterative generalised least squares (GLS).
4. It is important to note that whether you fix model parameters that were estimated during the initial fit or re-estimate them each time you refit the model to generate a new prediction, you should reflect the real-world production environment as much as possible. For instance, if a model parameter is re-estimated annually but employed monthly to produce one-step-ahead forecasts during the intervening months, the forecast evaluation should replicate this procedure as closely as possible. Conversely, if the model parameters remain fixed at their initial estimates, any additive outliers or level shifts that arise during the forecast horizon will not be incorporated, thereby diminishing the model's predictive accuracy in subsequent periods. More generally, as more and more observations are added to the end of the series, fixed model parameters will become less and less reflective of the true behaviour of the data. Thus, model parameters should be re-estimated as frequently as possible.

3.3.5 Calendar, Workday, Easter Effect Adjustments

3.3.5.1 Calendar Effects Adjustment

regARIMA allows for direct modelling of calendar effects (eg, weekday, holiday effects) by introducing regression variables:

$$Y_t = X_t\beta + ARIMA(p,d,q) + e_t \tag{3.46}$$

Among them, X_t there are calendar effect variables, such as the number of working days and holiday adjustments.

3.3.5.2 Weekday Effect Adjustment

regARIMA can more accurately correct the weekday effect for different periods.

$$Y_t = \beta_0 + \sum_{i=1}^{k} \beta_i X_{i,t} + e_t \tag{3.47}$$

Among them, the different working days X_{ij} variables are represented, and the β_i corresponding impact coefficients are represented. The following six regression variables define the most commonly used trading day effect:[16]

$$\begin{aligned} td_t^1 &= (\text{Monday to Sunday}) \\ td_t^2 &= (\text{Tuesday to Sunday}) \\ &\vdots \\ td_t^6 &= (\text{Saturday to Sunday}) \end{aligned} \tag{3.48}$$

It calculates the difference between the quantities for each day of the week (Monday, Tuesday, . . ., days on Sunday),[17] Different days of the week may affect potential economic activity by different magnitudes and directions. The weekday effect is captured by a regression variable that compares the weekday group (eg, Monday through Friday) to the weekend day group (eg, Saturday and Sunday) by the formula:

$$wd_t = \left(\text{weekday} - \frac{5}{2}\,\text{weekend} \right) \tag{3.49}$$

A 5/2 factor is required to make the weekday regression variable zero in the regular 7-day week combination. Any monthly deviation from the standard week will be reflected in the regression variable (eg, when greater than zero w_{et}.). This means that the month/quarter t has more working days than the standard week. This method assumes that weekdays have a similar effect and are different from weekend days. ARIMA can only be adjusted indirectly with the X-13ARIMA-SEATS.

3.3.5.3 Easter Adjustment

Easter dates move between March (Q1) and April (Q2).[18] The actual proportion of days before Easter was calculated using the Easter regression variables, specifically March (Q1) and April (Q2). After defining the length of the Easter effect, the regression amount is calculated as follows

$$e_t = \frac{W_t}{w} - \overline{W} \tag{3.50}$$

where W_t is the number of days in the month/quarter t; $\overline{W}$ is the long-term number of days for the month/quarter. Usually, $\overline{W}$ can be approximated 0.5 for March (q1) and April (q2),[19] that is, the number of days of the Easter effect is evenly distributed between the two periods. In X-13ARIMA-SEATS, the effect of the $\overline{W}$ length can be provided by the user (from 1 to 25) or automatically selected by the program (compare the lengths of 1, 8, and 15). Finally, the leap year effect is captured in the following way:

$$ly_t \begin{cases} 0.75 & \textit{if } t \textit{ is leap year Febury} \\ -0.25 & \textit{if } t \textit{ is non-leap year Fefury} \\ 0 & \textit{others} \end{cases} \tag{3.51}$$

The regression variables ly_t produce a deterministic four-year cycle, with a peak in February of a leap year. Over four years, the effect of leap years fully outweighs the negative effects of subsequent non-leap years.

Adjustments for calendar effects should be made only for those series for which there is both statistical evidence and economic explanations for calendar effects. This assessment should be based on the statistical and economic significance of its regression coefficients. Statistically speaking, the regression coefficient is considered to be significantly different from zero. When the associated *t*-statistic is above a certain threshold (usually an absolute value of 2), a lower threshold is acceptable. In addition, the sign of the regression coefficient should be economically interpretable. For example, the leap year effect should always be positive, the weekday effect of economic activities with a five-day production week should be positive, and the Easter effect should have a positive effect on the consumption of tourism-related services,[20] while other production activities should be negative, and so on. A series should not be provided for a calendar effect when the estimated coefficient for that effect is not statistically significant (ie, the *t*-statistic is below the selected threshold) or difficult to explain with economic significance (ie, the magnitude or sign of the coefficients is not credible).

It is important to note that the calendar effect is typically considered in monthly economic series rather than quarterly series, as calendar effects tend to cancel out in quarterly data, making them difficult to detect during estimation.

The X-13ARIMA-SEATS contains a process for automatically identifying additive outliers, temporary variation outliers, and level shifts.[21] The program includes the inclusion of dummy-type variables in the regression model for all possible periods within a specified period. The program calculates the regression coefficients for each specified type of outlier and adds all outliers whose statistically significant absolute value of *t* exceeds the critical value to the model.[22]

X-13ARIMA-SEATS allows users to incorporate specific calendar effects, such as Lunar New Year or Ramadan, with the capability to automatically select predefined intervention variables.[23] All regression effects should be based on the statistical significance of their regression coefficients (*t*-statistic value) and the economic and statistical significance of the results. For larger time series, an automatic selection process should always be used. In addition, as new observations are added to the series, the regression coefficients estimated to be associated with calendar effects should remain stable, and calendar effects that are not supported by economic justification should not be included in the adjustment.[24]

In addition, the X-13ARIMA-SEATS also allows user-defined regression variables in the regARIMA model,[25] such as Chinese New Year, Mid-Autumn Festival, Dragon Boat Festival, etc., which are related to calendar effects. These can be used in the regARIMA model as input to the program.

3.4 Summary and Conclusions

This chapter focuses on the general theoretical conceptual framework of seasonal adjustment.

1. A time series is a continuous observation of economic activity (eg, GDP, unemployment rate, etc.) recorded at fixed intervals (eg, monthly, quarterly), divided into stock (measurement at a certain point in time, eg, unemployment rate) and flow (cumulative over a certain period, eg, retail sales). Traffic series are more susceptible to calendar effects. Time series analysis aims to reveal patterns of their behaviour, which can be used to predict future trends.
2. The volatility of the original time series consists of implied trend, periodicity, seasonality, and irregular components. It is usually simplified into a three-part model $Y_t = T_t + S_t + I_t$, where the seasonal component covers repetitive effects such as weather and holidays, while the trend contains cyclicality. The amplitude of the seasonal component changes with the trend in the multiplicative model and remains stable in the additive model.
3. The choice of the time series decomposition model depends on the relationship between the components: if the components are independent and the seasonal amplitude is stable, the additive model is used; if the components are related to the trend level, the multiplicative model is used; if there is a series of multiplicative relations with zero values, the pseudo-additive model is chosen. The model is determined by fitting the data, and the multiplicative model can be converted to an additive form by logarithmic transformation. The seasonally adjusted series is the original series minus (or divided by) the seasonal component.
4. The identification of the model can be used to observe whether the seasonal changes with the trend graphically: if it is stable, it is an additive model, and if it is fluctuating, it is a multiplicative model. The X-13ARIMA-SEATS provides an automatic selection function to compare the fitting effect before and after logarithmic conversion based on the AICC value.
5. Seasonal factors include calendar effect, moving holiday effect, and outliers. These effects need to be corrected by pre-adjustment (regARIMA) to avoid distorting seasonal estimates.
6. Outliers fall into four categories: additive outliers (AO), level shifts (LS), transient-changes (TC), and slope outliers (RP). Outliers need to be detected and corrected by regression models (eg, dummy variables) or they will interfere with component estimation.
7. Pre-adjustment cleans the data through ARIMA and regARIMA models to handle outliers and calendar effects, where regARIMA is more flexible (supports user-defined regression variables such as Lunar New Year).

8. The ARIMA (Autoregressive Integrated Moving Average) model consists of three parts: autoregressive (AR), differential (I), and moving average (MA), which improves the model with historical prediction error. The model is expressed as (p, d, q) (P, D, Q), corresponding to non-seasonal and seasonal order, respectively. The order is determined by ACF/PACF or information criteria (eg, AIC), and the default 'airline model' (0,1,1) (0,1,1) is suitable for most economic series. ARIMA also detects outliers (eg, AO, LS).
9. The regARIMA model combines regression and ARIMA. The regression part attempts to model calendar effects and outliers and use dummy variables for working day and Easter variables; ARIMA part treats the time dependence of the residuals. The parameters are jointly optimised by maximum likelihood estimation and IGLS. The model needs to be statistically significant (eg, $|t|>2$) and economically justified (eg, the working day effect should be positive).
10. The core difference between ARIMA and regARIMA is that the latter directly models the calendar effect, while the former only adjusts indirectly through the difference.
11. Model diagnostics (eg, residual testing) and parameter stability are key to ensuring the quality of adjustments. Seasonally adjusted data are interpreted within the context of their economic significance and may be revised as new data become available.
12. This chapter lays the theoretical foundation for subsequent chapters (the X-11 method and the SEATS method) and provides a standard tool for economic data analysis. The X-11 moving average method will be discussed in Chapter, and the SEATS (parameter) method will be discussed in Chapter 7.

Notes

1 Findley, D. F., Monsell, B. C., Bell, W. R., Otto, M. C., & Chen, B.-C. (1998). "New Capabilities and Methods of the X-12-ARIMA Seasonal Adjustment Program". *Journal of Business & Economic Statistic*, 16(2), 127–152. https://doi.org/10.1080/07350015.1998.10524752.

2 In some cases, a hybrid model can be chosen. In particular, X-12-ARIMA includes a pseudo-addition model that $Y_t = T_t(S_t + I_t^{-1})$ shows sequences of multiplicative decomposition schemes, but whose value is zero in certain periods.

3 In time series analysis, a factor usually refers to the seasonal factor. In a multiplicative model, the seasonal component is usually expressed as a percentage or factor of the trend-cycle component. For example, if St=1.2, the current season's value is 20% above the trend level, and 0.8 means 20% below the trend.

4 We will give a detailed discussion of AIC-based and AICC-based options in Chapter 5.

5 The use of the regARIMA model for pre-tuning will be explained in more detail later.

6 For seasonality adjustment purposes, outliers that produce structural breaks in a sequence, such as horizontal shifts or seasonal outliers, can be assigned to trend or seasonal components. For more details on outlier assignments, see the Pre-Adjustment section.

7 If you can use the following RP model: $RP_t^{(t_0,t_1)} = \begin{cases} -1 \; for \; t \le t_0 \\ \dfrac{t-t_0}{t_1-t_0} - 1 \; for \; t \le t_0. \\ 0 \; for \; t \ge t_1 \end{cases}$

8 We are not prepared to give detailed proof here.
9 ARMA stands for Autoregressive Moving Average. In contrast to the ARIMA model, the ARMA process is smooth and does not require differentiation.
10 The X-13A-S uses the Bayesian Information Criterion (BIC) to select the ARMA order. The best model is the one with the smallest BIC value.
11 The X-13ARIMA-SEATS is software developed by the U.S. Census Bureau and supported by the U.S. Census Bureau. It is based on the X-11-ARIMA software developed by Statistics Canada, which was later developed by the U.S. Census Bureau as X-12-ARIMA and further developed as X13-ARIMA-SEATS. Similar software includes JDemetra+, developed by Eurostat. See also Findley, D. F., Monsell, B., Otto, M., Bell, W., & Pugh, M. (1988). *Toward X-12-ARIMA*. U.S. Census Bureau. Retrieved from http://www.census.gov.
12 If the observed time series is y_t a 'multiplicative' time series, it needs to be transformed by logs before regression can be performed.
13 See Findley, D. F., Monsell, B., Bell, W., Otto, M., & Chen, B. C. (1998). *"New Capabilities and Methods of the X-12-ARIMA Seasonal Adjustment Program"*. U.S. Census Bureau. Retrieved from.
14 See Harvey, A. C. (1990). *"The Econometric Analysis of Time Series"*. MIT Press. ISBN: 978-0262081897; Bell, R. W. (2004). "On Regime-Switching in Stochastic Volatility Models: Exact Likelihood Inference and Forecasting". *Journal of Econometrics*, 118.
15 For detailed estimation methods and procedures, please refer to the x13-arima-seats user guide and detailed x13-arima-seats outputs for estimation iterations are available. Under the program's default settings, for the AR and MA parameters, the identified model is initialized with an ARMA parameter value of 0.1. The default convergence tolerance is the change in the log-likelihood of the model that is less than absolute 0.00001 between IGLS iterations. Once this condition is met, the estimation routine stops and the final ARMA and regression parameter values are determined. See also U.S. Census Bureau (2007). *X-12-ARIMA Reference Manual*, version 0.3 (Beta), and X3-ARIMA-SEATS Reference Manual. Retrieved from http://www.census.gov.
16 The following describes the Monday-Friday work week (the default work week in X-13A-S) based on the implementation in most Western countries. However, the work week is different in other countries (especially Sunday to Thursday in many Muslim countries). The group of weekdays and weekends should be defined according to the statutory working week in the country.
17 National holidays shall be considered as non-working days. Therefore, the number of non-working days should be increased according to the number of national holidays, and the number of working days should be reduced accordingly. The same goes for weekday returners.
18 Only Catholic Easter is considered here. Orthodox Easter falls between April and May, so it does not affect the seasonal series.
19 The amount of Easter regression can also be non-zero in February, but this rarely happens.
20 In some countries, Easter has caused a spike in retail trade activity due to increased household spending; However, in other countries, most shops are closed during the Easter holidays.
21 In X13-ARIMA-SEATS, the specification name used to automatically identify outliers is OUTLIER.
22 The default cut-off value for the outlier t-statistic depends on the sequence length.
23 In X13-ARIMA-SEATS, the canonical name used to automatically recognize calendar effects is REGRESSION. The test is based on the Akaike Information Standard.
24 The X13-ARIMA-SEATS also provides another option for estimating trading day and Easter effects based on ordinary least squares (OLS) regression analysis of the final irregular components. The spec name for this option is X11 regressor. However, regARIMA models for estimating calendar effects are preferred because they generally provide better results than OLS regression.
25 In X13-ARIMA-SEATS, the canonical name for including the intervention variable in the regARIMA model is Regression Variable.

4

X-11 Method Based on Moving Averages

4.1 Introduction

Moving average methods constitute the core technique employed in the X-11 family[1] for time series decomposition. The X-11 family is based on a nonparametric modelling approach, relying on a set of predefined linear filters, specifically, moving averages, to adjust the series. The choice of filters is ultimately determined by the characteristics of the data series being processed.[2]

The moving average method does not rely on complex prior concepts or models. It is conceptually simple and highly flexible in application. It demonstrates strong performance in trend extraction, seasonal adjustment, and noise reduction. Moving averages have a long and rich history. According to Klein (2009),[3] the Bank of England's transformation of monetary data aimed to reveal patterns in gold and silver holdings while obscuring actual values. This process led not only to the first publication of a proportionally scaled index series in 1797 but also to the first publication of a moving average series in 1832. The use of a third-order asymmetric moving average allowed the Bank not only to conceal actual values but also to exploit the time lag introduced by the moving average to implement monetary policy.

Beginning with John Finlaison in 1829,[4] most of the early publications on moving averages appeared in actuarial journals in the United States, the United Kingdom, and Scotland. Finlaison (1829) constructed one of the earliest life tables based on age-graded individual observations and developed what was known as 'graduation by the summation method' into what we now recognise as *smoothing by moving averages*, particularly in the context of time series. Seeking a unified underlying trend that reflects the natural patterns in the data, Finlaison successfully derived a symmetric nine-point moving average formula after various experimental adjustments. This piecewise smoothing method was later extended by the Italian statistician Giovanni Schiaparelli (1866),[5] and the first systematic study of such averages was carried out by American mathematician E. L. De Forest (1873, 1875, 1876, 1877).[6]

The Dutch meteorologist Buys-Ballot[7] is generally recognised as one of the earliest practitioners of time series decomposition. In 1847, he analysed periodic temperature variation by modelling trends with polynomials and seasonal patterns with indicator variables, implicitly applying linear regression techniques for parameter estimation. The idea of decomposing time series gained further traction in the works of economists such as William Stanley Jevons (1862),[8] whose economic studies were often concerned with revealing *cycles*, a central concept in explaining and predicting economic crises.

In a 1919 paper presented to the Edinburgh Mathematical Society, E. T. Whittaker proposed an alternative graduation method[9] by deriving approximate solutions to different equations. He introduced a technique that minimised the sum of squares of third differences as a measure of irregularity, and the sum of squares of differences between fitted

DOI: 10.1201/9781003716655-4

and observed values as a measure of deviation from the observed data.[10] This method gained considerable popularity and is now known as the Whittaker–Henderson graduation.[11] The Whittaker–Henderson criterion was later used by Akaike and Ishiguro (1980)[12] as a smoothness prior in Bayesian models for seasonal adjustment, implemented in the computer program BAYSEA, and applied in spline smoothing.[13] The Whittaker–Henderson minimisation problem, which balances fidelity and smoothness, provides a comprehensive and globally convenient framework for understanding many properties of moving average methods proposed in the literature.

At the beginning of the 20th century, a significant portion of time series research focused on estimating seasonal variation. For example, Persons (1919)[14] developed the *link relative method*, which became one of the earliest major approaches to seasonal adjustment. Frederic R. Macaulay,[15] working at the National Bureau of Economic Research, improved the decomposition process in the 1920s by introducing the ratio-to-moving-average procedure, a variant of which has been widely used from the X-11 method in 1955 to the current X-13ARIMA-SEATS.

This chapter focuses on the moving average method. Following the introductory Section 4.1, Section 4.2 discusses time domain decomposition and moving averages, including brief definitions of simple moving averages, symmetric and asymmetric moving averages, weighted and composite moving averages, with particular emphasis on weighted symmetric moving averages and their desirable properties; Section 4.3 explores frequency domain decomposition, introducing important concepts and analytical tools such as gain and phase shift functions associated with moving averages; Section 4.4 details the X-11 algorithm, outlining its four processing steps and five iterative procedures; Section 4.5 analyses the properties and implications of different moving average filters; and finally, Section 4.6 presents a summary and conclusion.

4.2 Time Domain Analysis: Types and Characteristics of Moving Averages

The X-11 method uses moving averages (MAs) to estimate the unobservable components of a time series: trend, seasonal, and irregular. Time series can be analysed from two perspectives: (1) in the time domain and (2) in the frequency domain.

In the time domain, a time series is generally defined as a series of observed values $\{ Y_t \}$, where $t = 1,2,\ldots, T$. The core idea of the X-11 method is to replace the original data points with the average of a set of their neighbouring values. By applying various types of moving averages to smooth the time series, short-term fluctuations are reduced, allowing long-term trends and cyclical components to be extracted. At its core, the method decomposes a time series into its constituent components through smoothing techniques. Moving averages help eliminate irregular variations through smoothing and assist in identifying trends and seasonal patterns, making them one of the most commonly used techniques in seasonal adjustment of time series.

4.2.1 Simple Moving Average: Definition and Computation

Moving averages are fundamental to the X-12-ARIMA seasonal adjustment methodology, serving as essential tools for decomposing time series. Suppose $\{Y_t\}$ represents a time series. It is typically decomposed into three components: trend $(T)_t$, seasonal (S_t), and irregular (I_t), ie, $Y_t = T_t + S_t + I_t$. If the moving average with coefficients $\{\theta_t\}$ is regarded as an operator, it can be denoted by $M\{\theta_i\}$. Or simply M. When using a moving average M_θ to estimate the trend component, M_θ should ideally retain the trend while removing the seasonal component and minimising the influence of the irregular component.

The simple moving average is the most basic form of moving average. It smooths the data by taking the arithmetic mean of consecutive observations in the time series. For a time series y_t, the formula for the simple moving average is

$$M(y_t) = \sum_{k=-m}^{m} \theta_k y_{t+k} \tag{4.1}$$

Here, $k = 2m + 1$ represents the order of the moving average, and m is the number of observations before and after the central point. For example, a symmetric 3-term simple moving average is

$$M(y_t) = \frac{1}{3}(y_{t-1} + y_t + y_{t+1}) \tag{4.2}$$

In this symmetrical, simple 3-term moving average example, the weights of the moving averages are equal, and the $\frac{1}{3}$ weights of each observation can be expressed as $\left[\frac{1}{3}, \frac{1}{3}, \frac{1}{3}\right]$ or $\left[\frac{1}{3}, \underline{\frac{1}{3}}\right]$ or $\frac{1}{3}[1, \underline{1}]$. This moving average is symmetric, meaning the average is centred at period t: it equally considers one observation before (t–1) and one after (t+1). The weight applied to y_{t-1} is the same as that applied to y_{t+1}, which characterises the symmetry of the filter.

4.2.2 Symmetric versus Asymmetric Moving Averages: Characteristics and Phase Shifts

A Symmetric Moving Average (SMA) is usually a selection of past and future data equidistant around the current point, for example

$$M(y_t) = \frac{1}{3}(y_{t-1} + y_t + y_{t+1}) \tag{4.3}$$

Or in a more general form:

$$M(y_t) = \frac{1}{2m+1}(y_{t-m} + \cdots + y_t + \cdots + y_{t+m}) \tag{4.4}$$

The symmetrical moving average is centred on y_t, the data points on the left and right are evenly distributed, the coefficients can be expressed as $\left[\theta_{-m}, \cdots, \theta_{0,} \cdots \theta_{+m}\right]$, and the weight coefficients are θ_k. The advantages of the symmetrical moving average include better smoothing, the absence of lag effects and phase shifts, which together enable effective detection of turning points in time series. However, the symmetric moving average cannot be applied at the endpoints of a time series due to the lack of sufficient future observations at the beginning and end, resulting in data loss at these points. To avoid data loss, X-11 uses an asymmetric moving average, where the moving average calculation of the endpoint relies only on known past data, ie, an asymmetric moving average,

The Asymmetric Moving Average (AMA) uses only the current data (y_t), the past data $y(_{t-1}, y_{t-2})$, and does not use the future data, such as

$$M(y_t) = \frac{1}{3}(y_{t-2} + y_{t-1} + y_t) \tag{4.5}$$

The asymmetric moving average only uses past data points and uses different weights at the endpoints to solve the endpoint problem. However, the asymmetric moving average only uses the current and past data of t, t–1, and t–2, but ignores future data such as t+1 and t+2, which will introduce a certain phase shift due to the imbalance of weights, leading to a lag effect, especially near the endpoint of the time series, resulting in bias in trend estimation.

The X-11 method primarily relies on a symmetric moving average to decompose seasonal components. However, due to insufficient data at the endpoints of the series, symmetric filters cannot be applied directly at the boundaries, leading to the so-called endpoint data loss problem. To address this, asymmetric moving averages (AMA) are employed in certain cases to handle endpoint observations.

With the introduction of ARIMA and regARIMA models in X-12-ARIMA and X-13ARIMA-SEATS, it became possible to extend the time series using backcasting to fill in historical data and forecasting to project future values. This allows the use of symmetric moving averages throughout the series, including at the endpoints, effectively eliminating the limitations imposed by missing observations. In theory, this means that X-12-ARIMA and X-13ARIMA-SEATS can entirely avoid the need for asymmetric filters.

Nonetheless, in practical applications, asymmetric moving averages are still used when users choose not to enable ARIMA-based forecasts or when the fitted model lacks stability. In certain specific situations, even with ARIMA forecasts enabled, additional adjustments may be required, such as applying smoothed asymmetric filters with adjusted weights, to ensure reliable endpoint estimation.

4.2.3 Weighted Moving Average: Weight Design and Smoothing Effect

The symmetric moving average formula (4.3) and the asymmetric moving average formula (4.5) both assign equal weights of 1/3 to each point. These are known as simple moving averages where all data points contribute equally, with no weighting coefficients applied. This method is suitable for basic smoothing but may not effectively capture underlying trend changes.

In contrast, weighted moving averages assign different weights to different observations, thereby better capturing the dynamics of a time series. The general formula for a weighted moving average is

$$M(y_t) = \sum_{k=-m}^{m} \theta_k y_{t+k} \tag{4.6}$$

Where θ_k is the weight coefficient, and satisfies the following conditions: (1) the sum of the coefficients is 1, ie, $\sum_{k=-m}^{m} \theta_k = 1$ and (2) the weighted sum of the coefficients is 0, ie, $\sum_{k=-m}^{m} k\theta_k = 0$. The advantage of the weighted moving average is its ability to adjust weights based on the characteristics of the time series, enabling better capture of trend and seasonal components.

4.2.4 Henderson Moving Average: Weight Calculation and Trend Extraction

In the X-11 method, determining the weights is a crucial step. The choice of weights directly affects the smoothing performance of the moving average and its ability to capture the trend. One commonly used method for determining weights is the Henderson moving average.

The Henderson moving average is a key tool in the X-11 method for estimating the trend component. It effectively removes irregular fluctuations while maintaining the trend's smoothness. The X-11 method selects the length of the Henderson moving average automatically based on the ratio of the irregular to trend component (the I/C ratio). The Henderson moving average is most suitable for longer time series. For shorter series, its effectiveness may be limited, as there may not be enough observations to support a longer moving average.

The general formula for computing the Henderson moving average is

$$T_t = \sum_{i=-m}^{m} \theta_i y_t + i \tag{4.7}$$

where T_t is the trend component, $y_t + i$ is the observed value of the time series, and θ_i is the weight of the Henderson moving average.

The weights of the Henderson moving average are determined by minimising the sum of squares of their third-order differences, i.e. $min\sum\left(\nabla^3\theta_t\right)^2$, the weights of the Henderson moving average are calculated as

$$\theta_i = \frac{315}{8} \cdot \frac{(n+1)^2 - i^2}{(n+1)^2 - 1} \cdot \frac{(n+2)^2 - i^2}{(n+2)^2 - 4} \cdot \frac{(n+3)^2 - i^2}{(n+3)^2 - 9}$$

Where *n* is the order of the Henderson moving average, and *i* is the index of the weights. The weights of the Henderson moving average are symmetrical and satisfy $\sum_{i=-m}^{m} \theta_i = 1$. The advantage of the Henderson moving average is that it captures trends well and is highly reliable in estimating irregular fluctuations.

For certain more complex cases, such as calculating weights for a local polynomial of a specific order, the moving average requires trend modelling. For example, to perform a symmetric moving average based on a second-order polynomial, the trend is modelled as $y_t = \beta_0 + \beta_1 t + \beta_2 t^2$. To make a symmetrical moving average, you need to model the trend and calculate the weights of the moving average: for example, *m* = 3 and *k* = 2; *t* = –3, –2, –1, 0, 1, 2, 3; ($y_t = \beta_0 + \beta_1 t + \beta_2 t^2$, specific order: *k* = 2), we need to solve for $\hat{\beta}_0$ and get

$$\hat{\beta}_0 = \frac{1}{21}\left(-2y_{-3} + 3y_{-2} + 6y_{-1} + 7y_0 + 6y_1 + 3y_2 - 2y_3\right)$$

This is a symmetric moving average with weights $\frac{1}{21}[-2,3,6,\underline{7}]$.[16]

4.2.5 Composite Moving Average: Multi-Step Smoothing and Centring

The composite moving average is the symmetrical moving average used in X-11. When using a moving average of odd order, for example, for the simple 3-term moving average shown earlier, calculated as $\frac{1}{3}(y_{t-1}+y_t+y_{t+1})$, the average can be centred on the middle value (y_t). This is also known as the Centred Moving Average (CMA). When using a moving average of even order, for example, the simple 4-term moving average calculated as $\frac{1}{4}(y_{t-1}+y_t+y_{t+1}+y_{t+2})$, the centre of the moving average lies between the second and third values, that is, between y_t and y_{t+1}. Thus, a simple 2-term moving average is required for the original 4-term moving average smoothing value. This is known as the 2 × 4 moving average, also known as the 'compound' moving average.

For moving 2×4 averages, the centred moving average usually uses the two-step moving average method:

(1) Calculate 4 simple moving averages first: $M(y_4)=\frac{1}{4}(y_{t-1}+y_t+y_{t+1}+y_{t+2})$

(2) Calculate 2 more simple moving averages to get the median value: It can also be expressed as $CMA_t=\frac{1}{2}(My_4(t)+My_4(t+1))$

$$M(y_t)=\frac{1}{2}\left[\frac{1}{4}(y_{t-2}+y_{t-1}+y_t+y_{t+1})+\frac{1}{4}(y_{t-1}+y_t+y_{t+1}+y_{t+2})\right]=\frac{1}{8}y_{t-2}+\frac{2}{8}y_{t-1}+\frac{2}{8}y_t+\frac{2}{8}y_{t+1}+\frac{1}{8}y_{t+2}$$

so the weights are $\frac{1}{8}[1,2,\underline{2}]$.

4.2.6 Local Regression and Moving Average

Seasonal adjustment is, in essence, an almost nonlinear process. However, the most widely used seasonal adjustment methods are based on linear filters. These methods achieve progressive smoothing through moving averages and decompose time series into unobservable components via iterative procedures. All filters employed in these methods can be categorised as local polynomial regressions.

Local regression involves fitting a polynomial over a moving window (sliding one observation at a time), typically using least squares estimation, either weighted or unweighted. At the centre point of each window, the smoothed value of the series is taken as the value of the fitted polynomial at that time point. The smoothed value for the next time point is then obtained by fitting a polynomial to the next window.

It can be demonstrated that when observations are evenly spaced, local regression is equivalent to applying a specific type of moving average.

To obtain the weights for the moving average used to extract the trend from a local polynomial regression of degree d, such as in the case of a local quadratic (second-degree) polynomial, one proceeds as follows:

$$y_t=\beta_0+\beta_1 t+\beta_2 t^2+\varepsilon_t \tag{4.8}$$

Where y_t is the observation value of the time series at time point *t*; *t* is a time variable that represents the position of the point in time, eg, *t* = (-3, -2, -1, 0, 1, 2, 3); $\beta_{0,}\beta_1,\beta_2$ is the regression coefficient; β_0 is the intercept term of the model; y_t represents the predicted value when *t* = 0.

In local polynomial regression, β_0 represents the locally weighted average at time point *t* = 0, which corresponds to the central value of the moving average. β_1 is the coefficient of the linear term in the time variable *t*, indicating the linear trend of the time series. It reflects the rate of linear change in y_t over time. β_2 is the coefficient of the quadratic term in the time variable *t*, representing the quadratic trend (or curvature) of the time series. It captures the nonlinear variation in y_t to time. ε_t denotes the error term, representing random fluctuations in the data that the model does not explain.

Using a symmetric moving average because $m = 3$, particular order ($k = 2m$+1 = 7), quadratic ($d = 2$), so $t = (-3,-2,-1,0,1,2,3)$, a regression (4.8) to calculate the moving average (moving average), ie, in a given interval *t* = -*m*, . . ., *m*, the local quadratic regression (*d* = 2) is performed on y_t, and the intercept term $\hat{\beta}_0$, is obtained, which is equivalent to the local weighted average of the interval.

(1) In a given interval *t* = (-3,-2,-1,0,1,2,3), ie, $(2m+1=7)$

(2) Fit a quadratic polynomial: $y_t = \beta_0 + \beta_1 t + \beta_2 t^2 + \varepsilon_t$

where β_0 is the locally estimated value, ie, the centre point value of the moving average.

(3) Let the data point be *t* = (-3,-2,-1,0,1,2,3), the corresponding design matrix X is

$$X = \begin{bmatrix} 1 & -3 & 9 \\ 1 & -2 & 4 \\ 1 & -1 & 1 \\ 1 & 0 & 0 \\ 1 & 1 & 1 \\ 1 & 2 & 4 \\ 1 & 3 & 9 \end{bmatrix}$$

The dependent variable (*y*): $y = [y_{-3}, y_{-2}, y_{-1}, y_0, y_1, y_2, y_3]^T$. The regression parameter ($\hat{\beta}$) for estimating the local polynomial (4.8) using the least squares method given as

$$\hat{\beta} = (X^T X)^{-1} X^T y$$

Compute $X^T X$:

$$X^T X = \begin{bmatrix} 7 & 0 & 28 \\ 0 & 28 & 0 \\ 28 & 0 & 196 \end{bmatrix}$$

Its inverse matrix is

$$X^T X^{-1} = \begin{bmatrix} \frac{1}{7} & 0 & -\frac{1}{28} \\ 0 & \frac{1}{28} & 0 \\ -\frac{1}{28} & 0 & \frac{1}{196} \end{bmatrix}$$

To get $\hat{\beta}_0$, we focus on the formula

$$\hat{\beta}_0 = e_1^T \left(X^T X\right)^{-1} X^T y$$

where $e_1 = [1,0,0]^T$,
Solving $\hat{\beta}_0$, ie, we take only the first line, generated:

$$\hat{\beta}_0 = \frac{1}{7}\Sigma y_t - \frac{1}{28}\Sigma t^2 y_t$$

The matrix is used to calculate the specific weights, to obtain a corresponding moving average weight

$$\frac{1}{21}[-2,3,6,7,6,3,-2]$$[17]

$$\hat{\beta}_0 = \frac{1}{21}(-2y_{-3} + 3y_{-2} + 6y_{-1} + 7y_0 + 6y_1 + 3y_2 - 2y_3)$$

This is a symmetrical moving average with weights ($\frac{1}{21}[-2,3,6,\underline{7}]$ MA).[18]

The composite moving average better captures the trend of the time series by placing the centre point at the current moment, reducing the phase shift. More generally, we can express such a composite moving average as *n* × *m*. A composite moving average is constructed by combining multiple simple moving averages. The advantage of the compound moving average is that it can better eliminate seasonal fluctuations.

4.2.7 Trend-Cycle Decomposition: Application of 2×4 and 2×12 Moving Averages

The X-11 method employs the 2 × 4 (for quarterly data) and 2 × 12 (for monthly data) moving averages to decompose the trend-cycle component of a time series.

- The 2 × 4 moving average is used for quarterly data. It applies two successive 4-term moving averages to smooth the series and remove seasonal fluctuations. Its advantages include the following:

  (1) The 2 × 4 moving average is a symmetric 5-term moving average with weights {1, 2, 2, 2, 1}/8. The sum of the weights equals 1, which ensures that it preserves linear trends.

(2) The sum of squared weights equals 0.250, meaning it reduces the variance of white noise by 75%, offering high reliability in estimating the irregular component.

(3) The shape of the weighting function corresponds to a periodicity of 4, making it well-suited to quarterly series with stable seasonality.

Thus, the 2 × 4 moving average is effective at capturing the underlying trend in quarterly data.

- The 2 × 12 moving average is used for monthly data. It is computed by first applying a 12-term simple moving average, followed by a centred 2-term moving average. The resulting weights are: {1, 2, 2, 2, 2, 2, 2, 2, 2, 2, 2, 2, 2, 1}/24. This filter is also known as the centred 12-month moving average, preserves linear trends and effectively removes annual seasonality.[19] The sum of squared weights is 23/288, indicating that the variance of white noise is reduced by more than 90%. However, it does not perfectly recover all cycles shorter than one year. For instance, a function with a three-year cycle is only about 80% recovered. Despite this, the 2 × 12 moving average remains a robust tool for extracting trends in monthly data and suppressing irregular fluctuations.

4.2.8 Seasonal Component Decomposition: Application of 3×3, 3×5, and 3×9 Moving Averages

In the X-11 method, the seasonal component is decomposed using 3×3, 3×5, and 3×9 moving averages.

- The 3 × 3 moving average is used for the preliminary estimation of the seasonal component. Assuming that the smoothed series consists of seasonal and irregular components (in the additive decomposition model, $SI_t = S_t + I_t$), the weights of the moving average M_{3x3} are {1, 2, 3, 2, 1}/9. This filter smooths the seasonal-irregular component by applying a 3-term moving average, effectively reducing irregular fluctuations. Its advantage lies in its ability to capture seasonal variations while providing a relatively reliable estimate of the irregular component.
- The 3 × 5 moving average is applied for a refined estimation of the seasonal component. Under the same additive model assumption $SI_t = S_t + I_t$, the weights of the moving average M_{3x5} are {1, 2, 3, 3, 3, 2, 1}/15. This 5-term moving average further reduces irregular noise while enhancing the clarity of the seasonal structure. Its benefit lies in its improved ability to isolate seasonal patterns while maintaining strong reliability against irregular movements.
- The 3 × 9 moving average is used for the final estimation of the seasonal component. Again, assuming $SI_t = S_t + I_t$ the moving average M_{3x9} has weights {1, 2, 3, 3, 3, 3, 3, 3, 3, 2, 1}/27. It applies a 9-term smoothing filter to the seasonal-irregular component to further eliminate noise and emphasise the stable seasonal pattern. This final estimate provides a highly reliable seasonal signal and is particularly effective at removing short-term volatility.

It can be readily demonstrated that each of these symmetric moving averages preserves linear trends.

4.2.8.1 Use and Identification of the SI Ratio

The SI ratio is a key concept in the X-11 method. It is used to estimate the seasonal-irregular (SI) component by dividing the original time series by the trend component. The formula for calculating the SI ratio is

$$SI_t = \frac{y_t}{T_t} \tag{4.9}$$

where y_t represents the original series and T_t is the trend component. The primary purpose of the SI ratio is to help identify seasonal fluctuations and irregular variations. By analysing the SI ratio, one can assess the stability of the seasonal component and the magnitude of the irregular component.

In the X-11 procedure, the identification of the SI ratio is achieved through the computation of the Moving Seasonal Ratio (MSR). The formula for calculating MSR is

$$MSR = \frac{\sum n_j \bar{I}_j}{\sum n_j \bar{S}_j} \tag{4.10}$$

where $\bar{I}_j$ and $\bar{S}_i$ are the average annual variations of the seasonal and irregular components, respectively, and n_j is the number of months in the data. The higher the MSR value, the stronger the irregular fluctuations and the more unstable the estimation of the seasonal component.

4.3 Frequency Domain Analysis: Gain Function of Spectrum Decomposition and Moving Average

4.3.1 Basics of Frequency Domain Analysis: Fourier Transform and Periodic Recognition

In the frequency domain, time series can be expressed as a weighted sum of trigonometric basis functions (sine/cosine).[20] Spectral analysis is the process of identifying the cyclical components of a time series by transforming it into the frequency domain. A time series can be represented as a superposition of multiple sine waves of different frequencies. Each component of a time series has its specific frequency range in the frequency domain, and the core idea of frequency domain analysis is to measure the intensity of different frequency components by calculating the Power Spectral Density (PSD). In spectral decomposition, a time series is expressed as a sum of sine waves at different frequencies, facilitating the detection of trend, seasonal, and irregular components. The spectral method is a key approach for analysing the cyclical characteristics of time series.

Spectral analysis transforms a time series from the time domain to the frequency domain by means of a Fourier transform, expressed as a superposition of sine waves of different frequencies. The main purpose of spectrum analysis is to identify periodic components in a time series. For a time series y_t, its spectrum can be calculated using the Fourier transform $(S(f):)$

$$S(f) = \left|\sum_{t=1}^{T} y_t e^{-i2\pi ft}\right|^2 \tag{4.11}$$

Where f is the frequency and T is the length of the time series. Spectrum $S(f)$ represents the energy of a time series at f frequency. The cycle frequency of the 12-month year is associated with the corresponding time series component. For example, the zero frequency corresponds to the trend component. For monthly time series, the seasonal frequencies, $\frac{\pi}{6}, \frac{\pi}{3}, \frac{\pi}{2}, \frac{2\pi}{3}, \frac{5\pi}{6}$, are equivalent to 1, 2, . . ., 6 cycles per year. Specifically, the frequency $\frac{\pi}{3}$ corresponds to a six-month cycle, with 2 cycles completed per year. In the case of quarterly time series, there are two seasonal frequencies: $\frac{\pi}{2}$ (one cycle per year) and π (two cycles per year).

The trading day frequency is usually expressed as the length of the period. For monthly time series, the trading day frequency is associated with a 2.714 cycle. For the quarterly time series of the trading day, there are four frequencies: three of them correspond to length periods – 1.292, 1.850, and 2.128 – while the last one is associated depending on the length of the time series.

Spectral estimation identifies regular patterns in a time series by detecting peaks at a given frequency. Thus, for monthly series with strong seasonal effects, the spectrum will exhibit particularly high amplitudes at frequencies corresponding to annually recurring patterns, such as every 12 months, 6 months, or 3 months.

4.3.2 Spectral Decomposition: Frequency Characteristics of T, S, I Components

Spectrum is a theoretical concept, and in practice, seasonal adjustment software employs a number of spectrum estimators. Two spectrum estimators are implemented in JDemetra+: periodic plot and autoregressive spectrum. Typically, instead of a flat line, a spectrum estimator contains many peaks. The X-11 method divides the spectrum into different frequency ranges, corresponding to the components of the time series:

(1) The trend component corresponds to the low-frequency portion of the spectrum. The frequency of trend components is usually less than that of seasonal components. The trend component can be extracted using a low-pass filter.

(2) The seasonal component corresponds to the seasonal frequency portion of the spectrum. The frequency of seasonal components is usually fixed; for example, the seasonal frequency of monthly data is 1–12. Seasonal components can be extracted using a bandpass filter.

(3) The irregular components correspond to the high-frequency part of the spectrum. The frequency of irregular components is usually greater than that of seasonal components. Irregular components can be extracted by means of a high-pass filter. Users should focus on statistically significant peaks (p-values larger than 5%). Seasonal adjustment software often highlights these peaks. For example, JDemetra+ shows a yellow line indicating a 0.05 significance level, so peaks larger than this line (0.05) are considered significant.

If the time series y_t has a periodic component, y_t can be represented by a sine wave:

$$y_t = R\sin(\omega t + \varphi) \tag{4.12}$$

where, R = amplitude, ω = angular frequency, t = time, φ = phase angle.

4.3.2.1 Gain Function versus Phase Shift Function

The $G(\omega)$ function associated with ω is called the gain function of the moving average. The function $C(\omega)$, which is associated with ω, is called the phase shift function of the moving average. It is sometimes expressed as $C(\omega)/\omega$, which allows the phase shift to be measured in terms of the number of cycles. Using the Moving Average operator for $y_t = R\sin(\omega t + \varphi)$ gets:

$$M(y_t) = M(R\sin(\omega t + \varphi)) = G(\omega)R\sin(\omega t + \varphi + C(\omega)) \tag{4.13}$$

where,

- $M(y_t)$ denotes the right y_t some kinds of transformation.
- $G(\omega)$ is known as the gain function and is a function of angular frequency ω, and describes the amplitude variation of y_t at different frequencies ω transform. $G(\omega) = 1$ indicates that the period is completely preserved; $G(\omega) > 1$ indicates an increase in amplitude; if $G(\omega) < 1$, the amplitude decreases; $G(\omega) = 0$ indicates that the period is eliminated.
- $C(\omega)$ is called the phase offset function, and $C(\omega)$ depends on the angular frequency ω, which means that y_t is the phase during the transformation. It describes the phase change of y_t at different frequencies, usually expressed as $C(\omega)/\omega$ which denotes the phase shift in the number of periods, and shows the hysteresis introduced using an asymmetric moving average. For example, $C(\omega)/\omega = -1$ means that the phase is offset by one cycle, which means that the turning point may be detected one cycle later. If $C(\omega)/\omega > 0$, y_t is phase lagging, and if $C(\omega)/\omega < 0$, y_t is in phase ahead.
- After the transformation $M(y_t)$, the amplitude and phase vary with frequency ω. $G(\omega)$ and $C(\omega)$ describe these changes, respectively.
- $(\omega t + \varphi$ is measured in radians per unit of time, where π radians is equal to 180°, and ω is the 'angular frequency'. $(\omega t + \varphi)$ is usually represented by (1) at the frequency of $f = \dfrac{\omega}{2\pi}$ (number of periods per unit of time) and (2) periods or wavelengths denoted by $\dfrac{1}{f} = \dfrac{2\pi}{\omega}$ to measure.

4.3.3 Phase Shift of Asymmetric Moving Averages: Impact and Case Studies

Suppose we consider the series $y_t = \sin\left(\dfrac{\pi_t}{3}\right)$, and apply a 3-term moving average defined by the asymmetric operator $M(y_t) = \dfrac{1}{3}(y_{t-2} + y_{t-1} + y_t)$ to transform it:

$$\begin{aligned} M(y_t) &= \frac{1}{3}(y_{t-2} + y_{t-1} + y_t) \\ &= \frac{1}{3}R\left[\sin(\omega t - 2\omega + \varphi) + \sin(\omega t - \omega + \varphi) + sin(\omega t + \varphi)\right] \\ &= \frac{1}{3}R\left[1 + 2\cos(\omega)\sin(\omega t + \varphi - \omega)\right]^{21} \end{aligned} \tag{4.14}$$

Let $y_t = R\sin(\omega t + \varphi)$, the frequency (or period, $\frac{2\pi}{\omega}$), amplitude R, and phase of the series φ. Any transformation of the moving average to $\{y_t\}$ will also be a modified amplitude sinusoidal curve, showing a phase shift relative to the original series:

$$M(y_t) = M[\sin(\omega t + \varphi)] = G(\omega)\sin[\omega t + \varphi + C(\omega)] \tag{4.15}$$

By smoothing the series $y_t = \sin(\pi_t / 3)$ with the moving average $(y_{t-2} + y_{t-1} + y_t)/3$, Figure 4.1 shows the phase shift of the asymmetric moving average.

From (Figure 4.1), we can see $G(\omega) = \frac{1}{3}1 + 2\cos(\omega)$. $C(\omega) = -\omega \ or \ \frac{C(\omega)}{\omega} = -1$, it means a phase shift of one period (the turning point in the trend will be determined one period later). Figure 4.1 shows that the asymmetric 3-term moving average causes the sine wave phase to lag by one period $C(\omega)/\omega = -1$. The moving average introduced here induces a phase shift, which becomes evident when this asymmetry is applied to a simple straight line $y_t = at + b$:

$$M(y_t) = \frac{1}{3}y_{t-2} + y_{t-1} + y_t = \frac{1}{3}[\alpha(t-2) + b + \alpha(t-1) + b + \alpha t + b) = \alpha(t-1) + b = y_{t-1}$$

Thus, the gain function indicates the extent to which each frequency component of the series is attenuated or retained, while the phase-shift function reflects the lag (hysteresis) introduced by using an asymmetric moving average.

Although the asymmetric moving average introduces phase shifts, we see that the moving average eliminates certain frequencies and thus some of the seasonal components, because $G(\omega) = \frac{1}{3}(1 + 2\cos(\omega))$. The moving average has eliminated a period of three

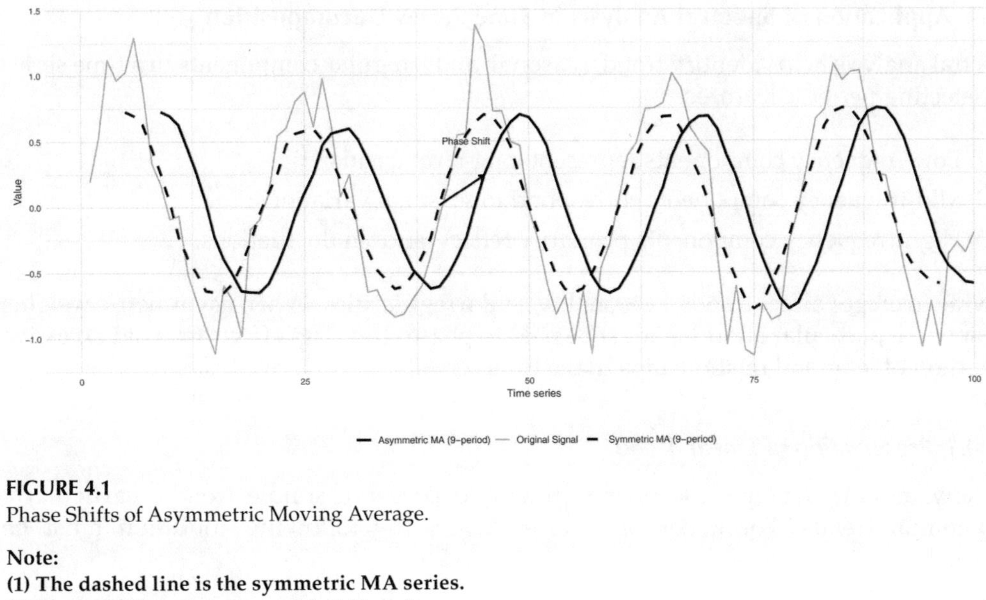

FIGURE 4.1
Phase Shifts of Asymmetric Moving Average.

Note:
(1) The dashed line is the symmetric MA series.
(2) The solid line: the asymmetric MA series.
Source: The simulated data.

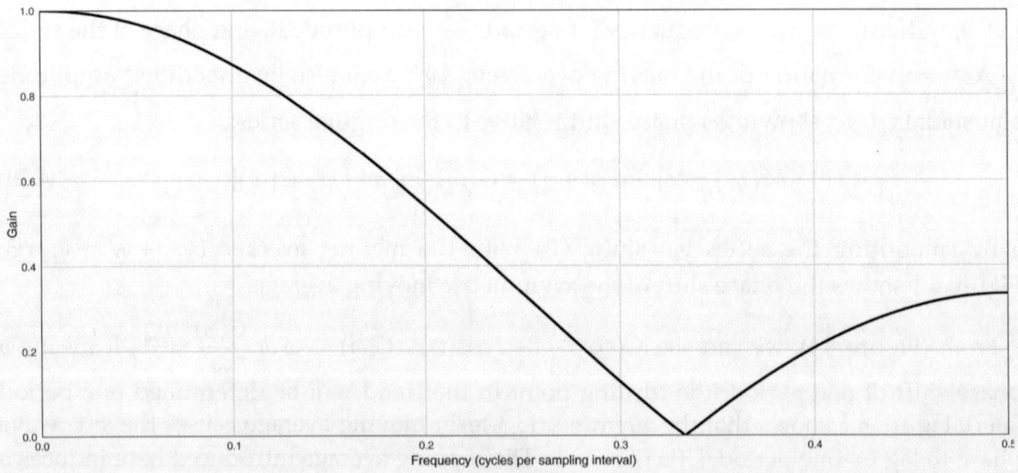

FIGURE 4.2
Asymmetric Moving Average $(y_{t-2}+y_{t-1}+y_t)/3$ Gain Function.
Source: The simulated data

months (indicating a quarterly change). Low-frequency variations (related to trend cycles) remain more or less constant.

For trend estimation, the ideal filter is a low-pass filter that retains low-frequency variations while eliminating high-frequency noise. For seasonal estimation, the ideal filter is a bandpass filter that retains variations at seasonal frequencies while suppressing both low- and high-frequency components. In general, a simple k-term moving average will eliminate cycles with a period length of k. However, many time series exhibit seasonal variations at more than one frequency.

4.3.4 Application of Spectral Analysis in Time Series Decomposition

Spectral analysis helps identify trend, seasonal, and irregular components in a time series by detecting periodic features:

(1) Low-frequency components represent long-term trends;
(2) Mid-frequency components correspond to seasonal variations;
(3) High-frequency components primarily reflect random fluctuations.

Moving averages affect trends, seasonality, and irregularities. When asymmetric weights are applied, particular attention must be paid to phase shift (lag) effects to avoid excessive distortion of temporal relationships in the time series.

4.3.4.1 Preservation of Linear Trend

For any moving average, the operation should preserve simple trends, particularly polynomial trends. For a constant series $y_t = a$, the following condition must be satisfied:

$$M(y_t)=\sum_{k=-m}^{m}\theta_k y_{t+k}=\sum_{k=-m}^{m}\theta_k a=a\sum_{k=-m}^{m}\theta_k=a \tag{4.16}$$

Thus, the sum of the moving average coefficients, $\sum_{k=-m}^{m}\theta_k$ must be equal to 1. To preserve a linear trend under any moving average, the following condition must hold for all *t*:

$$\begin{aligned} M(y_t) &= \sum_{k=-m}^{m}\theta_k y_{t+k}=\sum_{k=-m}^{m}\theta_k\left[a(t+k)+b\right] \\ &= at\sum_{k=-m}^{m}\theta_k + a\sum_{k=-m}^{m}k\theta_k + b\sum_{k=-m}^{m}\theta_k = at+b \end{aligned} \tag{4.17}$$

This results in, $\sum_{k=-m}^{m}\theta_k=1$ and $\sum_{k=-m}^{m}k\theta_k=0$, ie, the sum of the moving average coefficients is equal to 1, and the sum of the moving average weights is equal to 0.

In general, for a moving average that maintains a *d*-order polynomial, so that its coefficients are determined, its conditions must be satisfied: $\sum_{k=-m}^{m}\theta_k=1$ and $\sum_{k=-m}^{m}k_j\theta_k=0, j=1,\ldots d$, which is a necessary and sufficient condition. For the above asymmetric 3-term moving average,

$$M(y_t)=\frac{1}{3}(y_{t-2}+y_{t-1}+y_t)=\frac{1}{3}\left[a(t-2)+b+a(t-1)+b+at+b\right]=a(t-1)+b=y_{t-1} \tag{4.18}$$

What we get is

$$\sum_{K=2}^{0}\theta_K=\frac{1}{3}+\frac{1}{3}+\frac{1}{3}=1$$

and

$$\sum_{K=-2}^{0}K\theta_K=-2\times\frac{1}{3}-1\times\frac{1}{3}+0\times\frac{1}{3}=-1$$

Thus, the asymmetric moving average, although it remains constant, does not remain linear. By analysing the power spectral density map, it is possible to determine the statistical significance of the frequency and decide whether to remove a particular frequency component. The moving average can be viewed as a low-pass filter that removes high-frequency fluctuations and extracts underlying trends. Spectrum analysis, on the other hand, provides a more precise way to separate the individual components of a time series.

4.3.4.2 Elimination of Seasonality

As we saw when defining the gain function, moving averages can eliminate certain frequencies and thus some seasonal components. Fixed seasonality can be 'modelled' by the cycle function of period *k* (4 for the quarterly series and 12 for the monthly series). If we also assume that the seasonal coefficients sum to zero over a year, it can be shown that these functions form a $(k-1)$–dimensional subspace whose basis can be easily represented. For example, in the quarterly case, we find the subspace created by the row vector of the matrix:

$$\begin{bmatrix} 1 & -1 & 0 & 0 & 1 & -1 & 0 & 0 & \cdots \\ 1 & 0 & -1 & 0 & 1 & 0 & -1 & 0 & \cdots \\ 1 & 0 & 0 & -1 & 1 & 0 & 0 & -1 & \cdots \end{bmatrix}$$

Thus, the abolition of such a series introduces a constraint on the moving average coefficients, which are expressed in the form of a matrix, $C\Theta = \alpha$, where Θ is the column vector M of the moving average coefficients, C is a matrix of dimension (*k*-1, 2*m*+1), and its rows are made up of the above basis vectors (take k=4 as an example), and α is a zero matrix of dimension (*k*-1, 1). For example,

$$C\Theta = \begin{bmatrix} 1 & -1 & 0 & 0 \\ 1 & 0 & -1 & 0 \\ 1 & 0 & 0 & -1 \end{bmatrix} \begin{bmatrix} \theta_1 \\ \theta_1 \\ \theta_1 \\ \theta_1 \end{bmatrix} = \begin{bmatrix} 0 \\ 0 \\ 0 \end{bmatrix}$$

In this case, this gives

$$\begin{cases} \theta_1 = \theta_2 \\ \theta_1 = \theta_3 \\ \theta_1 = \theta_4 \end{cases}$$

so

$$\theta_1 = \theta_2 = \theta_3 = \theta_4$$

This moving average eliminates the seasonality of these periods 4, so its gain function will be cancelled out at frequency π/2 (=2π/4). The gain function is the easiest way to identify the frequency at which the moving average is eliminated. In general, a simple moving average of order *k* (all of which have a coefficient equal to 1/*k*) cancels out the fixed seasonality of period *k*, so its gain function is zero at 2π/*k*.

4.3.4.3 Reduction of Irregular Components

After accounting for trend and seasonality, it is still necessary to examine the impact of the moving average on the irregular component.

1) Reduction of variance in the irregular component is equivalent to reducing the irregular fluctuations. In the decomposition of an unadjusted series, the residuals are typically modelled as white noise, ie, a series of uncorrelated random variables e_t' with zero mean and constant variance. When subjected to a moving average, this white noise is transformed into a series of random variables e_t' with variance equal to $\sigma^2 \sum_{K=-\infty}^{\infty} \theta_K^2$, and reducing the variance of the irregular component corresponds to minimising $\sum_{k=-\infty}^{\infty} \theta_k^2$.

2) Appropriate selection of symmetric moving average coefficients can reduce the irregular component while preserving a linear trend. Taking a central three-term moving average as an example, the coefficient is $\{\theta_1, \theta_0, \theta_1\}$ the objective function is

$$\sum_{k=-\infty}^{\infty} \frac{\theta_k^2}{|k|} \tag{4.19}$$

We want to constrain $\sum_{k=-1}^{+1} \theta_k = 1$. and $\sum_{k=-1}^{+1} k\theta_k = 0$, the objective function (4.19) is minimised because it is a symmetric moving average, k = -1,0,1, and the objective function can be expanded as

$$\frac{\theta^2 - 1}{1} + \frac{\theta_0^2}{0} + \frac{\theta_1^2}{1},$$

Since $\frac{\theta_0^2}{0}$ is undefined, ignoring the k = 0 term, we get $\theta_{-1}^2 + \theta_1^2$. According to the second constraint $\sum_{k=-1}^{+1} k\theta_k = 0$, we have

$$-1 \cdot \theta_{-1} + 0 \cdot \theta_0 + 1 \cdot \theta_1 = 0 \Rightarrow \ -\theta_{-1} + \theta_1 = 0 \Rightarrow \theta_{-1} = \theta_1$$

Substituting this relation into the first constraint, $\sum_{k=-1}^{+1} \theta_k = 1$ gets

$$\theta_{-1} + \theta_0 + \theta_1 = 1 \Rightarrow \ \theta_1 + \theta_0 + \theta_1 = 1 \Rightarrow 2\theta_1 + \theta_0 = 1$$

So, $\theta_0 = 1 - 2\theta_1$.

Substituting $\theta_{-1} = \theta_1$ and $\theta_0 = 1 - 2\theta_1$ into the objective function,

$$\theta^2{}_{-1} + \theta_1^2 = \theta_1^2 + \theta_1^2 = 2\theta_1^2$$

At the same time, we have: $\theta_0 = 1 - 2\theta_1$.

Thus, the objective function (4.17) minimising the problem becomes $min\left[2\theta_1 + (1 - 2\theta_1)^2\right]$.

Expand and simplify $2\theta_1^2 + (1 - 2\theta_1)^2$:

$$2\theta_1^2 + \left(1 - 4\theta_1 + 4\theta_1^2\right) = 2\theta_1^2 + 1 - 4\theta_1 + 4\theta_1^2 = 6\theta_1^2 - 4\theta_1 + 1$$

Find the derivative of the function $6\theta_1^2 - 4\theta_1 + 1$ with respect to θ_1:

$$\frac{d}{d\theta_1}\left(6\theta_1^2 - 4\theta_1 + 1\right) = 12\theta_1 - 4$$

Set the derivative equal to zero and solve for θ_1:

$$12\theta_1 - 4 = 0 \Rightarrow 12\theta_1 = 4 \Rightarrow \theta_1 = \frac{4}{12} = \frac{1}{3}$$

Thus, we obtain the coefficients that minimise the function: $\theta_1 = \frac{1}{3}$. Based on our earlier derivation, $\theta_{-1} = \theta_1 = \frac{1}{3}$.

and $\theta_0 = 1 - 2\theta_1 = 1 - 2x\frac{1}{3} = \frac{1}{3}$.

Ultimately, we obtain a 3-term moving average where all coefficients are 1/3. This choice of coefficients for a centred 3-term moving average represents the optimal symmetric moving average (MA), as the simple three-term MA with uniform 1/3 coefficients eliminate third-order seasonality while effectively reducing irregular components and preserving linear trends in the data.

4.4 X-11 Algorithm Flow: Steps, Iterative, and Henderson Filters

4.4.1 Four Processing Steps and Three Stages of X-11

The X-11 method comprises four processing steps (A, B, C, D). Phase A involves preliminary pre-adjustment (data cleaning), while phases B through D perform component decomposition. Additionally, there are three stages (E, F, G) that provide statistics and charts.

Phase A is non-mandatory, allowing users to apply prior corrections to the series by adjusting factors. For example, introducing monthly (or quarterly) adjustment factors can correct for effects like public holidays or modify series levels (eg, strike impacts). Using these inputs, the program derives correction factors to adjust the original series. The adjusted series is presented in output Table B1, then proceeds to Phase B.

Phase B performs the first automated correction of the series, primarily consisting of initial estimation and adjustment of outliers and, where necessary, weekday effects. These estimates derive from applying the core algorithm detailed in Section 2.4. The resulting outputs include Table B19 (assessment of day-of-week or movable holiday effects) and Table B20 (values for correcting atypical points), which generate the corrected series in Table C1.

Phase C conducts the second automated correction. This step reapplies the core algorithm to refine estimates of weekday effects (Table C19) and atypical point corrections (Table C20). The finalised 'cleaned' series appears in output Table D1.

Phase D seasonal adjustment implements the X-11 core algorithm for the actual seasonal decomposition. It produces final estimates of:

- Seasonal component (Table D10)
- Seasonally adjusted series (Table D11)
- Trend-cycle component (Table D12)
- Irregular component (Table D13)

Phases E, F, and G generate statistics and charts. Phases E and F provide statistics for evaluating the quality of seasonal adjustment. Phase G delivers character-mode graphics.

4.4.2 Five-Step Iterative Algorithm: Estimation of Trend and Seasonal Components

The X-11 decomposition process is based on a five-step seasonal adjustment algorithm. This algorithm decomposes time series through iterative calculations, with each iteration refining component estimates:

1) Obtain an initial estimate of the trend cycle by applying a moving average to the original data.
2) Subtract this estimate from the original data to obtain the initial seasonal-irregular (SI) component. Then apply moving averages separately to the SI of each seasonal type to derive the initial seasonal components.
3) Subtract the initial seasonal component from the original data to obtain the initial seasonally adjusted series (trend-cycle/irregular). Apply a Henderson moving average to yield the second trend-cycle estimate.
4) Subtract the second trend-cycle estimate from the original data to obtain the second SI estimate. Apply moving averages separately to each seasonal type to derive the final seasonal component.
5) Subtract the seasonal component from the original data to obtain the final seasonally adjusted series. Apply a Henderson moving average to produce the final trend-cycle estimate.

4.4.3 Iterative Steps in Phase B: Trend-Seasonal Decomposition

More specifically, the X-11 cycle iteration consists of eight steps. Components are estimated twice, corresponding to two applications of the core algorithm with adjusted moving averages each time. Using Phase B (monthly data)[22] as an example,

Step 1: Adopt the 2×12 Moving Average Preliminary Estimation Trend (T) with a composite design that eliminates annual seasonality (see Section 4.2.6):

$$T_t^{(1)} = M_{2\times12}\left(y_t\right), \tag{4.20}$$

ie, calculate a centred 12-term moving average as a preliminary estimate of the trend:

$$T_t^{(1)} = \frac{1}{24}y_{t-6} + \frac{1}{12}y_{t-5} + \cdots + \frac{1}{12}y_t + \cdots + \frac{1}{12}y_{t+5} + \frac{1}{12}y_{t+6} \tag{4.21}$$

The 2 X 12 Moving Average used here is $\left\{\frac{1}{24},\frac{2}{24},\frac{2}{24},\frac{2}{24},\frac{2}{24},\frac{2}{24},\frac{2}{24},\frac{2}{24},\frac{2}{24},\frac{2}{24},\frac{2}{24},\frac{2}{24},\frac{1}{24}\right\}$,

Recreate the centre point of a linear trend, eliminate 12th-order constant seasonality, and minimise the variance of irregular components. The result is presented in Table B2.

Step 2: Estimation of the seasonality-irregular component

The first estimate of the seasonality-irregularity component is obtained by removing the trend period from the time series:

$$\left(S_t + I_t\right)^{(1)} = y_t - T_t^{(1)} \tag{4.22}$$

The result is presented in Table B3.

Step 3: Use a 3 × 3 moving average to preliminarily estimate the seasonal component

The estimation is based on the seasonal-irregular components of the previous step and corrected with extreme values:

$$S_t^{(1)} = M_{3\times3}\left((S_t + I_t)^{(1)}\right) \tag{4.23}$$

Using the 3 X 3 moving average is 5 coefficient terms, $\frac{1}{9},\frac{2}{9},\frac{3}{9},\frac{2}{9},\frac{1}{9}$, which is applied to the seasonality-irregularity ratio for each period over five years. The seasonality factor is then normalised using a centred 12-item moving average so that the seasonal effect is roughly offset over the entire 12-month period.

$$\tilde{S}_t^{(1)} = S_t^{(1)} - M_{2x12}\left(S_t^{(1)}\right) \tag{4.24}$$

The result is presented in Table B5.

Step 4: Estimate the seasonally adjusted series

The estimation of the seasonally adjusted series is done by removing the first estimate of the seasonal component (Table B5) from the starting series (Table B1):

$$SA_t^{(1)} = (T_t + I_t)_t^{(1)} = Y_t - S_t^{(1)} \tag{4.25}$$

Step 5: Estimate the trend period with 13 Henderson moving averages

The second estimate of the trend period (Table B7) was derived from a seasonally adjusted series smoothed using a 13-term Henderson filter (Table B6).

$$T_t^{(2)} = H_{13}\left(SA_2^{(1)}\right) \tag{4.26}$$

Step 6: Estimation of the seasonality-irregular component

The estimation of the seasonal-irregular component is achieved by subtracting the trend period from the original time series.

$$(S_t + I_t)^{(2)} = y_t - T_t^{(2)} \tag{4.27}$$

The result is presented in Table B8.

Step 7: Estimate the seasonal component via a 3 × 5 moving average

The second estimate of the seasonal-irregular component is obtained by removing the trend period from the time series:

$$S_t^{(2)} = M_{3\times5}\left((S_t + I_t)^{(2)}\right) \tag{4.28}$$

Here, a 3 × 5 moving average is used, over seven terms, with coefficients, $\frac{1}{15},\frac{2}{15},\frac{3}{15},\frac{3}{15},\frac{3}{15},\frac{2}{15},\frac{1}{15}$ and maintains a linear trend. The coefficients are then normalised so that their sum is roughly offset over the entire 12-month period.

$$\tilde{S}_t^{(2)} = S_t^{(2)} - M_{2x12}\left(S_t^{(2)}\right) \tag{4.29}$$

The result is presented in Table B10.

Step 8: Estimate the seasonally adjusted series

The estimation of the seasonally adjusted series is done by removing the second estimate of the seasonal component (Table B10) from the starting series (Table B1):

$$SA_t^{(2)} = (T_t + I_t)_t^{(2)} = Y_t - \tilde{S}_t^{(2)} \tag{4.30}$$

The output of this result is Table B11.

This completes the B-cycle of the X-11 algorithm. The C and D cycles are almost identical to the B cycles, but the starting point is the original series modified for extreme observations (determined in the previous iteration). It should be noted that, on the one hand, the estimation of the trend period in steps 1 and 5 needs to be extracted by smoothing the data, and on the other hand, the seasonal component estimated in steps 3 and 5 is the part of the periodic change in the time series, which usually needs to be extracted by removing the trend and irregular components, and the seasonal component moving average is used, which is a core difficulty in the X-11 algorithm, that is, how to reasonably select and use the moving average when estimating the trend period and seasonal component, Especially the Henderson filter. Choosing the right moving average, especially the Henderson filter, is a complex issue, as different moving averages have different effects on the estimation of trend and seasonal components. If the filter is not suitable, it can lead to an inaccurate estimation of trends or seasonal components.

4.4.4 Selection and Optimisation of Henderson Filters

The Henderson filter is a special moving average used to smooth time series data and extract trend components. Its design goal is to strike a balance between smoothness and responsiveness to trend changes. The length of the Henderson filter, such as 5, 13, 23, etc., affects the estimation of the trend. Shorter filters (eg, 5 terms) respond quickly to trend changes but are less smooth and may retain noise; longer filters (eg, 23 terms) provide smoother results but respond more slowly and may lag behind actual trend shifts. Selecting the appropriate filter is crucial to the algorithm, as it directly influences the quality of trend estimation.

Moving averages are also required for the seasonal component. In steps 3 and 5, the algorithm uses a moving average, such as a 3 × 3 or 3 × 5 moving average, to estimate the seasonal component. The role of these moving averages is to extract stable seasonal patterns from the seasonal-irregular component. The challenge lies in estimating the seasonal component such that the moving average effectively removes the irregular (noise) component while preserving the seasonal pattern. If the moving average is not chosen properly, it can lead to an inaccurate estimation of the seasonal component. Thus, the core challenge of the algorithm lies in balancing the estimate of trend and seasonal components. The most important thing to overcome these difficulties is to choose the right Henderson filter length based on the characteristics of the data (eg, noise level, how fast or slow the trend changes). Choose an appropriate moving average (such as 3 × 3 or 3 × 5) based on the stability of the seasonal pattern, pay attention to the changing seasonality, and adjust the choice of filters by observing the results of each iteration to capture these changes, thereby gradually refining the estimation of the trend and seasonal components through multiple iterations.

To evaluate the individual components of the series while accounting for potential outliers, X-11 is applied iteratively: estimating components, identifying outlier effects in the irregular component, re-estimating components on the adjusted series, detecting further interference effects, and so on.[23]

TABLE 4.1

Simplified Diagram of the X-11 Algorithm Flow

Part A: Preliminary Adjustments
• For significant known extremes
• For the trading day effect
Part B: First automatic correction of the sequence
• Estimation of irregular components
• Detection and automatic correction of extreme values
• Correction of the trading day effect
Part C: Second automatic correction of the sequence
• Estimation of irregular components
• Detection and automatic correction of extreme values
• Correction of the trading day effect
Part D: Seasonal Adjustments
• Preliminary seasonally adjusted sequences were calculated (Tables D1 to D6)
• Seasonally adjusted sequences were smoothed using the Henderson moving average, and seasonal factors were re-estimated (Tables D7 to D10)
• The final seasonally adjusted sequence (Table D11) was calculated, and the trend-cycle components (Table D12) and irregular components (Table D13) were extracted
Part E: Component corrections for large extreme values
Part F: Seasonally Adjusted Quality Indicators
Part G: Chart Analysis

4.5 Comparison of Characteristics of Moving Averages: Trend versus Seasonal Estimates

The composite moving average is a core technique in the X-11 seasonal adjustment method, used to smooth time series data and decompose it into trend, seasonal, and irregular components. The Henderson criterion is used to evaluate the smoothing effect of a moving average. The composite moving average coefficients used in X-11, together with their variance reduction ability and the Henderson criterion, are summarised in Table 4.2.

4.5.1 Trend Estimates: 2 × 4 and 2 × 12 Averages

When the following assumptions are met in the time series: (1) the trend component is linear or locally linear; (2) the seasonal pattern is relatively stable; (3) there is no significant autocorrelation between irregular fluctuations, then this kind of moving average has the best effect. In X-11 seasonal adjustment, the initial estimation of the trend component is obtained using composite moving averages: a 2 × 4 moving average (a 5-term filter) for quarterly data, and a 2 × 12 moving average (with 12 centred moving averages) for monthly data. The coefficient curve and gain function are shown in Figure 4.3.

From the coefficient curve of the 2 × 4 moving average in the upper part of the figure on the left, we can see that the coefficient is {1,2,2,2,1}/8, met: (1) symmetry ($\theta_k = \theta_{-k}$); (2) coefficients and $\sum\theta_k = 1$ (maintain linear trend), sum of squares, $\sum\theta_k^2 = 0.25$ (75% reduction in white noise variance). The weight distribution of the five items is symmetrical, with the centre point having the largest weight (2/8) and decreasing weight on both sides (1/8). Reflecting the way adjacent data is weighted, the centre weighting emphasises the current moment, and the symmetry guarantees no phase shift.

TABLE 4.2

The Composite Moving Average Coefficient Used in X-11 and Its Variance Reduction Ability and the Henderson Criterion

i	2 × 4	2 × 12	3 × 3	3 × 5	3 × 9
-6		1/24			
-5		1/12			1/27
-4		1/12			2/27
-3		1/12		1/15	3/27
-2	1/8	1/12	1/9	2/15	3/27
-1	1/4	1/12	2/9	3/15	3/27
0	1/4	1/12	3/9	3/15	3/27
1	1/4	1/12	2/9	3/15	3/27
2	1/8	1/12	1/9	2/15	3/27
3		1/12		1/15	3/27
4		1/12			2/27
5		1/12			1/27
6		1/24			
$\sum\theta_t^2$	0.2188	0.0799	0.2346	0.1644	0.1001
$\sum(\nabla^3\theta_i)^2$	0.1250	0.0139	0.1481	0.0356	0.0110

Note: (1) $\sum\theta_t^2$ is the composite moving average coefficient, and its ability to reduce Variance; (2) $\sum(\nabla^3\theta_t)^2$ is the Henderson criterion.

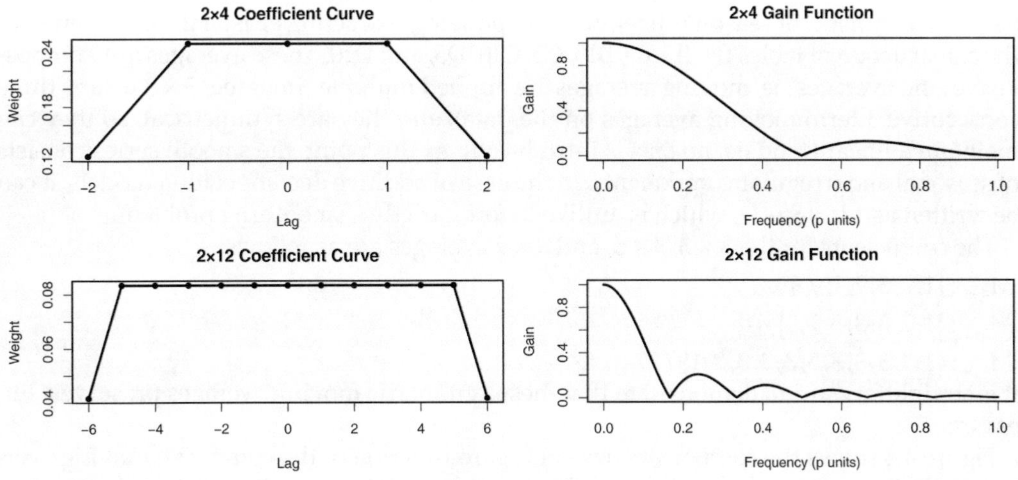

FIGURE 4.3

Composite Moving Averages of 2 × 4 and 2 × 12 used by X-11 for Coefficient Curves and Gain Functions.

Note: (1) Upper part: 2 × 4 moving average (quarterly data); Bottom half: 2 × 12 moving average (Monthly Data) (2) Sub-graph on the left: coefficient curve (time domain performance); Subplot on the right: Gain function (frequency domain performance)

Source: The simulated data

The gain function in the right figure is 0 at the frequency ω = π/2 (90^0 , corresponding to the quarterly period 4), which eliminates the seasonality. The low-band (ω→0) gain is close to 1 and retains the trend component. Falling edge: smooth attenuation from ω = 0 to π/2, showing the ability to suppress high-frequency noise. Retention for long periods (eg, three-year cycles) is about 80% (attenuation of low-frequency response). It is thus ideal for a quarterly series with constant seasonality.

From the lower part of the 2 × 12 moving average coefficient curve in the left figure, we can see that the average coefficient is $\left\{\frac{1}{24},\frac{2}{24},\frac{2}{24},\frac{2}{24},\frac{2}{24},\frac{2}{24},\frac{2}{24},\frac{2}{24},\frac{2}{24},\frac{2}{24},\frac{2}{24},\frac{2}{24},\frac{1}{24}\right\}$. The weights of the coefficients are $\frac{1}{24}[1,2,....,2,1]$, with (1) symmetry and linear trend maintenance; (2) sum of squares $\sum\theta_k^2 = 23/288$ (noise reduction >90%). The symmetrical weight distribution of 25 items is symmetrical, with the highest weight at the centre point and 1/24 at the edge. A wider window offers greater smoothness and is well-suited for suppressing high-frequency noise in monthly data. The gain function on the right shows that the gain is 0 at ω = π/6 (corresponding to annual cycle 12), which eliminates yearly seasonality. There is a small fluctuation at ω = π/3 (period 6), reflecting a partial inhibition of the semi-annual cycle; The passband characteristics show a low-band (ω < π/12) gain close to 1, perfectly preserving the long-term trend. The use of such moving averages works best when the trend period of our series is linear or locally linear, when seasonal factors are constant or vary little over time, and when the irregular components are unstructured and have limited amplitude. It is thus ideal for a monthly series with constant seasonality.

4.5.2 Seasonal Estimates: 3 × 3, 3 × 5, and 3 × 9 Averages

X-11 uses 3 × 3, 3 × 5, and 3 × 9 averages to smooth out seasonality-irregularity components based on estimates of seasonal-irregular components, extracting seasonal components for the construction of tables B4, B5, B9, B10, C5, C10, D5, and D10. These averages are 'composite moving averages', ie, moving averages are applied multiple times (eg, 3 × 3 means three consecutive 3-term moving averages on the data), and they are symmetrical, so they can maintain a linear trend (ie, no bias is introduced). At this point, the smooth series consists of seasonal and irregular components. In the case of additive decomposition models, it can be written as $SI_t = S_t + I_t$, which is, unlike before, strictly a smoothing problem.

The coefficients for the 3 × 3, 3 × 5, and 3 × 9 averages are as follows:

$M_{3x3} : \{1,2,3,2,1\}/9$

$M_{3x3} : \{1,2,3,3,3,2,1\}/15$

$M_{3x5} : \{1,2,3,3,3,3,3,3,3,2,1\}/27$

It is straightforward to demonstrate that these symmetric moving averages preserve a linear trend.

Figure 4.4 shows the coefficient curve and gain function of the composite moving average. From Figure 4.4, we can see some characteristics of the weight coefficient distribution of the 3 × 3, 3 × 5, and 3 × 9 moving averages shown by the coefficient curve on the left (time domain performance): (1) Symmetry: the weight of the centre point is the largest, and the two sides gradually decrease; All coefficient curves are symmetrically distributed, eg, [–3, –2, –1, 0, 1, 2, 3], ensuring that the linear trend is unbiased (no phase shift is introduced); (2) Smoothing strength: 3 × 3 coefficient distribution is narrow (the smoothing range is small), local smoothness is strong, and it is suitable for high-frequency noise

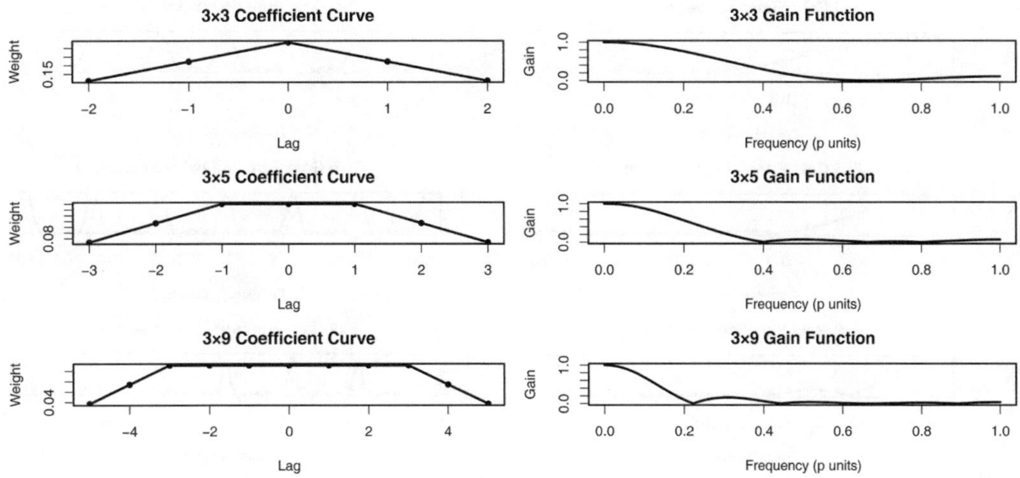

FIGURE 4.4
Coefficient Curves and Gain Functions of the Composite Moving Average Used in X-11 Seasonality Factor Estimation.

Note: (1) The left side is the coefficient curve (time domain performance), and the right side is the gain function (frequency domain performance); (2) 3 × 3 is shown in the upper part, 3 × 5 is shown in the middle part, and 3 × 9 is shown in the lower part.

Source: The simulated data

suppression; The 3 × 9 coefficient has a wide distribution and a larger smoothing range, which is suitable for low-frequency noise suppression.

The gain function on the right (frequency domain representation) reflects how well the filter retains or attenuates signals at different frequencies. It can be seen that the main lobes of 3 × 3 are wider, and the high-frequency attenuation is faster; the side lobes are obvious. The 3 × 9 main lobe is narrower, and the side lobes are better inhibited but may be over-smoothed to non-seasonal low-frequency components. Ideally, seasonal filtering should completely preserve the annual cycle (eg, 12 months) and its harmonics, attenuating other frequencies. However, the actual gain function is not ideal (there are side lobes), and the effect of non-ideality is limited because the seasonal component is concentrated in a specific frequency (such as 1/12, 2/12, etc.).

As shown in Figure 4.4, this non-ideality does not affect the decomposition effect due to the cyclical nature of the seasonal component, as the seasonal-irregular component itself should not contain any cyclic component with a period of three to six years. For example, applying the 3 × 3 filter to each month of the seasonal-irregular component estimate is equivalent to applying a moving average with 49[24] coefficients to that component:

{1,0,0,0,0,0,0,0,0,0,0,2,0,0,0,0,0,0,0,0,0,0,0,
3,0,0,0,0,0,0,0,0,0,0,0,2,0,0,0,0,0,0,0,0,0,0,0,1}/9

This operation is a typical sub-series smoothing technique in X-11 seasonal adjustment, which achieves cross-period noise suppression through a high-dimensional moving average while maintaining the structural characteristics of seasonal patterns.

The same goes for the 3 × 5 and 3 × 9 filters, which are the associated 73[25] and 121[26] moving averages, respectively! The gain function for these moving averages is shown in Figure 4.5. It can be seen that these moving averages retain the annual seasonality since they recover precisely the octave of $30^0 = \frac{2\pi}{12} = \pi / 6$.

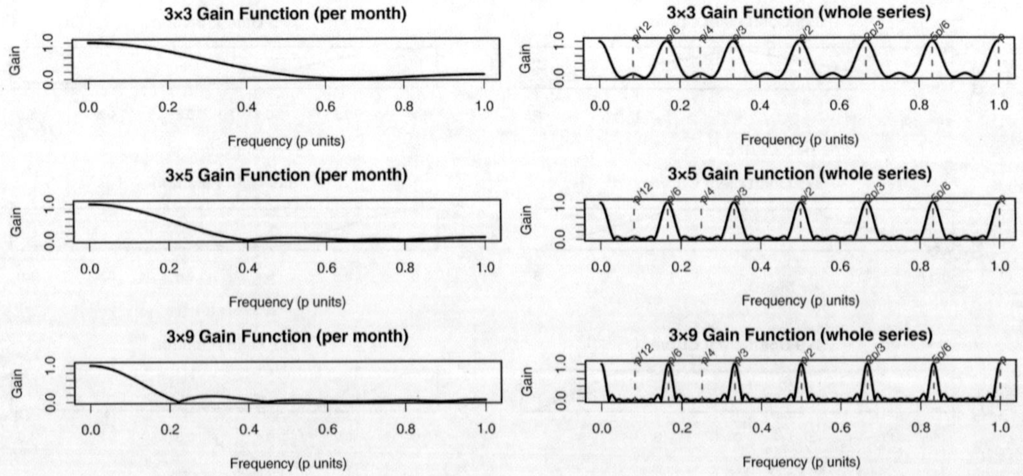

FIGURE 4.5
Gain Function for 3 × 3, 3 × 5 and 3 × 9 Averages.

Note: (1) 3 × 3 is shown in the upper part, 3 × 5 is shown in the middle part, and 3 × 9 is shown in the lower part; (2) The left side is the growth function for each month, and the right side is the gain function for the whole series.

Source: The simulated data

Figure 4.5 shows a detailed analysis of the gain functions, 3 × 3, 3 × 5, and 3 × 9. On the left side is the monthly gain function, showing the frequency response of the filter to the 'sub-series per month'. Monthly seasonality (eg, January, February, etc.) is retained, and irregular noise is attenuated. It can be seen that the gain of 3 × 9 is close to 1 at the annual frequency (eg, 1/12), indicating perfect preservation of seasonality; Other frequencies are more attenuated. On the right side is the full range of gain functions, reflecting the filter's comprehensive frequency response to the 'entire time series'. The gain of 'annual periodicity' at a frequency of k/12 (k = 1,2, . . .) is high to ensure that seasonal components are not filtered out. Noise suppression gains close to 0 at non-seasonal frequencies such as trend or random fluctuations.

4.5.3 Henderson Moving Average

The Henderson moving average is used in X-11 to extract trends from estimates of seasonally adjusted series. The coefficients of the Henderson moving average used in X-11 are represented in Table 4.2[27] as well as the associated criteria, variance reduction, and smoothing values. The Henderson moving average is the core tool of the X-11/X-12-ARIMA, which is primarily used for the extraction of trend components. The core features of the Henderson moving average include

(1) Symmetry and unbiasedness, and the weight coefficient of central symmetry, eg, the coefficient of the 13-term Henderson is (−0.019, 0.028, 0.109, . . ., 0.28, 0.109, . . .) to ensure that the linear trend is unbiased. At the endpoint of the series, the 'Musgrave Asymmetric Moving Average' (Figure 4.6) is used to avoid information loss.

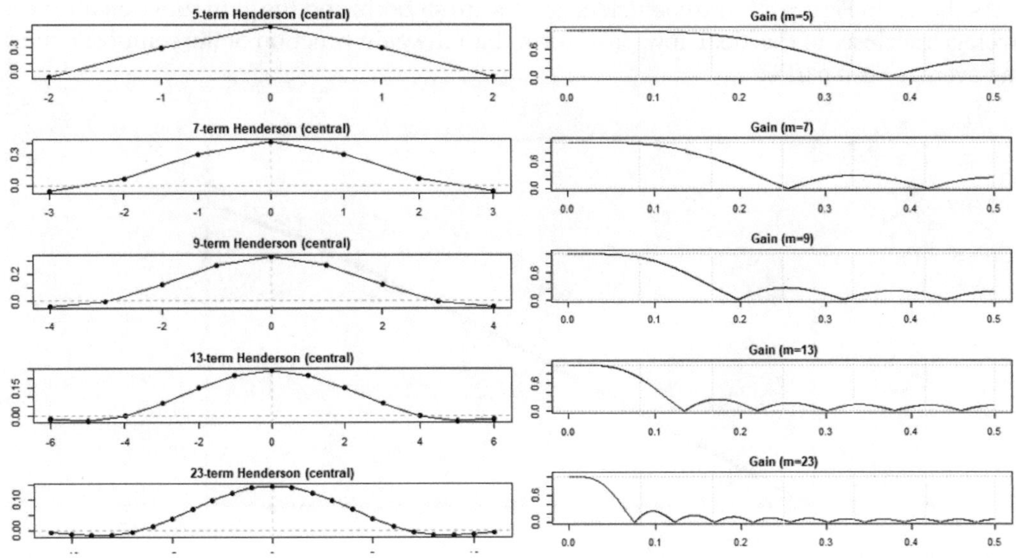

FIGURE 4.6
Coefficient Curve and Gain Function for the X-11 Henderson Moving Average.

Note: (1) From top to bottom are the coefficient curves and gain functions. (2) The coefficient curve is on the left side, and the gain function is on the right side.

Source: The simulated data.

(2) Smoothness and frequency response: the higher order (eg, 23 terms) is smoother, which is suitable for low-frequency trend extraction, and the lower order (eg, 5 terms) retains more short-term fluctuations. The gain function is close to that of an 'ideal low-pass filter' (Figure 4.6, right) and effectively separates trend (low frequency) from noise (high frequency).

(3) Flexibility: Odd order (5 to 101 items) is supported, and users can choose to balance smoothness and timeliness according to data characteristics.

Figure 4.6 shows the coefficient curve versus the gain function. From Figure 4.6, we can see the (time domain) coefficient curves on the left: (1) the weight distribution of the 5-term, 7-term, 9-term, 13-term, and 23-term Henderson moving average; (2) all curves are symmetrical with the centre point, and the weight decreases from the centre to both sides (eg, the centre coefficient of 13 terms is the largest, and the two ends are close to 0); (3) 5 narrow windows, sensitive to local fluctuations; 23 wide windows, strong ability to extract global trends.

Right (frequency domain) gain function (1) low-pass characteristic: the gain function is close to 1 at low frequency (trend component), and the high frequency (noise) is rapidly attenuated, close to the ideal low-pass filter; (2) high-order comparison: the 23-item main lobe was narrower, and the side lobe inhibition was better (high-frequency attenuation was more complete), 5 main lobes are wide, which may retain some high-frequency noise; (3) seasonal inhibition: effectively filter out seasonal (such as 12-month cycle) and irregular components to highlight trends.

As shown in Figure 4.6, the coefficient curves are smooth, and the gain function for these averages is closer to the ideal 'low-pass' form than the gain function of the composite moving average seen earlier.

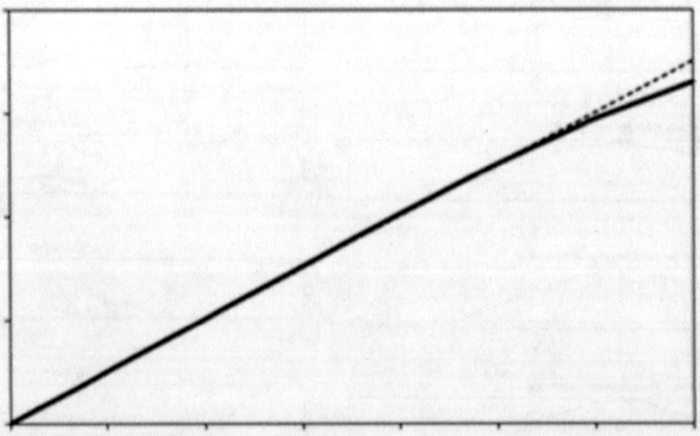

FIGURE 4.7
Uses the Linear Smoothing of the 13-term Henderson Moving Average Supplemented by the Musgrave Asymmetric Moving Average.

Source: The simulated data.

Figure 4.7 shows that the symmetric weights used in the middle of the series are 13-term Henderson, and when a symmetric moving average cannot be used for the latest or oldest data (the endpoint of the series), the Musgrave asymmetric weights (eg, the forward weights are used for the last period) to avoid truncation bias, thus maintaining overall smooth continuity and reducing the estimated variance at the endpoints (no mutation at the tail in Figure 4.6).

4.5.3.1 Comparison of the Henderson Moving Average and the Composite Moving Average

- Henderson: extracting 'trends' (low-frequency components); composite moving averages (3 × 3, etc.), extracting 'seasonality' (specific cycle components).
- The Henderson gain function is closer to the ideal low-pass (Figure 4.7, right), and the composite moving average needs to match the seasonal frequency (Figure 4.5, right). The Henderson coefficient is smoother (Figure 4.7, left), and the composite moving average is multiple average overlays (Figure 4.4, left).

The Henderson moving average is the gold standard for trend extraction with 'symmetric weighting' and 'optimised frequency response'. The high-level (eg, 23-item) is suitable for long-term trend analysis, and the low-level (eg, 5-item) is suitable for short-term volatility monitoring. Combined with Musgrave's asymmetrical processing, it effectively solves endpoint problems and is an indispensable tool in X-11 seasonal adjustments.

4.6 Summary and Conclusions

The moving average is fundamental to the 'X-11' family of seasonal adjustment programs and serves as a core technique in their time series decomposition. This chapter focuses on the time- and frequency domain characteristics of the moving average, as well as the iterative procedure of the X-11 algorithm.

1. The moving average is the core technology of the X-11 seasonal adjustment method and is used to break down the trend, seasonal, and irregular components of a time series. Moving averages have a long history and were used in early applications for financial and meteorological data, such as those of the Bank of England (1797) and Buys-Ballot (1847). The X-11 method is flexible for nonparametric modelling with predefined linear filters (moving averages) and is suitable for trend maintenance and noise reduction.
2. Time domain analysis and moving average: The simple moving average smooths the data through the arithmetic average, and the symmetrical form (such as the 3-term moving average) has no phase shift, but the endpoints need to be asymmetrical. The symmetric moving average maintains a linear trend, while the asymmetric moving average introduces a phase lag. The weighted moving average is given different weights to optimise the smoothing effect, which needs to be met. $\sum_{k=-m}^{m}\theta_k = 1$ and $\sum_{k=-m}^{m}k\theta_k = 0$. The Henderson moving average determines the weights by minimising the sum of the three differences of squares, which is suitable for trend estimation, and the length is automatically selected by the I/C ratio. Composite moving averages, such as 2 × 4 moving averages (quarterly data) and 2 × 12 moving averages (monthly data), eliminate seasonality and preserve trends through two-step averaging. Local regression is equivalent to a specific moving average, and the weighted centre point value is calculated by local polynomial fitting, such as quadratic regression.
3. Frequency domain analysis identifies periodic components through the Fourier transform, and trend (low frequency), seasonal (medium frequency), and irregular fluctuation (high frequency) correspond to different frequency ranges. The gain function $G(\omega)$ measures the amplitude retention of the frequency ($G = 1$ is completely preserved, $G = 0$ is eliminated); The asymmetric moving average introduces a lag, which can be seen from the phase shift function $C(\omega)$, for example, $\frac{C(\omega)}{\omega} = -1$, indicating a lag of one period.
4. Seasonality elimination: Eliminate fixed periodicity (eg, quarterly data period 4) using a moving average and a constraint factor. Irregularities are reduced through symmetrical moving averages to minimise variance (eg, $\frac{1}{3}[1,1,1]$ 75% reduction in white noise variance).
5. X-11 algorithm flow: Four processing steps include A (pre-adjustment): extreme value and trading day effect correction; B/C (iterative correction): estimates trends and seasonality through moving averages, detects outliers; D (seasonally adjusted): Eventually decomposed into trend, seasonal, and irregular components.
6. Five-step iterations: initial trend estimation (eg, 2 × 12 moving average), → seasonal-irregular (SI) ratio calculation→ seasonal estimation (eg, 3 × 3 moving average), → seasonally adjusted series, → final trend (Henderson moving average).
7. The Henderson filter balances smoothness and response speed (eg, 13-term moving average), and length affects the quality of trend estimation. Characteristics of different moving averages, trend estimation using 2 × 4 moving averages (quarterly): eliminate period 4, retain linear trends, and reduce noise variance by 75%;

2×12 moving average (monthly) eliminates annual seasonality and reduces noise variance by 90%; seasonal estimates use 3×3/3×5/3×9 moving averages: corresponding to 5/7/9 term coefficients respectively, and retain the annual seasonal pattern; The Henderson moving average coefficient is smooth (eg, 13 terms), and the gain function is close to the ideal low-pass filter, which is suitable for trend extraction.

8. The moving average is the core of X-11, which is nonparametric and flexible, but it is necessary to pay attention to the phase shift of the asymmetric moving average; Composite moving averages (eg, 2×12) and Henderson moving averages are key tools that are used for seasonal and trend decomposition respectively; frequency domain analysis assists in identifying components, and spectral peaks correspond to trend, seasonal and noise.
9. X-13ARIMA-SEATS introduces SEATS model decomposition as an option; this chapter focuses on the time/frequency domain characteristics of the moving average and the iterative process of X-11, and the subsequent chapter (Chapter 7) will discuss the SEATS parameterised method.

Notes

1 The X11 series contains X11, X-11-ARIMA, X-12-ARIMA 和 X-13ARIMA-SEATS.

2 By default, the selection of the moving average is done in a limited number of predefined filters but is done depending on the I/C and I/S ratios of the series characteristics.

3 Klein, J. L. (2009). "*Statistical Visions in Time: A History of Time Series Analysis 1662–1938*". Cambridge University Press.

4 Finlaison, John. (1829). "Report on the Mortality of the Government Life Annuitants". In *Appendix B: On the Method of Constructing Mortality Tables*, pp. 235–239. House of Commons Parliamentary Papers, London.

5 Schiaparelli, Giovanni Virginio. (1866). "Sul modo di ricavare la vera espressione delle leggi della natura dalle curve empiriche" ("On the True Expression of the Laws of Nature Derived from Empirical Curves"). *Memorie dell'Accademia delle Scienze di Torino*, 21(2), 1–50.

6 De Forest, E. L. (1873). "On Some Methods of Interpolation Applicable to the Graduation of Irregular Series, Such as Tables of Mortality". *Smithsonian Report 1871*, 275–339; De Forest, E. L. (1875). "Additions to a Memoir on Methods of Interpolation Applicable to the Graduation of Irregular Series". *Smithsonian Report 1873*, 319–349; De Forest, E. L. (1876). *Additions to a Memoir on Methods of Interpolation Applicable to the Graduation or Adjustment of Irregular Series of Observed Numbers*. Tuttle, Morehouse and Taylor Co., New Haven; De Forest, E. L. (1877). "On Adjustment Formulas". *The Analyst*, 4, 79–86, 107–113.

7 Buys-Ballot, C. (1847). "*Les changements périodiques de température*". Kemink et Fils, Utrecht.

8 Jevons, W. S. (1862). "On the Study of Periodic Commercial Fluctuations". In *Investigations in Currency and Finance*. Macmillan, London, 1884.

9 Whittaker, E. T. (1923). "On a New Method of Graduation". *Proceedings of the Edinburgh Mathematical Association*, 41, 63–75.

10 Henderson, R. (1924). "A New Method of Graduation". *Transactions of the Actuarial Society of America*, 25, 29–40. Thus, his difference equation is the condition for the minimum value $\Sigma(\Delta^3 u_x)^2 + k\Sigma\left(u_x - u_x''\right)^2$, where ux is the graded value of the function and u_x'' is the ungraded value of the function $\Sigma(\Delta^3 u_x)^2 + k\Sigma\left(u_x - u_x''\right)^2$.

11 It is worth pointing out that Bohlmann (1899) proposed a very similar criterion, using the first difference instead of the third, known as the Bohlmann criterion. Bohlmann criterion is the

Hodrick-Prescott local version of the filter, Bohlmann, G. (1899). "Ein Ausgleichungsproblem". *Nachrichten von der Gesellschaft der Wissenschaften zu Gottingen, Mathematisch-Physikalische Klasse*, 1899, 260–271. Retrieved from http://eudml.org/doc/58455.

12 Akaike, H., & Ishiguro, M. (1980). "Bayesian Seasonal Adjustment". *Biometrika*, 67(1), 221–233.

13 Härdle, W. (1990). "Applied Nonparametric Regression". Econometric Society Monographs. Cambridge University Press, Cambridge; Akaike, H., & Ishiguro, M. (1980). "BAYSEA, a Bayesian Seasonal Adjustment Program". Computer Science Monographs, 13. Institute of Statistical Mathematics, Tokyo.

14 Persons, W. M. (1919). "Indices of Business Conditions". *Review of Economics and Statistics*, 1, 1. Harvard University Press. Persons' method was further expanded in his 1923 work "The Construction of Index Numbers".

15 Macaulay, F. R. (1931). "*The Smoothing of Time Series*". NBER Book Series. National Bureau of Economic Research, New York, 1938. 年, Macaulay further expanded his research and published a monograph "The Movements of Interest Rates, Bond Yields, and Stock Prices in the United States since 1856". NBER.

16 Kendall, S. M., & Keith Ord, J. (1990). "*Time Series*", 3rd ed., chapter 3, Edward Arnold, London.

17 Kendall, S. M., & Keith Ord, J. (1990). "*Time Series*", 3rd ed., chapter 3, Edward Arnold, London.

18 In local polynomial regression, β_0 does not directly represent a weight. Instead, it is the intercept term estimated via weighted least squares, representing the locally weighted average at the time point $t = 0$. However, the computation of β_0 involves weights because, in local polynomial regression, each observed value y_t at time t is assigned a different weight. The magnitude of this weight depends on the distance between t and the centre point ($t = 0$). Points closer to the centre receive larger weights, while points farther away receive smaller weights. Weighted least squares: When fitting a local polynomial regression model via weighted least squares, the calculation formula for $\hat{\beta}_0$ is $\hat{\beta}_0 = \frac{1}{21}(-2y_{-3} + 3y_{-2} + 6y_{-1} + 7y_0 + 6y_1 + 3y_2 - 2y_3)$. The coefficients $\frac{1}{21}[-2,3,6,\underline{7}]$ represent the *weights*, indicating the contribution of each time point's observed value to $\hat{\beta}_0$. Since $\hat{\beta}_0$ is the locally weighted average at $t = 0$, it is essentially computed as a weighted average. Thus, $\hat{\beta}_0$ can be regarded as the 'weighted average centred at $t = 0$', where the weights are determined by the local polynomial regression fitting process.

19 In this context, when listing moving average coefficients, it suffices to specify the order of the moving average and the first (k+1) coefficients according to Kendall (1973). For example, {1,2, 2,2,2,2,2,2,2,2,2,2,2,1}/24 can be compactly written as [13]: 1/24 {1,2,2,2,2,2,2}. See Kendall, M. (1973). "*Time Series*". Griffin, London/Hafner Press, New York, Often co-published); Kendall, M., & Stuart, A. (1978). "*The Advanced Theory of Statistics*", vol. 3, chapter 46. Griffin, London/ Macmillan, New York, later editions.

20 Jean-Baptiste Fourier, in his Théorie Analytique de la Chaleur, published in 1807, determined that any mathematical function can be decomposed into the sum of sine and cosine functions. This theorem gave rise to the initial harmonic analysis, which was later generalised to spectrum analysis. Cf. Fourier, J.-B. (1822). "*Théorie analytique de la chaleur*". Firmin Didot, Paris.

21 Follow the $\sin(A \pm B) = \sin A \cos B \pm \cos A \sin B$ rules of application, and then simplify the cause: $M(y_t) = \frac{1}{3}R(1 + 2\cos(\omega)R\sin(\omega t + \varphi - \omega)]$

22 Mark [(1)] refers to the first estimate and Mark [(2)] refers to the second estimate.

23 A detailed approach to the decomposition of the X-11, See Ladiray, D., & Quenneville, B. (2001). "*Seasonal Adjustment with the X-11 Method*". Springer; Dagum, E. B. (1980). "*The X-11-ARIMA Seasonal Adjustment Method*". Springer; Dagum, E. B., et al. (1999). "*X11ARIMA Version 2000: Foundations and User's Manual*". Springer; Gray, A. (1990). "*How the X-11 Program Implements a Trend-Seasonal Decomposition*". Springer.

24 A 3×3 moving average calculation is applied consecutively (meaning: first a 3-term average, then a 3-term average of *those results*, followed by *another* 3-term average). This is equivalent to a single 49-term moving average (since 3×3×3+2x(3x3)+2x3=49 terms, but the consecutive application expands the effective window). The centre point receives the maximum weight,

with weights symmetrically decreasing toward both ends. The specific coefficients (simplified example; actual X-11 weights may differ slightly) are: Weights = [1/27, 3/27, 6/27, 7/27, 6/27, 3/27, 1/27]. Actual 3×3 moving averages in X-11 may involve more complex weighting schemes, but their core characteristics remain: symmetry, smoothing capability, and preservation of linear trends.

25 The 3×5 moving average calculation is applied consecutively (meaning three successive applications of a 5-term moving average). This is equivalent to a single 73-term moving average (calculated as: (5×5×5) + 2×(5×5) + 2×5 = 73). Compared to the 3×3 moving average, this provides greater smoothing and covers a wider time window. Example coefficients (centrally symmetric): Weights = [1/125, 5/125, 15/125, 30/125, 45/125, 51/125, 45/125,. . .]. Actual 3×5 moving averages in X-11 may involve more terms, but their core characteristic is stronger noise suppression capability.

26 The 3×9 moving average calculation involves consecutively applying three successive 9-term moving averages. This is equivalent to a single 121-term moving average (calculated as: (9×9×9) + 2×(9×9) + 2×9 = 729 + 162 + 18 = 121). This configuration provides (1) the widest coverage in the time domain and (2) the strongest smoothing effect. Example coefficients (exhibiting high symmetry): Weights = [1/729, 9/729, 45/729, 165/729, 495/729,. . .] (symmetric continuation toward centre and ends). While exceptionally effective for extreme noise suppression, this filter may excessively smooth seasonal details due to its aggressive weighting profile.

27 With X-12-ARIMA, any odd Henderson moving average of order less than 101 can be used.

5

Application of X-13ARIMA-SEATS Software

5.1 Introduction

Seasonal adjustments are widely used in official statistics and financial sector (banking) data. The X-13ARIMA-SEATS software, a widely used tool for time series decomposition and seasonal adjustment, has been the standard method for official statistics in the UK since 2001, following its endorsement by the UK National Program Committee for Statistical Quality and Methods, in alignment with European best practices and the approach of the Bank of England. This method is widely adopted by statistical agencies in major economies around the world.

This chapter primarily describes the use of the X-13ARIMA-SEATS software, including its operating environment and, in particular, the *spec* command file used by X-13ARIMA-SEATS. It covers topics such as ARIMA model selection for decomposition and, especially, the use of the *regression* spec (regARIMA). The procedures and diagnostics for the seasonal adjustment of X-13ARIMA-SEATS are also described and discussed in detail. Chapter 6 will delve deeper into other advanced topics related to seasonal adjustments.

Following the introduction in Section 5.1, Section 5.2 provides a general description of how to use X-13ARIMA-SEATS, including the data file format used to run X-13ARIMA-SEATS, the X-13ARIMA-SEATS command file (spec) and save file; Section 5.3 describes the operating environment of the X-13ARIMA-SEATS – DOS, OxEdit (interactive file editor), and Win-X13 (direct operation in Window); Section 5.4 describes the various command (spec) files used by X-13ARIMA-SEATS; Section 5.5 describes the procedures used by the X-13ARIMA-SEATS; Section 5.6 describes how the X-13ARIMA-SEATS run standard output; Section 5.7 discusses the inspection and diagnosis of seasonal adjustments; Finally, Section 5.8 gives a summary and conclusions.

5.2 Basis of X-13ARIMA-SEATS Application

There are two files to run X-13ARIMA-SEATS: (1) the executable file x13a.exe and (2) the model list file x13a.mdl, both of which must be built in the same directory on the computer. You must also create two additional files: (1) a data file that uses X-13ARIMA-SEATS for seasoning, such as the name myfile.dat, and (2) a specification file that specifies the program options for seasoning a data series, called a spec file. For example, name myfile.spc. Both files must be in text format and can be easily created using Notepad (a text editor) or OxEdit (a user-friendly editor), or Win X–13. The X-13ARIMA-SEATS can use these files to run seasonal adjustments.

DOI: 10.1201/9781003716655-5

5.2.1 Data Files

A data file is a list that contains the raw data needed for seasonal adjustments. The data can be in free-form, date-value format, or X13-save format. Free-form is the more commonly used format.

5.2.1.1 '.dat' Free-Format

If the data is free-form, include a single column of data saved with the '.dat' extension, as in the following example:

89814.18

85582.54

85059.04

85880.88

97711.15

78266.99

84479.21

79941.65

In the 'series' spec file, you need to have a 'start' statement with the start date of the series, for example:

```
series{title="name of file"
   start= 1991.1
   period=4
   file="c:\research\test.dat"
          }
```

5.2.1.2 'Date-Value' Format

If the data contains a date that is a date-value (date-value format), the separator separating each field is a tab or a space, as in the example shown below:

1991 1 89814.18

1991 2 85582.54

1991 3 85059.04

1991 4 85880.88

1991 5 97711.15

1991 6 78266.99

1991 7 84479.21

1991 8 79941.65

In this case, the 'start' parameter in the 'series' spec is not required. Tags are not required in the data file, but they are allowed, such as the following:

```
series{title="name of file"
    period=4
    file='c:\research\test.dat'
        }
```

5.2.1.3 'X13-Save' Format

The file in the X-13ARIMA-SEATS save format contains the date, and the file can only be generated in the save format of X-13ARIMA-SEATS. Here's an example of X13-save data:

date----- mydata.d11---------------
199101 8.98E+04
199102 8.56E+04
199103 8.51E+04
199104 8.59E+04
199105 9.77E+04
199106 7.83E+04
199107 8.45E+04
199108 7.99E+04

5.2.2 Command (Spec) File

To run X-13ARIMA-SEATS for any series, you must create a command file, also known as a 'spec' file or input file. A 'spec' file is a text file that is used to specify program options. The name of the 'spec' file must end with the '.spc' extension. A 'spec' file contains a set of functional units called 'spec'. The X-13ARIMA-SEATS reads these 'specs' to obtain the information needed for the time series, such as the decomposition model used, the analysis performed, and the desired output. Each spec includes options to control a specific function. For example, the 'series' spec defines the raw data to be used and the time range for analysis, while the 'transform' spec specifies any transformations to apply, including prior adjustments.

The X-13ARIMA-SEATS includes four types of spec files: (1) data entry and conversion; (2) regARIMA modelling; (3) seasonal adjustment; and (4) seasonally adjusted diagnosis. The X-13ARIMA-SEATS spec file must start with a 'series' or 'composite' spec, while other specs can be entered in any order. The general input syntax for a spec file looks like this:

```
specification_name{
        argument1=value
        argument2=(value1 value2 value3)
        argument3="A string value"
        argument4=2003.9        # date can be written as '2003.9' or
   #'2003.Sep'.
        }
```

Where: '#' after the line of the symbol is considered a comment, the date is in the format yyyy.month period (for example, 2003.9 or 2003.SEP), if the parameter has two or more values, these values must be enclosed in parentheses, for example, (value1 value2 value3), character values, such as titles and file names, must be enclosed in quotation marks, for example, 'A string value', parameters can be set in each spec; parameters specify the functions that the X-13ARIMA-SEATS program will execute; if no parameters are provided (for example, by using empty parentheses ()), default values are typically selected automatically. Parameters can be written in any order in the spec, and the X-13ARIMA-SEATS is case-insensitive, ie, uppercase, lowercase, or mixed-case. A spec file can include the name and path of any optional file, including data for the modelled time series, data for user-defined and predefined regressions, the value of any user-defined prior adjustments, and the type of model to be attempted using the automated model process. The simplest spec file looks like this:

```
Series{
        Title="Example of Specification for default seasonal adjustment"
        Start-1994.1
        Period=4
        Name="default"
        Fime="c:\research\test.dat"
        }

Transform{
        Function=log
        File="c:\research\testtp.tpp"
        Type=tempory
        }

arima{
    model=(0,1,1)(0,1,1)
    }

Regression{
        Aictesy=(EASTER)
        }
X11{}
```

The above spec file includes the following:

(1) series: X-13ARIMA-SEATS file name used, start time, data type (quarterly data), data name, and storage location of the original data (path of data storage);

(2) transform: takes the logarithm of the series, the path where the temporary prior file name is stored, and the file type to adjust the series a priori;

(3) arima: specifies the ARIMA model as (0,1,1) (0,1,1);

(4) regression model: contains Easter variables, tests the Easter effect of known or suspected influence series, tests the significance of regression variables, 'aictest=(EASTER)';

(5) X-11 is a method chosen for a seasonal adjustment.

5.3 Runtime Environment Configuration

The X-13ARIMA-SEATS can be run (1) under the DOS system, using the DOS prompt; (2) using an interactive file editor, such as OxEdit; (3) using Win-X13 to operate directly on Windows.

5.3.1 Operation in DOS Environment

```
The X-13ARIMA-SEATS program can be run in the DOS environment using the
following command: C:\X13a>X13a path\myfile¶
```

The symbol¶ is the 'return' or 'enter' key, which means that the enter key is pressed to run the program. In this statement, path\myfile.spec is the primary input (spec) file used by X-13ARIMA-SEATS. If the spec file is not in the same directory as the X-13ARIMA-SEATS program, you will need to enter the path to the file. The standard output of the run will be written to the file path\myfile.out. The optional 'print' parameter in each spec controls what is added (or removed) to the standard output. Any input errors are printed in the corresponding error messages, which are stored in a file named path\myfile.err.

If the program can locate the error, the spec file will mark the line containing the error with a caret (^) below the error message; if the program is unable to locate the error, it will only print an error message; if the error is fatal, 'ERROR' will be displayed before the error message, sometimes accompanied by a suggestion of what to change; for non-fatal errors, 'WARNING' is printed. WARNING is also sometimes used to draw attention, but caution is required. The X-13ARIMA-SEATS first reads the entire spec file and reports all input errors found. This enables the user to correct multiple input errors per run. If any fatal error is detected, the program will stop. The warning does not stop the program, but the user should be reminded to double-check the inputs and outputs to verify that the desired results are being produced.

To examine the effect of different modelling and tuning options on a given series, it is sometimes necessary to use different names for the output and input. The general format for specifying an alternate file name for the output file is

```
C:\X13a>X13a path\myfile path\newname¶
```

In this case, X-13ARIMA-SEATS runs using the spec file 'myfile.spc', the output will be stored in the '.out' file, and any error messages will be stored in the '.err' file. In production situations, it is often necessary to run multiple series in a single execution.

5.3.1.1 Run the Command spec.mta *File*

The X-13ARIMA-SEATS allows running (a) a single spec mode, where each series will use the options detailed in a single input spec file; (b) multiple spec modes, where each series has a specified, individually input spec file.

When running multiple specs for X-13ARIMA-SEATS, it must first create an input *spec.mta* file containing the name of the individual file to be run. The input spec file can have up to two entries per line: the spec input file name for a given series (and path information if necessary), and an optional output file name for that series. If the user does not provide an output file name, the spec input file name and path will be used to generate the output file. The *spec.mta* input files are processed in the order that their individual spec input file names are entered. For example, the input spec files are series 1.spc, series2.spc, and series3.spc:

```
series1
series2
series3
```

The *spec.mta* file will also run in this order. And let's say all these spec files are stored in the x13a directory. If they are stored in the 'c:\research\specs' DOS directory, the spec file should be:

```
c:\ research\specs\series1
c:\ research\specs\series2
c:\ research\specs\series3
```

To run X-13ARIMA-SEATS using the input *spec.mta* file, you can use the following command line:

```
X13a -m mygroup¶
```

Where -m is a flag, *mygroup* is the name of the *spec.mta* input file. *'.mta'* is the extension, but the extension *'.mta'* does not need to be typed as in the command line above. Just enter mygroup, not mygroup.mta. If necessary, the path information should be included in the input spec file name. The name of the X-13ARIMA-SEATS used to generate the output file is taken from the individual spec file listed in the spec file, not the filename of *the spec.mta*. The example above will generate output files named series1. out, series2.out, and series3. out, which correspond to a single spec file given in *mygroup.mta* instead of a comprehensive output file named *'mygroup' out*. To specify an alternate output filename, simply add the desired output filename to each line of the input spec file as follows:

```
c:\research\specs\series1 c:\research\alternative\series1
c:\research\specs\series2 c:\research\alternative\series2
c:\research\specs\series3 c:\research\alternative\series3
```

5.3.1.2 Data Metafile

The data metafile (.dta) is a single file used by X-13ARIMA-SEATS for batch processing of multiple time series. Each line of the .dta file may contain up to two entries: the full filename of a data series (including path if needed) and an optional output filename for that series. If no output filename is provided, X-13 generates output using the path and filename of the data file. If the data file is not in the current working directory (often x13a), the full path and filename must be specified.

X-13ARIMA-SEATS processes the series in the order they appear in the .dta file. All data files listed must be in the same format, since they are processed with a single spec file. For example, to process three data files (series1.dat, series2.dat, and series3.dat), the data metafile mygroup.dta would contain

```
Series1.dat
Series2.dat
Series3.dat
```

If these files are stored in c:\research\data\, the lines must include the full paths:

```
c:\research\data\series1.dat
c:\research\data\series1.dat
c:\research\data\series1.dat
```

To run X-13ARIMA-SEATS on the data metafile, use the following command line:

```
X13a myspec -d mygroup¶
```

where myspec is the spec file to be applied, mygroup is the data metafile, and the -d option specifies that a data metafile is being used. Note that the input metafile file must be saved with the '*.dta*' extension (eg, *mygroup.dta*). However, when providing the name of the data spec file in the command line, as shown earlier, only the file name should be included, not the extension. If appropriate, the path information should be provided along with the data spec file name. The filename used by X-13ARIMA-SEATS to generate the output file is derived from the data filename listed in the spec file, not from the spec filename. The example given earlier will produce output files named series1. out, series2.out, and series3. out, which correspond to the individual data file names given in the original file *mygroup. dta*, rather than the comprehensive output file names of *mygroup. out*. If you want to specify an alternate output filename, simply add the desired output filename to each line of the data spec file as follows:

```
c:\research\data\series1.dat c:\research\alternative\series1
c:\research\data\series2.dat c:\research\alternative\series2
c:\research\data\series3.dat c:\research\alternative\series3
```

5.3.1.3 Log File

Each time X-13ARIMA-SEATS is run, a log file is generated. This includes a summary of modelling and seasonally adjusted diagnostics for each series or spec file processed. When X-13ARIMA-SEATS runs multiple spec modes, the log file will be stored in the same name and directory as the spec file. Conversely, when X-13ARIMA-SEATS runs single spec mode, the log file will be stored with the same name and directory as the data spec file. In both cases, the log file is stored with the extension '.log'; for example, the command,

```
X13a -m mygroup¶
```

Run each spec file stored in mygroup.mta and store the user-selected diagnostic information in the log file mygroup.log. If a single series is processed, the output directory and file name are used with the '.log' file extension to form the name of the log file (myseries.log). Users can specify which diagnostics are stored in the log file using the 'savelog' parameters in Series, Composite, Transform, X11, Regression, Pickmdl, Estimate, Check, Slidingspan, and History Spec. If an error is found in one of the spec files where the spec file runs, all input files containing the error (.err) are provided in the log file.[1]

5.3.2 OxEdit Interactive Editor

Create a spec file in OxEdit, run it in OxEdit, and display the analysis results. OxEdit is a reliable, freeware text editor that operates based on file extensions.

5.3.2.1 Installation of OxEdit

OxEdit can be downloaded from the http://www.oxedit.com website: 'macro keys'[2] have been created for OxEdit to call X-13ARIMA-SEATS directly from the editor. The installation of OxEdit and description of the X-13ARIMA-SEATS 'macro keys' is as follows:

(1) Let OxEdit be installed in the directory 'c:\x13a'. If not, the following file needs to be modified to reflect the path of the installed X-13ARIMA-SEATS.

(2) The OxEdit setting runs with OxEdit 0.3 version. If you're using version 0.3, the installation process will be simpler; just follow these instructions from step 5 onwards.

(3) Once OxEdit is installed, copy 'x13a.tool'[3] and 'x13.def' into the installation directory of OxEdit.

(4) Copy the remaining files (runx13.bat, runx13v.bat, runx13vb.bat, runx13vl.bat, runxall.bat, viewlog.bat, and viewout.bat) to the installation directory of X-13ARIMA-SEATS. Usually 'c:\x13a'.

(5) Activate OxEdit 'macro keys'. Let's assume that X-13ARIMA-SEATS is installed in the 'C:\x13a' directory. If not, the following file must be edited to reflect the correct path.

(6) To do this, select 'View (Figure 5.1) -> Preferences -> Add/Remove Modules'. Click on 'From . . . Load', then navigate to the installation directory of OxEdit (c:\program files\oxedit) and double-click x13a.tool.

(7) Click 'OK' and then close. If the installation is successful, the contents of a.spc file should appear colour-coded when you open it.

(8) Clicking on 'Module-> Run X-13ARIMA-SEATS (single.spc)' will bring up the option to run X-13ARIMA-SEATS.

After installing OxEdit, copy 'x13a.tool' and 'x13.def' to the directory where OxEdit is installed. Click on the directory 'tools' and select 'add/remove modules' (see Figure 5.1). Then select 'load from . . .', find 'X-13-tools' from the installation directory, and copy it (runx13.bat, runx13v.bat, runx13vb.bat, runx13vl.bat, runxall.bat, viewlog.bat, and viewout.bat) to the 'list of modules'. Note that if the path information for these files differs from your settings, you will need to update it to match your paths.

In the UK ONS NBS mainframe, an interface will be programmed on the central ONS data repository (CORD) of the X-13ARIMA-SEATS program. This release will access data stored in a central database and enable users to apply time series analysis methods using X-13ARIMA-SEATS.

5.3.3 Win-X13 GUI

Win-X13 is the Windows interface for the X-13ARIMA-SEATS program, a powerful tool for seasonal adjustment and time series analysis. Win-X13 is an interface program that runs X-13ARIMA-SEATS in Windows. This interface will create the input spec file required to use the X-13ARIMA-SEATS seasonal adjustment; run X-13ARIMA-SEATS and display the output file; create a datasheet containing model information and diagnostic information for running series in X-13ARIMA-SEATS; and output graphics from X-13ARIMA-SEATS.

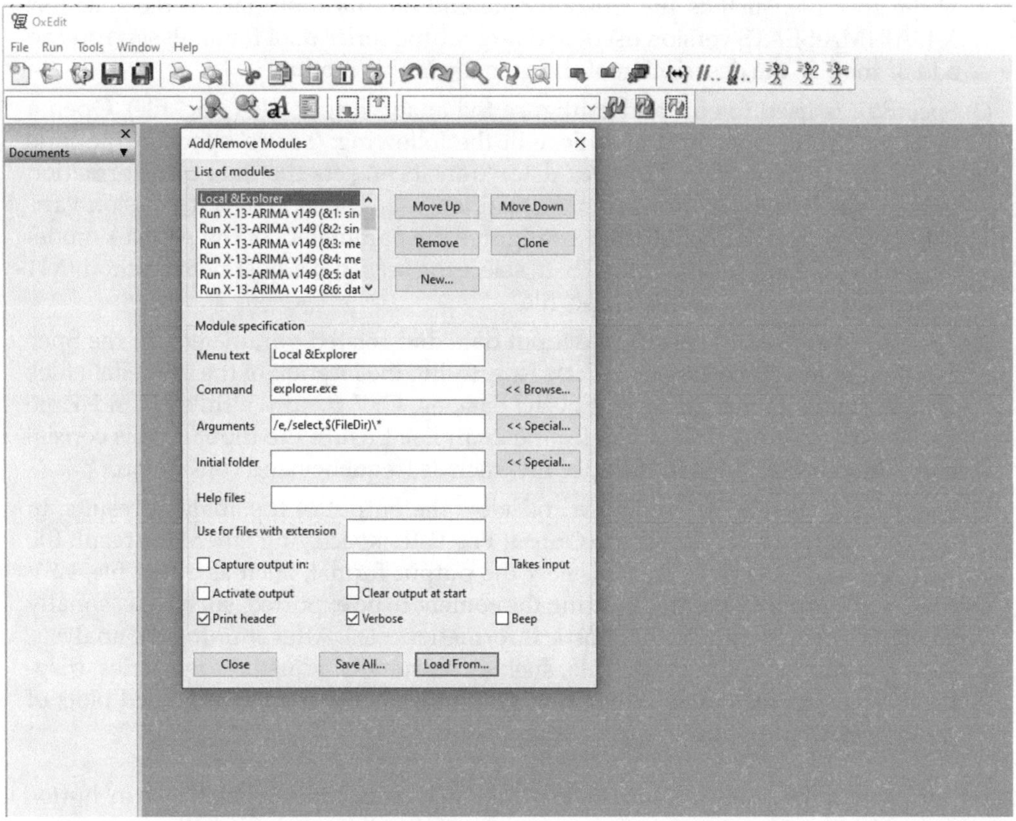

FIGURE 5.1
OxEdit Settings.

Win-X13 is designed to run X-13ARIMA-SEATS and X-12-ARIMA version 0.3. If you are using an earlier version of X-12-ARIMA, some features may not be available. The X-13ARIMA-SEATS does not support the HTML converter used in the X-12-ARIMA program to convert output files to HTML.

5.3.3.1 The Main Functions of Win-X13 and How to Use It

The main functional modules of Win-X13 include *'data dta'*; *'spec mta'*; *'spec edit'*; *'spec input'*; and *'spec output'*.

(1) data meda (original data file) is used to manage and view the raw data information of the time series. Open Win-X13 and load time series data files (such as '.dat' or '.csv' files). In the 'Data Meta' module, view the details of the data, such as the time range of the data (start date and end date); frequency of data (monthly, quarterly, or annual); and the source and description of the data. If needed, you can edit the original data information to supplement or correct the data.

(2) spec meta (spec file) is used to manage and view information about the analysis spec file ('.spc' file). Open a '.spc' file. In the 'Spec MTA' module, view the details

of the spec file, such as the creation date and modification date of the spec file; X-13ARIMA-SEATS version used; and target time series data for analysis. You can add or modify the description of the spec file in this module

(3) Spec Edit is used to edit and configure the analysis spec file ('.spc' file). Open a '.spc' file. In the Spec Edit module, edit the following: 'Series' specifies the time range and frequency of the input data; 'transform' sets the data transformation options (such as logarithmic transformation); 'regression' adds a regression variable (eg, holiday effect); 'ARIMA' configures the parameters of the ARIMA model (eg, order (p, d, q)); 'X11' or 'SEATS' to select the seasonal adjustment method (X11 or SEATS). Save the modified '.spc' file.

(4) Spec Input is used to specify the input data and related parameters. In the Spec Input module, set the 'Data File Path' (specifying the location of the input data file); Data Format (defines the format of the data, eg, CSV, fixed width, etc.); and Time Range (specifies the time range for the analysis). Ensure the input data is consistent with other settings in the Spec file, such as frequency and conversion.

(5) Spec Output is used to configure and view the output of the analysis results. In the Spec Output module, set the Output File Path (specifying where the result file will be saved); Output Format (select the output format, such as a text file, CSV file, etc.); 'Output Content' (selecting the content to be exported, such as seasonally adjusted series, residuals, diagnostic information, etc.). After running the analysis, view the results in the output file, such as seasonally adjusted time series, diagnostic statistics for model fitting, and graphical results such as exploded plots of trends, seasonality, and irregular components.

After completing the configuration of 'Spec Edit' and 'Spec Input', click the 'Run' button to start the analysis. When the analysis is complete, review the result file in the 'Spec Output' folder. If needed, you can adjust the specification file and rerun the analysis.[4]

5.3.3.2 Win-X13 Installation

(1) Find the official Win-X13 installation website or specify the download link to get the installation package of Win-X13. Make sure the downloaded version is compatible with your operating system (for example, Windows 10 or 11).

(2) Download the installation file (usually an '.exe' file) and double-click to run. If you are prompted for User Account Control, click Yes to proceed with the installation.

(3) During the installation process, select the installation directory. The default path is usually 'C:\Program Files\WinX13'; you can change the path as needed. Click 'Next' to continue.

(4) Once the installation is complete, set up the software.

(5) If Win-X13 needs to be integrated with other software such as OxEdit, '.tools', or '.def' files, please make sure that these relevant configuration files have been copied to the correct directory. For example, copy 'x13a.tool' and 'x13.def' to the installation directory of OxEdit.

5.3.3.3 Win-X13 Settings

Double-click Win-X13.exe to launch the Windows interface. The first time you run this program (and when you can't find the previously specified X-13ARIMA-SEATS

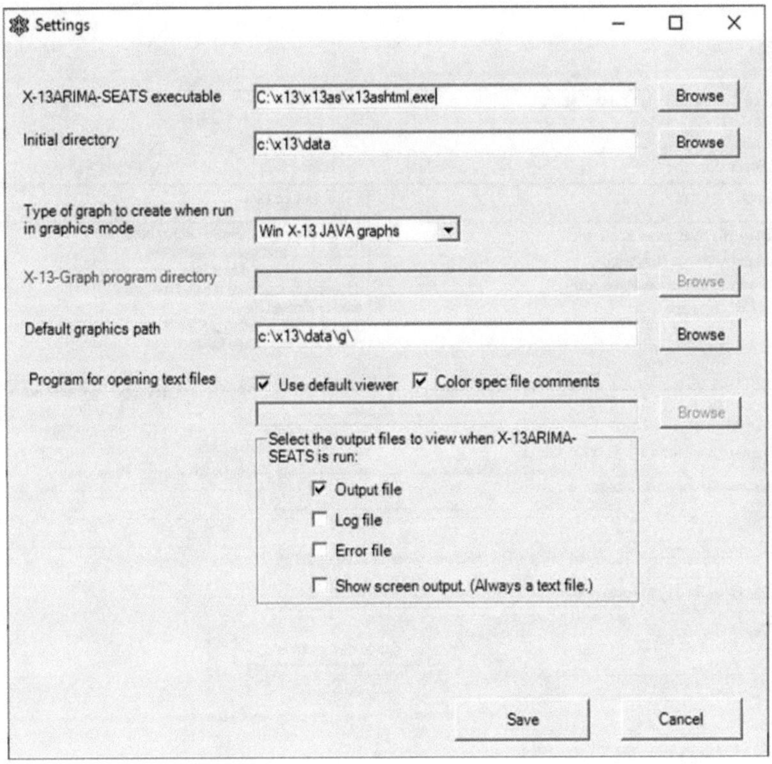

FIGURE 5.2
Win-X13 Setting.

executable when you start Win-X13), a Settings window appears. If a valid X-13ARIMA-SEATS executable has been provided, Win-X13 will open on the home screen. From here, if any changes are needed, you can access the Settings screen from the File menu. The X-13ARIMA-SEATS executable must be set before the interface can be run. All other settings are optional, such as (See Figure 5.2)

(1) setting the path for X-13ARIMA-SEATS to perform software installation;
(2) the path of the original data;
(3) the type of chart created when running in graph mode;
(4) X-13-Graph program directory;
(5) default graphics path;
(6) programs for opening text files: (1) default settings, (2) colour files;
(7) selecting the output file you want to view when running X-13ARIMA-SEATS: (a) Output File, (b) Log File, (c) Error File, (d) Display Screen Output (See Figure 5.3).

Using Win-X13, open and run the X-13ARIMA-SEATS input file, open the output file, set the options to run X-13ARIMA-SEATS, and select the output you want to view when the spec file is running. Other options in the menu include creating spec files, spec.mta files and data.dta files, as well as creating graphs and diagnostics from previously run spec files (See Figure 5.4).

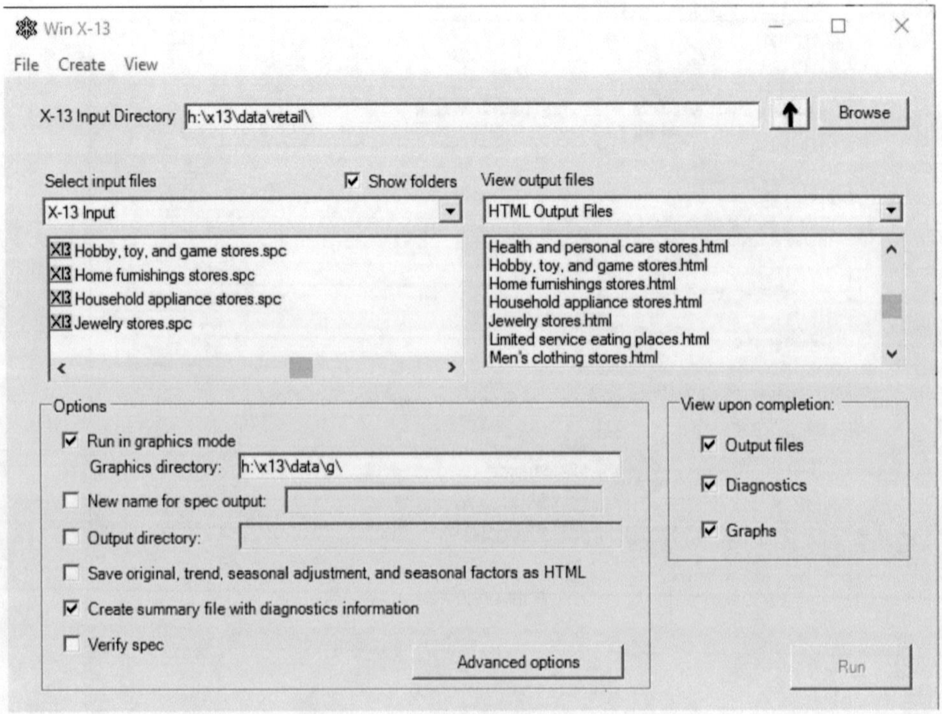

FIGURE 5.3
Win-X13 Main Interface.

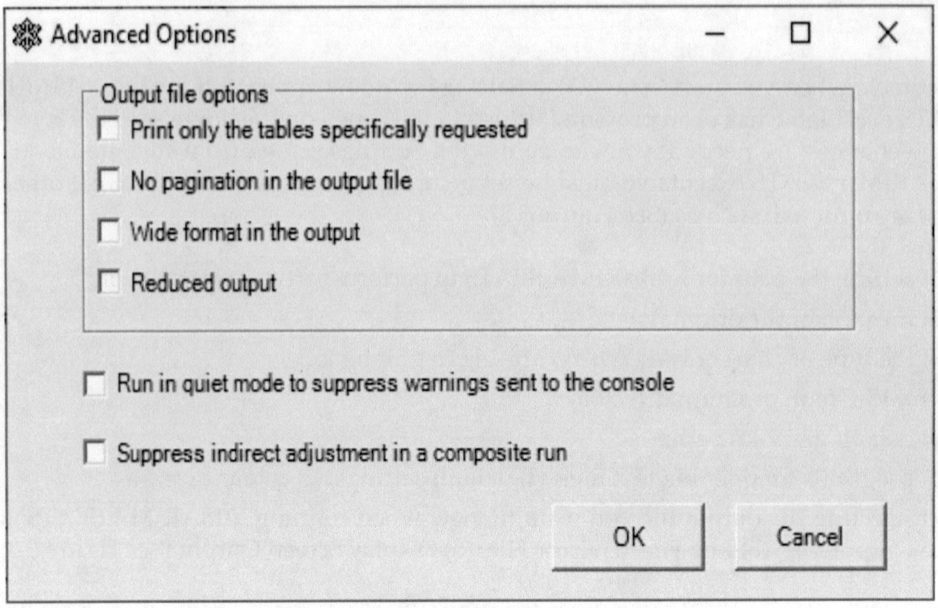

FIGURE 5.4
Other More Advanced Options.

Press Advanced Options to get more X-13ARIMA-SEATS options. In the window that appears, you can

1. Suppress all output except for the table explicitly requested in the spec file
2. Turn off pagination in the output file
3. Request a wide format in the output
4. Specify requirements for reduced output
5. Run the X-13ARIMA-SEATS in quiet mode, suppressing warnings sent to the console and printed to the Screen Output tab
6. Indirect adjustments in compound operations are prohibited.

The X-13ARIMA-SEATS has a built-in file viewer that opens spec files, MTA files, and other text files. Use the viewer to edit text files; Save them (using the Save button or Alt+S); and save them as new files (using the 'Save As' button or Alt+A). Among other editing features, the viewer allows searching for text (using the Find text box and button, Alt+F or Ctrl+F) and alternative text (using the Replace button, Alt+P or Ctrl+H) (See Figure 5.5).

FIGURE 5.5
View of the Spec File.

A. Text output

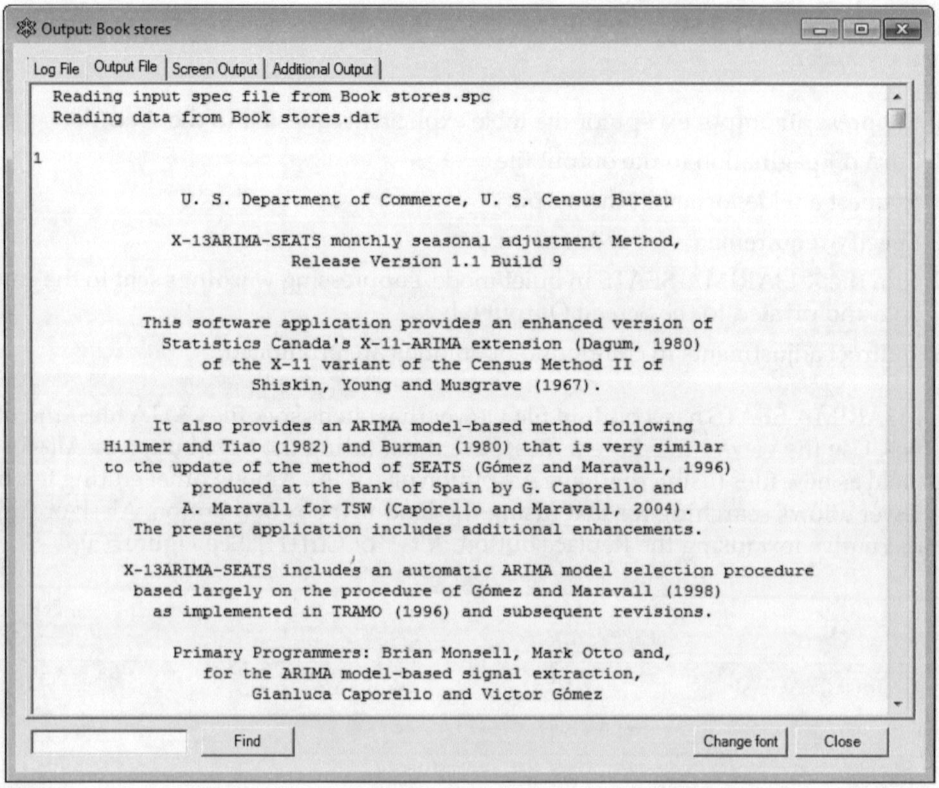

FIGURE 5.6
Text Output.

B. HTML output

When a spec file with the output name OutputName is run with an accessible HTML version of X-13ARIMA SEATS, the following HTML file is created:

- OutputName.html (Output File)
- OutputName_log.html (log file)
- OutputName_err.html (Error file)
- Running either a .spec.mta or .data.dta file results in the generation of an additional file: MetaFileName_mta.html or DataMetaFileName_dta.html (an index linked to all outputs, logs, and error files for the series in the run).

The specific file that X-13ARIMA-SEATS displays at runtime depends on the selection in the Settings screen. Screen output can also be displayed. But this is always a text file and will open in Win-X13's default viewer or the alternative text viewer selected in the Settings window.

C. Diagnostic form

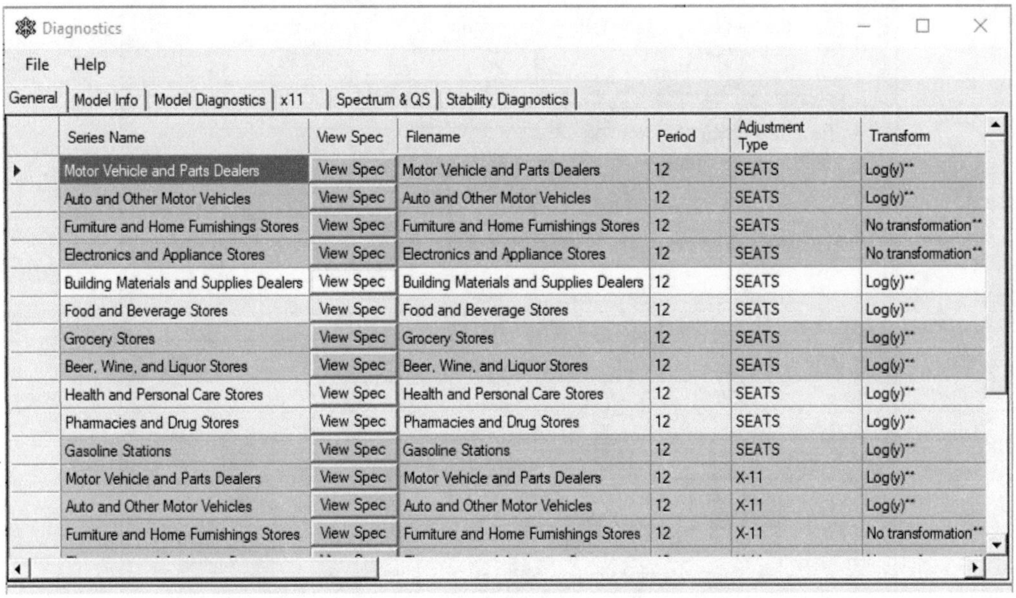

FIGURE 5.7
Stability Diagnostics Table.

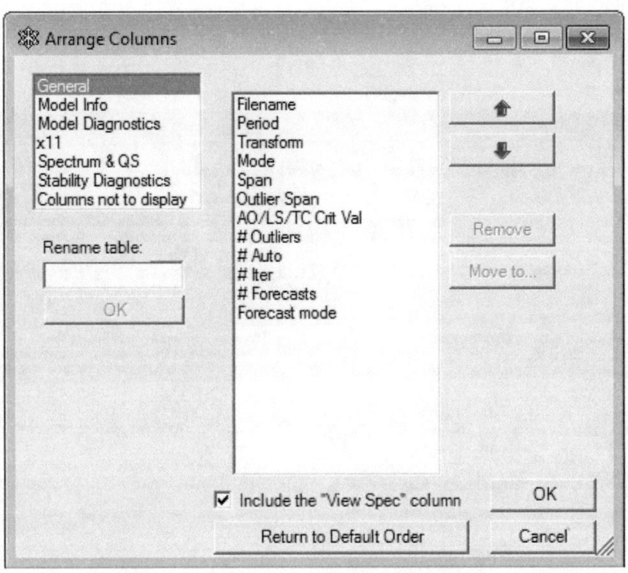

FIGURE 5.8
Set up to Selectively Display or Print the Results.

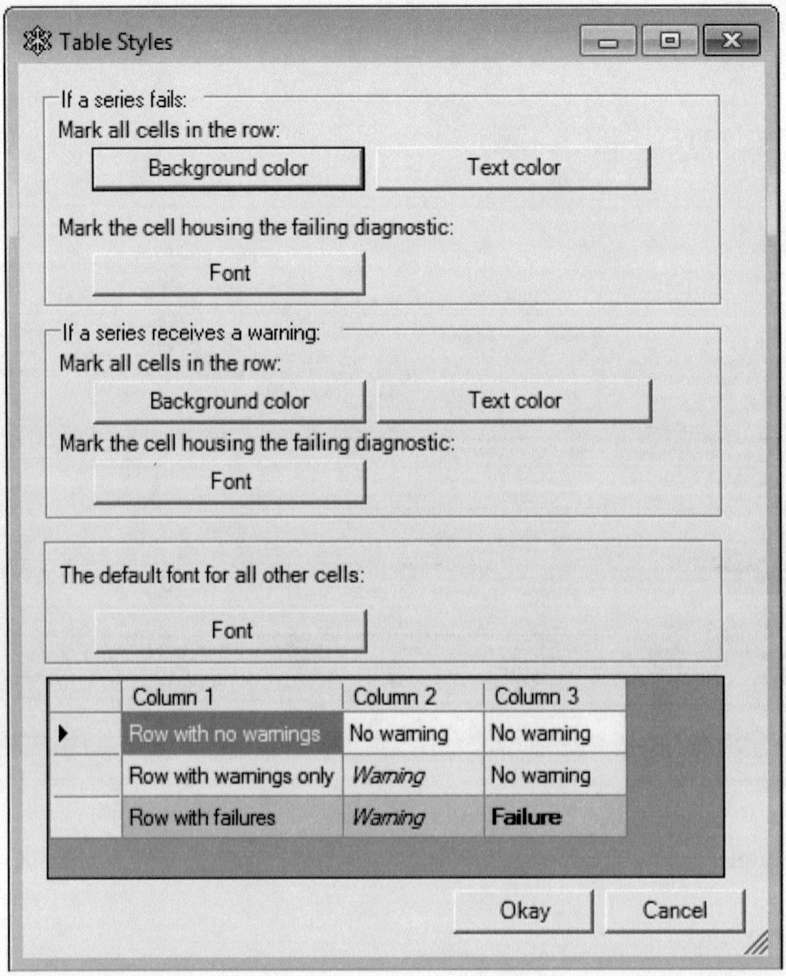

FIGURE 5.9
Set the Type of the Table to Be Printed.

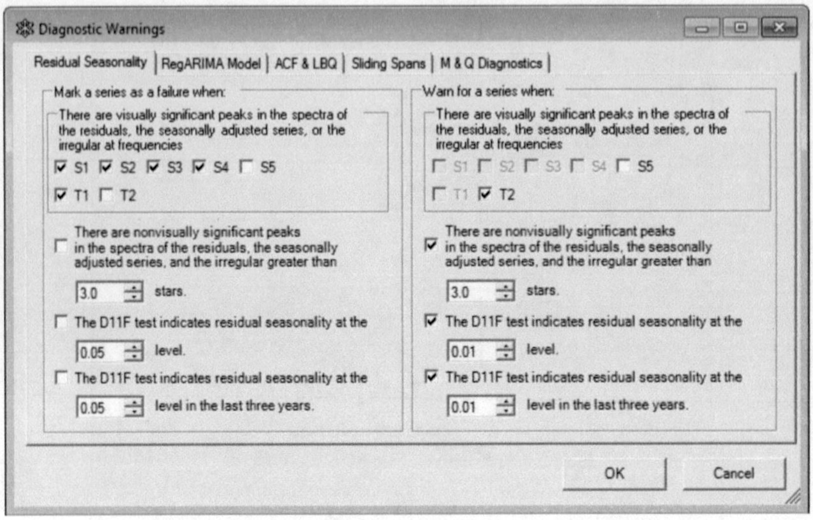

FIGURE 5.10
Select the Diagnostics for Failures/Warnings from the File Menu.

Check the installation result.

If Win-x13 is installed and set up correctly according to the above steps, you can open a '.spc' sample file and check whether it is loaded and running correctly. If the file contents are colour-coded and you can run the X-13ARIMA-SEATS via the menu options, the installation is successful. If you can't start Win-x13, try running the program as an administrator. If the file cannot be loaded, please verify that the path to the .spc file is correct. Check that the relevant 'macro keys' or configuration files (such as OxEdit) are installed correctly and that the profile path is correct.[5]

5.4 Detailed Explanation of Command Files (Spec Files)

The command '.spc' file used by X-13ARIMA-SEATS Seasonal Adjustment requires three basic functions:

(1) 'series' spec (identifier of the data file);

(2) 'arima' spec (select ARIMA model);

(3) 'x11' spec (run the X-11 decomposition method). The spec file used for prediction does not require the 'X11' function. Each function is also called a 'spec'.

The other 'spec' functions that need to be used are

(1) 'transform' spec;

(2) 'regression' spec (regression function);

(3) 'Outlier' spec (Outlier).

The name of the .spc file will be in the saved table.

5.4.1 Series Spec

The 'series' spec identifies the dataset (data file), for example

```
series{
      start=1992.04
      file="LFSone.dat "
      }
```

The key options are: 'period=4' (4 is quarterly, 12 is monthly). If it does not include 'period= ', the default value is used, which is = 12. 'span=(year.month, year.month)' is the data range used to fit seasonally adjusted (ie, long series). The first parameter is the start time, and the second parameter is the end time. If the first parameter is left blank, separated by a comma, ie, span=(, year.month), it means starting from the second parameter series to the end of the second series. If the second parameter is left blank, ie, span=(year. Month,), it means starting from the first parameter series to the end of the first series.

5.4.2 Transform Spec

The 'transform' spec is a transformation of the series, depending on whether the series is addition or multiplication, and is the first step in fitting the series to the ARIMA model. When the series is additive, the regARIMA model fits directly to the original series; When the series is multiplicative, the model fits to the original series through a number transformation. A logarithmic transformation is a change in the scale of the fitting model series; any predictions or previous adjustments estimated by the model are converted back to the original scale, and the multiplication effect is converted into an additive effect.

To decide which transformation to use and the corresponding decomposition model, X-13ARIMA-SEATS fits a regARIMA model to both untransformed and transformed series and selects a logarithmic transformation unless the following conditions are met:

$$AICC_{nolog} - AICC_{log} < \Delta_{AICC}$$

Where $AICC_{nolog}$ is the AICC value obtained by fitting the regARIMA model to the unconverted series, $AICC_{log}$ is the AICC value obtained by fitting the regARIMA model to the transformed series, and Δ_{AICC} is the value entered for the 'aicdiff' parameter, and the default value is –2.[6]

5.4.3 ARIMA Spec

The 'arima' spec is to review historical structural features of a series, assume that these structures are unchanged, and predict the future based on this structure. There are many ARIMA modelling methods, which differ primarily in their systematic approaches to feature detection and identification. In the simple case of a non-seasonal series, the ARIMA model takes the form of decomposing the current value into the previous value plus a random quantity called the 'error'. The error is considered to be an independent sample of the normal distribution with a zero mean and constant variance. For example,

(1) The first-order autoregressive model is $X_t = \varnothing_1 X_{t-1} + \varepsilon_1$, where X_t denotes the series value of time t, and ε_1 denotes the error value;

(2) The first-order differential moving average model is $X_t - \varnothing_1 X_{t-1} = \varepsilon_1 - \theta_1 \varepsilon_{1t-1}$;

(3) The second-order autoregressive and second-order moving average models are $\varnothing_1 X_{t-1} - \varnothing_2 X_{t-2} = \varepsilon_1 - \theta_1 \varepsilon_{1_{t-1}} \theta_2 \varepsilon_{1_{t-2}}$

where autoregression depends on some linear combination of the previous values of the series. The moving average relies on a linear combination of previous errors. The difference is that autoregressive and moving average effects apply to the difference value of the series, not to the series value itself.

Within this general framework, the model order is denoted (p, d, q), where p represents the autoregressive order, d the degree of differencing, and q the moving average order. It is possible to distinguish between the different types of effects of the ARIMA model; for example, the ARIMA model can be written as (1,0,0), (1,0,1), and (2,1,2). If a seasonal component is present, it indicates an effect that recurs annually. This is modelled in the same way as traditional effects, ie, relying on one year, two years. Values that appear before, not values that were a period or two ago. For a monthly series, this means that the seasonal autoregressive component will involve X_t and X_{t-12}, X_{t-24}, . . ., and so on, reflecting the relationship between the current value and its values from previous years.

For a quarterly series, the same is true for seasonal moving averages or differences. Seasonal series typically have a regular component, defined by (p, d, q) as described earlier, and a seasonal component of a parameter defined by itself, written (P, D, Q). The combination of these two components is indicated by putting two parentheses together.

Due to the effects of sampling variations or irregularities in the observed series, it is often possible to find different ARIMA models that fit a given time series. A basic principle of model selection is to give preference to the simplest model, ie, the one with the fewest parameters, such as the 'airline' (0,1,1) (0,1,1) model. The process of searching for a model usually involves adding parameters, during which you can remove some of the previously selected parameters until a satisfactory fit is found. The X-13ARIMA-SEATS provides a tool for automatic model selection by testing for autocorrelation in the residuals, ie, an estimate of the error.

There are two methods of ARIMA model selection: (1) automatic model selection, and (2) selecting a fixed model.

Automatic selection includes : 'automdl' and pickmdl'. You can choose 'automdl' :

```
Automdl{
        }

Or select 'pickmdl'
Pickmdl{
        File="c:\x12av135a\extended.mdl"
        }

(2 A fixed model selection:
        arima {
        model =(0,1,1) (0,1,1)
        }
```

'automdl' selection is based on the Ljung–Box Q statistic and the standard error of the residuals. The Q statistic is

$$Q = n(n+2)\sum_{j=1}^{h}\frac{\hat{\rho}_j^2}{(n-j)} c.f.\chi^2_{1-\alpha.h}$$

where n is the sample size, ie, the total number of observations in the time series; h is the lag order, ie, the maximum lag of the residual autocorrelation coefficient used for testing; $\hat{\rho}_j$ is the autocorrelation coefficient of the residuals at the lag j order; $\chi^2_{1-\alpha.h}$ is the critical value of the Chi-square distribution with degrees of freedom h at the significance level α; c.f. stands for 'compare with', ie, the Q-value needs to be compared to the Chi-square cut-off to conclude.[7]

The 'automdl' spec starts with a simple model, (0,1,1) (0,1,1), and allows the program to add items in succession, continuing the selection by continuously improving the first simple model. It will automatically specify limits for model selection based on the model complexity. The complexity limit is the maximum default value specified according to the parameters p, q, which are built into the program. The maximum values of 'd and D' are defined by the *maxdiff* parameter. The maximum values of 'p and q' are specified by the *maxorder* parameter. The default values are d ≤ 2, D ≤ 1, (p, q) ≤ 2, (P, Q) ≤ 1, which can be expressed as *maxdiff*=(2 1), *maxorder*=(2 1), until the fit satisfies some predetermined goodness-of-fit. For normal use, these defaults are sufficient, and they should not be overridden

TABLE 5.1
Alternative Values for 'maxorder'

Series length	Monthly series	Quarter series
<6 years	2, 1	2, 1
6–10 years	3, 1	3, 1
10–15 years	4, 1	3, 1
>15 years	4, 2	4, 2

unless the program is unable to produce a satisfactory model with them. If it does not generate a satisfactory model with the defaults, you can try something else, but you shouldn't do it until you've exhausted all other possibilities. The only option for *'maxdiff'* is (2 2), which shows a higher override value for *maxorder.*

Although the program automatically selects the ARIMA model in most cases, there are times when it does not. If the most suitable ARIMA model cannot be found automatically, the system will select 'airline',[8] ie,(0,1,1)(0,1,1)model, which is the simplest and most suitable ARIMA model for all data.

The 'pickmdl' spec searches for predetermined candidate models, usually in an order of increasing complexity, to find a satisfactory fit. The five default models in 'pickmdl' were specified by Statistics Canada when it developed the X-11-ARIMA and are retained in the X-13ARIMA-SEATS:

```
(0,1,1) (0,1,1)*
(0,1,2) (0,1,1)
(2,1,0) (0,1,1)
(0,2,2) (0,1,1)
(2,1,2) (0,1,1)
```

The X-13ARIMA-SEATS are selected from a limited number of 5=five ARIMA models in the 'pickmdl'. In the .mdl file, the model ending with * is considered the default model. The X-13ARIMA-SEATS selects ARIMA models from this limited set of 'pickmdl' based on the following three criteria:

(1) If the absolute mean percentage error of the sample prediction for extrapolated values in the past three years is <15%;
(2) If the residual series-dependent test (ie, the p-value of the residuals that are not correlated with Ljung–Box) Q is > 5%, and
(3) if there is evidence of excessive differential in the MA portion of the test model, ie, $\sum_{ns} \beta_{MA} < 0.9$.

If the model does not pass all the tests, no predictions will be generated. In this case, you can specify that you choose to use the first model by setting 'method=first': (0,1,1)(0,1,1). The alternative is 'method=best', which means that all models are tested and the one with the lowest prediction error is accepted.

If the program also performs automatic outlier identification (using the outlier spec), the estimates from the first ARIMA model tested are used as default estimates for the other models. This can also be used for outlier identification of each test model. It can use the optional parameter 'identify=all', which is the recommended option as it requires both model and regression estimation, improving the efficiency of both estimates.[9]

In addition to the ARIMA model capturing relationships in the time series, the series can also be affected by external deterministic factors, such as calendar effects, trading day variables, 'additive outliers', 'level shifts', 'temporary changes', 'slopes', trends or seasonal disruptions, and often the exact magnitude of these factor effects is unknown; the most effective way to do this is to use the regARIMA model, which removes these effects from the original series to estimate the 'true' seasonality.

5.4.4 Regression Spec

The 'regression' spec is the function of regARIMA in the X-13ARIMA-SEATS. It removes all factors that affect the seasonal analysis from the original series to estimate the 'true' seasonality. It is also a pre-adjustment (data cleansing) before seasonal adjustment, which directly affects the quality of seasonal adjustment. The regARIMA model includes both deterministic effects and ARIMA components, estimating them simultaneously. The 'regression' spec includes two types of regression variables: (1) program-specified (default) regression variables; and (2) user-defined regression variables.

The specified (default) regression variables include the following:

- Seasonal constant: This variable is rarely needed because its existence implies a deterministic trend with an order equal to the total number of differences, which is difficult to justify in most cases.
- Deterministic seasonality (fixed seasonal effects) can be in the form of seasonality constants or triangular regression variables,[10] but not both. The advantage of triangular regression variables is that fewer parameters are used. Although many series exhibit deterministic or near-deterministic seasonality, the use of seasonal constants is not particularly recommended. This is because deterministic seasonality is equivalent to the ARIMA model with a seasonal moving average coefficient of 1. [11]In addition, if there is a need to reduce the seasonal over-differential to a stable seasonal ARMA with a seasonal mean, this must be done manually, as the X-13ARIMA-SEATS will not be done automatically in the same way that a non-seasonal over-differential is handled. It is important to note that constant and deterministic seasonality variables, unlike other regression variables, are not adjusted out of the original series before estimating seasonality; These variables exist solely to enable the prediction of the series.
- Easter and trading day (TD) effect variables, X-13ARIMA-SEATS has a built-in series of variables to deal with the different effects of Easter on economic activity. If you choose to automatically test the 'Easter effect' and 'trading day effect', you can set it as

```
regression{
       aictest=(easter td)
    }
```

If, according to this test, the regression variables of 'Easter[15]' and 'td' are selected, you need to turn off the automatic test for the Easter and trading day effects, replace the 'aictest=(easter td)' row in 'regression' with the 'variables=(easter[15] td)' row, ie, fit the adjustment and save the previous adjustment, using the fixed choice of 'easter[15]' and 'td':

```
regression{# aictest=(east td)
        variable=(Easter[15] td)
        save(hol td)
        }
```

- Outliers (AO), level shifts (LS), temporary changes (TC), and slopes (RP) variables can be automatically detected using the 'Outlier' spec. If 'additive outliers', 'level shifts', 'temporary changes', 'slopes', or one-time events (TCs) are automatically detected, these can all be included in the regression variables.

The automatic detection of outliers and one-time events is

```
Outlier{
        types = (ao, tc)
        method =addone
        # span =()
    }
```

Where 'addone' is the forward selection, and then the backwards elimination. Identified and detected outliers and one-time events can be fixed by 'regression': for example, the identified outlier value 'AO2009.2' can be directly written to the 'variables' of 'regression' spec as:

```
Regression{
        # aictest=(td)
        Variables=(ao2008.02)
        Save=(otl)
        }
```

or the identified temporary change value 'TC1997.06'.

```
Regression{
        # aictest=(td)
        Variables=(TC1997.06)
        Save=(otl)
        }
```

This output is generated as a summary:

Regression Model			
Variable	Parameter Estimate	Standard Error	t-value
TC1997.06	0.0918	0.03038	3. 02

Automatic Test Level Offset (LS):

```
Outlier{
      Type=(ls)
      Method=addone
      }
```

or check for all 'outliers' (AO, TC, LS) as:

```
Outlier{
        Type=(all)
        Method=addone
        }
```

The level offset is then fitted, and a fixed selection model is adopted in the Regression spec:

```
Regression{
        # aictest=(td)
        Variables=(ls1995.12)
        Save=(ot1)
        }
```

Generate a summary of this output.

Regression Model			
Variable	Parameter Estimate	Standard Error	t-value
LS1995.12	47056.1906	13 S 67.98995	3.39

To test the ramp as well, you need to specify a start date and an end date. Returns unadjusted data for five months of testing ramp:

```
Regression{
        # aictest=(td)
        Variables=(rp1995.08-1995.12)
        Save=(ot1)
        }
```

Generate a summary of this output:

Regression Model			
Variable	Parameter Estimate	Standard Error	t-value
Rpl995.08–1995.12	12645.6602	11905.51222	1.06

We will focus on the regression of the 'regression' spec user-defined variable in Chapter 6.

5.4.5 X11 Spec

Using Automatic selection

(1) The 'seasonalma=sma' in X11 spec defines how the X-13ARIMA-SEATS estimates the seasonal component from a series that has been de-trended, and it specifies what type of the seasonal moving average 3x3, 3x5, or whatever average usedfor the estimation from the global I/S ratio:

TABLE 5.2
SMA Selection to Global I/S Ratio

I/S<2.5	sma=3x3
2.5<I/S<3.5	Delete the last year and recalculate
3.5<I/S<5.5	sma=3x5
5.5<I/S<6.5	Delete the last year and recalculate
I/S>6.5	sma=3x9

If you run the 'X11{}' spec, it means that the seasonal moving average and the trend moving average are running without any options, as follows:

```
x11{
        }
```

X11 does not specify 'sma', the seasonal moving average is automatically estimated by X-13ARIMA-SEATS based on the global I/S ratio estimate. If the recognised 'sma' is 3x3, the x11 fixed-parameter model can be written as follows:

```
x11{
        seasonalma=s3x3
        }
```

S3x3 is the same moving average set for different months/quarters. If the I/S value of a month is not less than 2.5, you can also set a different 'sma' for different months/quarters, for example,

```
x11{
    seasonalma=(s3x3 s3x3 s3x3 s3x5 s3x3 s3x3 s3x3 s3x3 s3x3 s3x3 s3x3 s3x3)
    }
```

Note: (1) The moving average is applied across years, ie, the 3x3 moving average for May 2002 is (May2000 + 2May2001 + 3May2002 + 2May2003 + May2004)/9.

The Trend Moving Average (trendma/tma) defines the Henderson Moving Average X-13ARIMA-SEATS for estimating the length of the trend, and the parameter 'trendma=n' in the 'X11{}' spec specifies the length of the Henderson Trend Moving Average ($0 < n < 102$); n must be an odd number. Note that moving averages are applied within the year. If no TMA is specified, X-13ARIMA-SEATS automatically estimates the TMA based on the global I/C ratio; the results are shown in Table D12.

If the identified TMA is estimated from the global I/C ratio, if it is 9, then the 'x11{}' fixed-parameter model can be written as follows:

```
x11{
        trendma=9
        }
```

TABLE 5.3
TMA Selection to Global I/C Ratio

I/C<0.99	tma=9(monthly), tma=5(quarterly)
1<I/C<3.49	tma=13(monthly),tma=5(quarterly)
I/C>3.50	tma=23(monthly),tma=7(quarterly)

TABLE 5.4

Replacement Criteria

$SoI_t < 1.5\sigma_t$	No substitutions $\omega_t = 100$
$1.5\sigma_t < SoI_t < 2.5\sigma_t$	*Partial replacement* $\omega_t = 100*(2.5- SoI_t)$
$SoI_t > 2.5\sigma_t$	Replace completely $\omega_t = 0$

5.4.6 Other Key Options

(a) appendfcst = yes, the table contains the predicted value;
(b) 'Sigmalim=(1.5, 2.5)' is to reset automatic outlier detection/handling;
(c) 'save=(dn)' is the key output, provided in tabular form.
   (1) The default values for 'sigmalim' in the 'x11{}' spec are 1.5 and 2.5;
   (2) During the X-11 algorithm, the outliers in the 'SoI' are identified and replaced to improve the estimation of S. For example, in the B loop, outlier identification and substitution occur twice: from tables B3 to B4, and from B8 to B9;
   (3) Determine whether 'SoI' is an outlier, which is calculated as the standard deviation (St) of the five-year irregularity around year t (ie, the two years before t, the t itself, and the two years after);
   (4) Replace the weighted average of the value itself and its four closest non-outlier seasonal neighbours by the following criteria.

Assuming non-anomalous neighbours, for the monthly series, the substitution formula is

$$SoI_{t*} = \frac{\sum \omega_{t+i} SoI_{t+i}}{\sum \omega_{t+i}}, \text{ i=-24,-12,0,12,24}$$

For volatility series, you can specify upper and lower bounds, which may rise or fall in response to peak/volatility changes:

```
x11{
      sigamalim =(2.8,3.8)
      }
```

The final weights are shown in Table C17

C 17 Final weights for irregular component
From 1992.Apr to 2010.Feb
Observations 215
Lower sigma limit 2.80
Upper sigma limit 3.80

	Jan May Sep	Feb Jun Oct	Mar Jul Nov	Apr Aug Dec	S.D.
1992				100.0	
	100.0	100.0	100.0	100.0	0.6
	100.0	100.0	100.0	100.0	0.6

Note: Only one value has been replaced in this example

1997	100.0	100.0	100.0	100.0	
	100.0	0.0	100.0	100.0	
	100.0	100.0	100.0	100.0	0.6
1998	100.0	100.0	100.0	100.0	
	100.0	100.0	100.0	100.0	
	100.0	100.0	100.0	100.0	0.7
1999	100.0	100.0	100.0	100.0	
	100.0	100.0	100.0	100.0	
	100.0	100.0	100.0	100.0	0.8

'save=dn' in the 'X11{}' spec is a different output table for X-13ARIMA-SEATS, which needs to be 'saved' in a different part of the spec file. The key tables include: A1 is the unadjusted data (save as '.a1'), saved in 'series'; B1 is the pre-adjusted data, saved in the 'series'; The tables '.td', '.hol', '.outl', and '.usr' are td, easter, outliers, and user-defined regressors, which are stored in 'regression'; Table '.D8', '.D10', '.D11', '.D12', '.D13' is saved in 'x11'.

5.5 Seasonal Adjustment Procedures

In general, there are two steps to seasonal adjustment using the X-13ARIMA-SEATS program: (1) run the auto-selected default spec file; (2) rerun the spec file with fixed parameters, and the new spec file with fixed parameters will be created according to the parameters automatically identified and selected by the default spec file.

5.5.1 Automatically Select the Default Spec File

The X-13ARIMA-SEATS program has a strong automatic identification and selection function, almost all function parameters in the spec file, such as the 'ARIMA' decomposition model, the selection of regression variables in the 'regression' model, the calendar effect (trading day effect), the moving holiday (such as the Easter effect), various additive outliers and level shifts, all have the function of automatic identification, testing and replacement. In the default spec file, all function items are automatically identified and tested. The default spec file that is automatically selected in the following example is based on monthly data:[12]

```
series{
        title="Monthly Seasonal Adjustment Spec file with default
        settings"
        file="default.dat"
        format="datevalue"
        period=12
        decinals=2
        }
```

```
transform{
function=auto
}

automdl{  }
regression{
        aictest=(easter, td)
        }

estimate{
        maxiter=5000
        }

outlier{
        types=(ao Is)
        }

forecast{maxlead=12
        save=(fct)}

Force{type=denton
     rho=1
     lambda=1
     round=no
     usefcst=no
     mode=ratio
     save=saa
     }

x11{
        appendfcst=yes
        save=( d11 d12 d13)
        }
```

Among them, the functions in the spec, eg, transform{}, automdl{}, regression{}, outlier{}, x11{}, do not specify parameter values, indicating that all of these functions use the automatic recognition and selection function. We summarise the functions and related parameters of the spec file in detail in Table 5.5.

5.5.2 Non-Seasonal Series Processing

The purpose of seasonality adjustment is to remove seasonality from the observed time series. To do this, one must first determine whether the series is seasonal or non-seasonal, that is, whether it should be seasonally adjusted. There are three possibilities for raw data: (1) there is a recognisable seasonality; (2) there is no identifiable seasonality; another possibility is somewhere between (1) and (2), ie, with slight seasonality. These are given after Table D8 of the output. If the series is identified as non-seasonal, it generally does not require adjustment; however, it is important to evaluate, using historical data, whether the adjustment framework should still be retained. This is also the first and most important thing to do when running the default spec file. The easiest and fastest method to determine

TABLE 5.5

Spec Function and Example Value Parameters, and Usage Descriptions

spec	Parameter descriptions with example values	Usage notes
series	title='Series xxx'	The label of the program's output
	start=1992.1	The start date of the data specified by 'file'
	period=12	Data period specified by 'file' (12 for Month, 4 for Quarter)
	decimals=2	The number of decimal places displayed in the output table
	file='xxx.dat'	File name (with extension) for the input data
transform	function=auto	Automatic selection of logarithmic transformations (for multiplicative components) or no transformations (for additive components) based on AICC
	or: function=log	Or: Fixed Conversion Function (log)
		Or: Fixed without conversion (none)
	or: function=none	
pickmdl	*Leave blank and use the default values.*	Use the 'pickmdl' program to automatically select ARIMA models. Cannot be used with 'automdl' or 'arima'
automdl	*Leave blank and use the default values*	Use the 'automdl' program to automatically select ARIMA models. Cannot be used with 'pickmdl' or 'arima'
arima	model=(0,1,1)(0,1,1)	Fixed ARIMA model order. Cannot be used with 'pickmdl' or 'automdl'
outlier	*Leave blank and use the default values*	Automatically identify and replace additive outliers and level shifts based on statistical significance
regression	aictest=(easter td)	Based on AICC's automatic detection and modelling of regression effects, such as Easter or trading days, 'aictest' cannot be used in conjunction with the 'variable' parameter below
	variables=(easter[8] ao1996.7)	Fixed regression variables to model the variables identified by the 'aictest' and 'outlier' specs. In this example, a priori adjustments will be made for the Easter effect with an accumulation period of 8 days and the additive outliers in 1996.7
	save=(hol ao)	Save the output table associated with the 'regression' spec. In this example, the holiday (Easter) and additive outliers are saved a priori adjustment factors
forecast	maxlead=24	Append the prediction to the fitted ARIMA model. Usually set to 2 years: 'maxlead=24' (monthly data) or 'maxlead=8' (quarterly data)
x11	seasonalma=s3x5	Specifying the seasonal moving average that will be used in the X11 algorithm, here we are using a 3x5 moving average to estimate the seasonality factor. If this parameter is omitted, the moving average is automatically selected based on the global I/S ratio
	trendma=13	Specifies that the X11 algorithm will use the Henderson Moving Average to estimate the trend. If this parameter is omitted, the moving average is automatically selected based on the global I/C ratio.
	save=(d10 d11 d12 d13)	Save the output table associated with the 'x11' spec. In this example, the seasonality factor (d10), seasonally adjusted series (d11), trend component (d12), and irregular component (d13) are saved

seasonality in a series is by referring to Table F3. Empirical studies have established the following general criteria for seasonality: if $F_s > 7$, $F_M < 2$, and M7 < 1, the series can be classified as seasonal; otherwise, it is considered non-seasonal. F_s, F_M, and M7 statistics are calculated as follows:

$$F_s = \frac{S_A^2/(k-1)}{S_R^2/(N-k)} \quad F_m = \frac{S_B^2/(N-1)}{S_R^2/(N-1)(k-1)} \quad \text{M7} = \sqrt{\frac{1}{2}\left(\frac{7}{F_s} + \frac{3F_m}{F_s}\right)}$$

where F_s captures the F-statistic of stable seasonality; F_m captures the F-statistic of moving seasonality, S_R^2 defined as the variance arising from the seasonal factor. S_B^2 represents the total of squared annual values, reflecting the overall variation across years. S_R^2 is the residual variance; k represents the seasonal period; N is the number of observations; M7 compares the relative contributions of stable seasonality (F_S) and moving seasonality (F_M) based on the F-test statistic (from Table F2. I).

M7 is used to determine the seasonal pattern of the original series, and low values indicate that X-13ARIMA-SEATS has identified stable seasonality. The closer the M7 value is to 0, the better the quality of the seasonal adjustment. If the value is around 1, it indicates mild seasonality in the series; if it is close to 3, the series is considered non-seasonal. The M7 statistic is calculated by X-3ARIMA-SEATS, a combination test that checks for the presence of stable and moving seasonality, and for the presence of identifiable seasonality. This information is in Table D8. A and F3.

Although an M7 statistic of less than 1 indicates seasonality, there are two other situations to be aware of:

(1) A series is 'non-seasonal' when it has only slight seasonality, but as new data become available, the series may change from 'non-seasonal' to 'seasonal'. This may result in revisions to published seasonally adjusted series. It is recommended that when a series is considered slightly seasonal, it should still be adjusted, as the adjusted series differs from the original.

(2) If the current test shows 'no seasonality', but the history shows that the series has been adjusted as a seasonal series, it is recommended to maintain the adjustment unless the seasonality test confirms no seasonality and the M7 value exceeds the cut-off (1.250 monthly, 1.050 quarterly), and if there is no adjustment in the previous year's analysis, continue to leave the adjustment unless the seasonality test shows seasonality and the M7 is below 1.150 (monthly series) or 0.900 (quarterly series). For 'non-seasonal' series, ie, series that do not require any adjustments, the setting 'Adjust data equals original.' start data' (NSA=SA), you can run the 'type=summary' directive specific to the 'x11' spec, for example:

```
series{
      title="DefaultTitle"
      period="DefaultPeriod"
      decimals=2
      format= "datevalue"
      file="Filename.dat"
      save=(al)
      }
```

```
transform{function=none}

automdL{}

forecast{
        Maxlead=12
        save=fct
        }


force{
     type=denton
     rho=1
     lambda=1
     round=no
     usefcst=no
     mode=ratio
     save=saa
     }

X11{
   type=summary
   appendfcst=yes
   save=(d10 d11 d12 d13)
   }
```

Specifying the command 'type=summary' in the 'x11' spec will cause the X-13ARIMA-SEATS call to perform the seasonal adjustment process regularly, but only for summary purposes. If you check the output of 'summary', you will find that the seasonally adjusted value d11 (SA) and the unseasonally adjusted value a1 (NSA) are exactly equal (NSA=SA). All non-seasonal series can be set up using this type of Spec file.

5.5.3 Fixed Model Spec File Configuration

If the series is identified as 'seasonal' after running the auto-selected default spec file, you should update the spec file to use a fixed model. This means that the parameters selected during automatic model selection should be explicitly specified and fixed in the spec file. For example, if 'arima' spec automatically selects a (0,1,1) (0,1,1) model, you should write it directly as 'model = (0,1,1) (0,1,1)', then delete the original 'function=auto' line. If 'regression' spec automatically selects 'aictest=(easter td)' and '(td easter[1])', and 'outlier{}' automatically selects outliers as (ao1998.3 ls1993.2), then the 'regression' spec needs to be changed to a fixed regression variable, which can be directly written as 'variables=(td easter[1] ao1998.3 ls1993.2)', and deletes the original 'aictest=(easter td)' and 'outlier{}' rows.

Similarly, if the 'X11' Spec automatically selects values with a seasonal moving average of '3x5' and a trend moving average of '5', then the 'X11' Spec fixed selection model can be written directly as '(seasonalma=s3x5 trendma=5)'. The spec file of a fixed model is as follows:

```
series{
   title="DefaultTitle"
   period="DefaultPeriod"
   decimals=2
```

```
    format='datevalue'
    file='Filename.dat'
    save=(a1)
    }
Transform{
    function=log     # If 'log' is the parameter value that the default
    # spec recognises and selects
    }
arima{
    model=(0,1,1)(0,1,1)  # If (0,1,1) (0,1,1) is the parameter value
    # that the default Spec recognises and selects
    }
regression{
    variables=(td easter[1] ao1998.3 ls1993.2)

# if (td easter[1] ao1998.3 ls1993.2) is the parameter value recognised
# and selected by the default Spec

    save=(hol ao)
          }
x11{
    appendfcst=yes
    seasonalma=s3x5 # If s3x5 is the default parameter value that Spec
                    # recognises and selects
    trendma=5       # If 5 is the parameter value that the default Spec
                    # recognises and selects
    save=(d11 d10 d12 d13)
    }
```

5.6 Output File and Result Interpretation

5.6.1 Generated Standards Files

- '.out' is the output file for each running series; '.err' is the error file for each run series; '.log' is a log file for each run series that stores certain diagnostics requested by the user. The most important of these is the output file, which provides the seasonal adjustment results as well as useful quality diagnostics. The output file consists of a fairly long list of tables identified by capital letters followed by numbers.
- Table A shows the prior adjustments for the series, including the adjustments specified by the user in the 'transform', the regression effects of the regARIMA model, and the automatically identified outliers or seasonal breaks, or the data is identified as non-seasonal.
- Tables 'B', 'C', and 'D' are the output of the seasonally adjusted run, consisting of three iterations of the 'X11 method'. The output of the first iteration is saved in Table B, the output of the second iteration is saved in Table C, and the output of the last iteration is saved in Table D, and Table D is final.

- Tables 'E' and 'F' provide the diagnosis of seasonal adjustment, and Table 'G' is a graphic.
- Table 'R' shows the revision history; Table 'S' shows the sliding span diagnostics.

Most forms will not appear in the output unless explicitly requested by the user. The user can selectively print the output, or specify the desired output length and options with the 'print' parameter in the 'x11' spec:

- With 'print= none', only the estimation of the regARIMA model is provided, in particular, the A table, regARIMA coefficients, residual checks (if needed), and predictions are given. No seasonal adjustment-related output will be printed. Output with 'print=brief' includes everything printed with the 'none' option, as well as basic seasonal adjustment tables and diagnostics.
- Using 'print=default' gives the same output as above, only with more seasoning tables and diagnostics.
- Use 'print=alltables' to print all output tables, including the 'intermediate' tables in the first and second iterations of the X11 method.
- Use 'print=all' to give all the resulting graphics. It is for this reason that no one needs this option, because the quality of the charts is very low. The default options are more than enough for most seasonal adjustment audits.

5.6.2 Key Seasonal Adjustment Tables

Of all the output tables, some are more important than others:

- Tables A6, A7, A8, and A9 show trading days, holidays, outliers, and user-defined regression components, respectively, as they are estimated according to the regARIMA model. These components include the effects of programs and user-specified variables. The effects given in Tables A6–A9 can be used not only for analysis but also as a priori adjustments to production runs, as an alternative to re-estimating these effects each time a new data point becomes available.
- Table B1 shows all a priori-adjusted original series, including regression models and automatically detected outliers, and Table B1a shows predicted values for a priori-adjusted original series.
- Table D10 shows the final seasonal components. The Seasonal Moving Average filter used is shown at the top of the table, which should be fixed and used until the next reanalysis.
- Table D11 shows the seasonally adjusted series.
- Table D12 shows the final trend components, and at the top of the table, the Henderson Moving Average filter is used. This should be fixed and used for production runs until the next reanalysis.
- Table D18 shows the composite factor of the trading days and holidays used in the seasonal adjustment, which is equal to the sum (or product of the multiplication series) of Tables A6 and A7. Like these tables, it includes the default calendar variable effect and the effect of user-defined variables.
- Tables C17, D8, D8a, D9, D9a, E5, E6, and F3 are also important as useful diagnoses.

5.6.3 Secondary Output Table

These are forms that users rarely need to use, including the following:

- Table D13 shows the final irregular components.
- Table D16 is the sum (or product) of Tables D10 and D18, showing the total number of calendar and seasonal adjustments for the series.
- Table E4 shows the annual total for the original series divided by the annual total for the seasonally adjusted series (in the case of additive seasonal, it is subtraction rather than division); The second column of Table E4 shows the same ratio (or difference) and is the original and seasonally adjusted series for extreme adjustments. This table can be used as a diagnostic of adjustment quality, particularly when annual totals are constrained. In the multiplicative model, the further the ratio deviates from 100 (or from 0 in the additive model), the greater the impact of the annual constraint on the seasonal adjustment. Table E7 shows the month-over-month changes in the components of the trend cycle (Table D12).
- Table F2A – F2I shows some of the diagnoses in more detail. These changes involve variations in specific components of the series, run durations, or ANOVAs. Most of these diagnoses are used to derive the single-valued M diagnosis of Table F3.
- In the F2 table, Table F2E is considered useful because it gives the months of cyclical dominance (MCD), ie, the number of months it takes for the change in the trend period to become greater than the change in the irregular components. If a span period shorter than the MCD shows a change, it will be considered irregularly dominant; if the MCD is greater than 6, it indicates that the series is extremely unstable and the quality of the seasonal adjustment is unlikely to be good.

Not all output Tables are printed by default:

Users can print only the specifically required tables for in-depth analysis.

- Tables C1 and D1 show the series of 'X11 boxes' in the second and third iterations of X11. These series are the original series for previous adjustments, regression effects, and extreme adjustments identified in the previous X11 iteration.
- Tables D2, D4, D5, D6, and D7 show preliminary estimates of the series components in the final X11 iteration.
- Tables B2–B13 and C2–C13 are essentially the same as the corresponding D-Tables, but apply to the first and second iterations of X11.
- Table B17 is the same as C17, but applies to the first X11 iteration. Tables B20 and C20 show the coefficients on which the extremums determined in the first and second X11 iterations were adjusted before the next (second or third) iteration.
- Tables C15 and C16 are generated only when running irregular regressions. C15 provides the regression output, while C16 presents the regression results of components such as trading days and holidays.
- Table D8B, like D8, marks any extrema based on whether they are identified during the regression/automatic outliers or X11 iterations and is useful information for analytical purposes.

- Tables D12B and D13B show the trend and irregular components of the model regression portion after the determined extremums (the user is reminded that additive outliers and temporary changes are assigned to the irregular components, while level shifts and slopes are assigned to trend periods).
- Tables E1, E2, and E3 show the raw series, seasonally adjusted series, and irregular components, corrected for extreme values.
- Table E11 shows a robust estimate of the final seasonally adjusted series. It is equivalent to Table E2, except for those points that are considered extreme, ie, those points that have been assigned zero weight in Table C17.
- Table E16 shows the final adjustment ratios.
- Table F1 is a 'smoothed' seasonal adjustment that is derived from the smoothing of the original seasonal adjustment by a simple moving average, the length of which depends on the 'cyclically advantageous month' diagnosis.
- Table F4 is generated only if the model for the monthly series already includes the trading day component. It displays the impact of the trading day component on the monthly adjusted data, based on the month's length and the weekday on which the month begins. It can serve as a secondary check on the reasonableness of the trading day impact. Users should consider whether known patterns in weekly activities might explain why some values are high while others are low.

5.6.4 Error and Log Files

Each series, after X-13ARIMA-SEATS runs, will generate an error file. This file stores the following information:

- Errors in input files: including syntax errors, internal inconsistencies in input files, or problems reading data files.
- Problems encountered during operation that cause the program to stop. The singularity and non-convergence of the regression matrix.
- Problems encountered during operation, although automatically fixed, must still warn the user, including changes to certain programs due to insufficient data that cannot be run with user-specified options, such as shorter spans for historical analysis.
- Series attributes identified during processing for which no specific action was taken, such as trading days or seasonal peaks identified in the spectrum.

When a data input spec file runs multiple series, if you want to keep some information about all series in a single file, rather than checking a large number of output files, you can use the 'savelog' parameter of the corresponding spec to specify a log file to be generated at the time of the spec file runtime:

```
series{start=1991.1
       period=12}
transform{function=auto}
automdl{maxorder=(4 1)
       maxdiff=(2 1)
       savelog=(AMD MU)}
```

```
regression{aictest=(easter td)
          savelog=aictest
         }
X11{savelog=(IDS M7)}
```

The log file summarises the information for all series, including the automatic ARIMA model selected for this spec file, the AIC test results for Easter and trading days, and the two diagnostics of seasonality adjustment (recognisable seasonality and M7).

5.7 Inspection and Diagnosis of Seasonal Adjustments

Before using the results of seasonal adjustment, the diagnosis should be checked to ensure the quality of the adjustment. The X13 output first checks the spec and then the original series (Table A1). After these basic checks, you'll review the output diagnostics. These can be divided into (1) a priori-adjusted diagnosis and (2) a seasonally adjusted diagnosis. Since a priori adjustment is a prerequisite for good seasonal adjustment, the user should first be satisfied with the quality of the a priori adjustment.

5.7.1 Inspection and Diagnosis of A Priori Adjustments

A good prior adjustment means that any internal and external factors affecting the series are appropriately adjusted, including all outliers, level changes, and seasonal interruptions [13]without missing any significant variables and proper fitting of the regression model. According to the three test checks below Table D11, it was proved that there were no remaining seasonal remnants. The M and Q statistics in Table F3 are in the correct range.[14]

5.7.1.1 A Priori-Adjusted Test

The test of prior adjustment includes the t-value, Chi-square value, and AIC and AICC tests. The t-value is used to test the significance of a single variable, and the Chi-square test is used to test the joint significance of a set of regression variables. Therefore, the Chi-square statistic is generated only when a set of regression variables is used, as in the following example:

Regression Model			
variable	**Parameter Estimate**	**Standard Error**	**t-value**
Trading Day			
Mon	0.0089	0.00581	1. 54
Tue	0.0013	0.00591	0.22
wed	0.0256	0.00601	4.26
Thu	0.0109	0.00590	1.84
Fri	−0.0012	0.00609	−0.19
Sat	−0.0177	0.00605	−2.92
*Sun (derived)	−0.0278	0.00594	−4.68

Regression Model			
variable	**Parameter Estimate**	**Standard Error**	**t-value**
Easter[1]	0.0281	0.01160	2.42

* For full trading-day and stable seasonal effects, the derived parameter estimate is obtained indirectly as minus the sum of the directly estimated parameters that define the effect.

Chi-squared Tests for Groups of Regressors			
Regression Effect	df	Chi-square	P-value
Trading Day	6	167.19	0.00

As you can see in the example above, although some variables of 'days' (such as Monday, Tuesday, . . .) are not significant, ie, the coefficient is not significantly different from zero. However, the Chi-square test for the trading day effect gives a p-value of 0.00, which is within the recommended 5% significance level (ie, less than or equal to 0.05), indicating that the trading day effect is significant. If a variable is found to be significant, it should generally be included in the model. If all trading day variables are significant, the AICC test and diagnostic tools for assessing the quality of seasonal adjustments should be used to determine which variable should be chosen.

The AIC test evaluates whether the 'regression' spec should include a particular regression variable. The AICC statistic is a key diagnostic tool to consider when selecting regression variables, especially those found to be significant, as it helps assess the performance of the seasonal adjustment. For example, you can specify the regression variable 'tdlcoef':

```
regression {aictest=(td1coef)}
```

This will generate the following likelihood statistics:

Likelihood statistics for model without tdlcoef Likelihood Statistics	
Effective number of observations (nefobs)	138
Number of parameters estimated (np)	3
Log likelihood	130.8304
Transformation Adjustment	–980.3013
Adjusted Log likelihood (L)	–849.4709
AIC	1704.9417
AICC (F-corrected-AIC)	**1705.1209**
Hannan Quinn	1708.5104
BIC	1713.7235

Likelihood statistics for model with tdlcoef Likelihood Statistics	
Effective number of observations (nefobs)	138
Number of parameters estimated (np)	4
Log likelihood	130.6730
Transformation Adjustment	–980.3013
Adjusted Log likelihood (L)	–849.6283
AIC	1707.2565

AICC (F-corrected-AIC)	**1707.5573**
Hannan Quinn	1712.0148
BIC	1718.9656
***** AICC (with aicdiff=0.00) prefers model without tdlcoef	*****

The output example above gives the AICC statistic of 1707.5573 for the model with the 'tdlcoef' regression variable and 1705.1209 for the model without the 'tdlcoef' regression variable. Therefore, a model without 'tdlcoef' is preferred.

When checking for any significant variables, you can use the automatic outliers in the 'outlier' spec to identify any unexplained outliers or level shifts. If an outlier is automatically detected in the same month or quarter, it may be evidence of a seasonal outage.

5.7.1.2 Diagnosis of a Priori Adjustments

Diagnosis of a priori adjustment includes the following:

- In Table D8 (Seasonal-Irregular Ratio, SI), a sudden increase or decrease in the value of a month/quarter may indicate a seasonal interruption.
- Month-to-month or quarter-to-quarter changes in the unadjusted series in Table E5 can mean seasonal interruptions, such as the following:

E 5 Quarter-to-quarter percent change in the original series
From 1991.2 to 2002.4
Observations 47

	1st	2nd	3rd	4th	AVGE
1991		21.3	4.6	8.3	11.4
1992	−24.3	24.7	6.1	8.1	3.7
1993	−22.3	18.6	5.6	10.1	3.0
1994	−24.7	24.1	4.3	8.3	3.0
1995	−24.8	28.9	6.0	8.2	4.6
1996	−26.6	26.9	9.7	5.3	3.8
1997	−18.2	7.9	2.7	−0.5	−2.0
1998	−8.7	10.5	1.7	−1.9	0.4
1999	−4.4	11.7	1.7	−2.6	1.6
2000	−6.5	9.5	3.9	−4.6	0.6
2001	−2.6	11.9	−1.6	−1.9	1.5
2002	−2.1	12.5	−0.6	−5.7	1.0

As shown in Table E5, the signs for the fourth quarter were positive before 1996 and negative after 1996. In addition, although the signs of other quarters have not changed, there have been some significant changes in the size of the numbers: the first quarter rose from less than −20 to single digits, the second quarter from 30 to 20 to about 10, while the third quarter was also affected, suggesting that the seasonal disruption may have occurred sometime between 1996.3 and 1997.1.

In addition, Table E5 can be used to identify outliers or level shifts. Outliers differ significantly from subsequent quarterly values, exhibiting larger magnitudes or more pronounced changes in the following month. These changes are typically unbalanced, not followed by opposite adjustments in subsequent periods. Regression diagnosis is a key basis for evaluating whether the regARIMA model needs to be adjusted. Once the user is satisfied with the quality of the prior adjustment, the next step is to check the quality of the seasonal adjustment itself.

5.7.2 Output Diagnostics for Seasonal Adjustments

The first thing to check is the seasonal test for M7 diagnosis in Table D8a and Table F3. The M7 value measures the degree of variation relative to stable seasonality and helps determine whether the series is seasonal or not. If the series is not seasonal, then it should not be seasonally adjusted. If the series is seasonal, you should continue to check for other diagnostics of seasonal adjustment, including the following:

- M1 is the ratio of irregular components to seasonal components. If M1>1, it indicates that the irregular component is too large, which may affect the accuracy of the seasonal component. The M6 can also measure irregularities, but only when using a 3x5 seasonal filter. M6 failure means that a filter shorter than 3x5 should be used.
- The M8 to M11 diagnosis shows fluctuations in seasonal composition. These diagnostics can be used to show potential problems in seasonal components, such as seasonal interruptions. M10 and M11 are similar to M8 and M9, but are only used at the end of the series. Therefore, comparing M10 and M11 with M8 and M9 can also help identify problems at the end of the series. Note that even if there are no issues with seasonal adjustments, M10 and M11 may fail, for example, if the fitted predictions used to generate the ARIMA model are poor.
- It must be mentioned that if the M7 is high, then the M8–M11 may also be high.
- M4 is used to measure the autocorrelation of irregular components. This diagnosis is not a high priority due to the low correlation requirements of seasonally adjusted quality to irregular components.
- M2 measures the number of irregularities compared to a straight trend. If the trend in the series cannot be adequately approximated by a straight line, the M2 statistic may produce misleading results. M3 and M5 measure irregularities compared to trends and are not important diagnoses.

5.7.3 Other Diagnostic Tools

- The test for residual seasonality is shown after Table D11: if the test shows 'residual seasonality present', it may be due to (a) that the series is too long, and the first few years can be considered removed; (b) if an inappropriate length of the seasonal filter is used, consider changing the length of the seasonal filter or using a different filter for different months/quarters.
- Table E6 can be used to check for outliers and seasonal breaks, and it can also be used to check for residual seasonality. In particular, seasonally adjusted series should not contain any seasonal elements. Therefore, if the monthly variation consistently shows the same sign in certain months, this may indicate residual seasonality.

- Table D9a (Moving Seasonality Ratio) gives the annual change in (preliminary) seasonality and the irregular components of month/quarter. It is common to use the appropriate seasonal moving average (using program auto-selection) unless the user specifies a specific filter. However, when the seasonal composition changes too rapidly in a particular month/quarter, or when the irregular composition fluctuates too much, it may sometimes be necessary to use different seasonal filters in certain months or quarters, in which Table D9a can be used to identify such problems.
- The SI ratios in Table D8 and the corresponding charts can be used to detect heteroskedasticity. If they fluctuate too much in a month, a shorter filter may be needed.
- The spectral graphical output for some of the components immediately following the F-table, labelled G.0 through G.2, is the only part of the G-output that is generated by default and shows seasonal or transactional spikes in a seasonally adjusted or irregular series. A peak is defined if there are at least six more '*' values on the graph than adjacent values.

5.8 Summary and Conclusions

This chapter provides a comprehensive introduction to the X-13ARIMA-SEATS seasonal adjustment software, addressing its operating environment, the structure of the specification (spec) file, the procedural steps of seasonal adjustment, the interpretation of outputs, and the application of key diagnostic tools. As a tool widely used in the official statistics and financial industry, the core functions of the X-13ARIMA-SEATS include (1) data preprocessing; (eg, outlier detection, calendar effect adjustment); (2) model selection (ARIMA, regARIMA); (3) seasonal adjustment (X-11 or SEATS method); (4) diagnostic validation (eg, M7 statistic, residual seasonality test). This chapter aims to equip the reader with the skills to carry out the full seasonal adjustment process, from data entry to result interpretation, with a focus on maintaining adjustment quality through diagnostic evaluation.

1. Presentation of the background of the X-13ARIMA-SEATS and their importance in official statistics. Clarify the objectives of this chapter: explain the basic operation of the software, command file configuration, and diagnostic methods.
2. General description of the use of X-13ARIMA-SEATS. (1) Data file format: support free format, date-value format and X13-save format; (2) Spec file structure: core hives (eg, 'series', 'transform', 'x11'), syntax rules (parameter order, annotation writing); (3) Running process: Prepare a data file ('.dat') and a spec file ('.spc') and execute it through the command line or interface.
3. Operating Environment: (1) DOS system: run through the command line, generate output ('.out'), error ('.err') and log ('.log') files: (2) OxEdit: interactive editor, support syntax highlighting and direct call X-13ARIMA-SEATS; (3) Win-X13: Graphical interface, providing data management, spec editing, and result visualisation functions.
4. The command (spec) file is the core configuration file of the X-13ARIMA-SEATS. The core command (spec) files include 'series': define the data source and time

range; 'transform': select an addition/multiplication model (eg, logarithmic transformation); 'arima'/'automdl': configure the ARIMA model (automatic or manual); 'regression': deals with calendar effects (trading day, Easter) and outliers; – 'x11': specifies a seasonal/trending moving average (eg, 's3x5', 'Henderson 13 terms'). Advanced options: eg, 'sigmalim' controls the outlier substitution threshold, and 'save' saves the key output table.

5. This chapter focuses on the use and methods of program-specified regression variables (built-in variables) and user-defined regression variables in the 'Regression' spec, which is critical for series seasonality.
6. Seasonal Adjustment Procedure: (1) Automatically select the default spec: identify seasonality (M7 statistic), ARIMA model, and regression variables; (2) Fixed model spec: based on the automatic selection of curing parameters (eg, 'model = (0,1,1) (0,1,1)'; (3) The non-seasonal series processing: skip the adjustment directly ('type=summary').
7. Run standard outputs, key output tables: 'A1–A9': a priori adjustment results (eg, outliers, calendar effects); 'D10–D13': final seasonal, trend, irregular components. 'F3': Seasonal diagnosis (M7 value). File types: '.out' (result), '.err' (error), '.log' (summary diagnostic).
8. The seasonal adjustment consists of two tests and diagnoses: (1) the examination and diagnosis of a priori adjustment, and (2) the inspection and diagnosis of seasonal adjustment. Prior adjustment refers to the appropriate treatment of internal and external factors affecting the series, including all outliers, level shifts, and seasonal breaks, ensuring that no significant variables are omitted. It also involves the proper specification and fitting of the regression model to optimise adjustment results based on the characteristics of the data. Good prior adjustment is essential for reliable seasonal adjustment. However, to achieve the expected seasonal adjustment, the user needs to be very familiar with the key diagnostic tools of the X-3ARIMA-SEATS, especially the significance of regression variables (t-test, AICC) and Outlier/Seasonal Interruption Detection (Tables D8, E5). The M7 value indicated seasonality (a value <1 is significant); residual seasonality was assessed using the test following Table D11; irregular component analysis was conducted using the M1 and M4 diagnostics.
9. In conclusion, the X-13ARIMA-SEATS is a powerful seasonal adjustment tool, and readers need to learn and master its basic operation and advanced functions to lay a solid foundation for practical application. The next section will further explore some other important topics related to seasonal adjustment.

APPENDIX 5.1: References for Key Parameters and Associated Notes in the X-13ARIMA-SEATS Menu

The X-13ARIMA-SEATS' Menu Reference is a collection of instructions in its core Spec file that is used to configure the analysis process. These menu items define all the steps from data readout, preprocessing, model recognition (ARIMA), seasonality adjustment (X-11 or SEATS), diagnostics, to result output. The following descriptions are grouped by functional logic:

1. Core process control
    1. 'series' Function: Define the time series data to be analysed.

        Key parameters:

        `title = '...'`: Set a title for the analysis.

        `file = 'filename'`: Specify the path to the external file that contains the data.

        `name = variablename`: Specify the name of the variable (such as the CSV column name) that contains the data in the file.

        `period = n`: Define the seasonal period of the data (eg, 12 for monthly data and 4 for quarterly

        data).

        `start = c(year, period)`: Define the start date of the data (eg,'c(1990, 1)' for January 1990).

        `data = (value1, value2, . . .)`: Embed data directly in the SPEC file (for short series).

        `span = (start, end)`: Specify the range of data subsets that will actually be used in the analysis

        (eg,'span = (2000, 1, 2023, 12)').

        Importance: Required. All analysis starts with defining a 'series'.
    2. `transform` Function: Perform a preliminary conversion of the original series.

        Key parameters:

        `function = log/none/auto`: `log` takes the natural logarithm, 'none' does not convert, and 'auto'

        lets the program automatically select based on the characteristics of the data (usually based

        on variance stability). 'log' or 'auto' are most commonly used.

        `power = n`: Specify the exponent of the power transformation (rarely used).

        `adjust = (const | leap | none)`: Deal with calendar effects (eg, leap years), 'const' adjusts for trading days/leap years (recommended), 'leap' only adjusts for leap years, and 'none' does.

        Importance: Very commonly used. Affects the subsequent modelling and adjustment effects.

3. `regression` Function: Define regression variables for modelling and removing identifiable effects before ARIMA modelling or during seasonal adjustments.

   Key parameters:

   `variables = (effect1, effect2, . . .)`: Specify the type of effect to include.

   Common Effects:

   `td` / `tradingdays`: Trading day effect (the difference in activity from Monday to Sunday).

   `easter` / `holiday`: Easter/Mobile Holiday Effect (specify the number of days of influence

   'easter[days]', eg,'easter[8]' for 7 days before Easter + Easter day).

   `lpyear` / `leapyear`: Leap year effect.

   `ao` / `ao.date`: Additive Outlier.

   `ls` / `ls.date`: Level Shift.

   `tc` / `tc.date`: Temporary Change.

   `ramp` / `ramp.date1 date2`: Ramp.

   `user` / `user.varname`: User-defined regression variables (to be provided in the data file).

   `save = (b1, b2, . . .)`: Save the regression coefficient results for a specific effect (eg,' save = (td, ao)').

   `aicdiff = n`: Set the AIC variance threshold for variables to enter the model.

   Importance: Very commonly used and critical. Used to model calendar effects and outliers.

4. `outlier` Function: Automatically detect and process outliers (AO, LS, TC) in time series.

   Key Parameters:

   `types = (ao, ls, tc)`: Specify the type of outlier to detect.

   `critical = n`: Set a threshold for the t-statistic (the default value is 3.5 or 4.0), the lower the value, the more sensitive the detection.

   `span = (start, end)`: Specifies the range of subsets of detected outliers.

   `method = (addone/all)`: `addone` (default) to add the most significant outliers one by one; 'all' takes into account all candidates at the same time (computationally intensive).

   Importance: Very commonly used. Identify and correct for the impact of special events in the data to improve model stability.

5. `automdl` function: Automatically identify the order (p, d, q) of the ARIMA model (P, D, Q). This is one of the core strengths of the X-13ARIMA-SEATS.

   Key Parameters:

   `maxorder = (n1, n2)`: Sets the maximum order (usually '(2,1)') for the regular difference ('n1') and the seasonal difference ('n2').

   `acceptdefault = (yes/no)`: Whether to accept the default model '(0 1 1) (0 1 1)' ('aero model') as an alternative (usually `yes`).

`mixed = (yes/no)`: Whether hybrid models are allowed (including both AR and MA items, usually 'yes').

`savelog = (aic/bic)`: Save the AIC or BIC value of the candidate model in the log.

`diff = (nd, nsd)`: Manually fix the regular difference order ('nd') and seasonal difference order ('nsd'). Usually do not set, let the process perform Automatic recognition of the series.

Importance: Extremely commonly used. Most analyses rely on this feature to automatically select

ARIMA models. Parameters typically use default values or fine-tune 'maxorder' and 'acceptdefault'.

6. 'X11' Function: Perform seasonal adjustments using the classic X-11 method.

   Key Parameters:

   `mode = (add/mult/pseudoadd/logadd)`: Specifies the mode of seasonal adjustment. 'mult' (multiplicative model) and 'add' are the most commonly used. 'auto' is usually determined by 'transform'.

   `seasonalma = 'filter'`: Specify the type of seasonal moving average filter (eg, 's3x3', 's3x5', 's3x9', 'stable').. "auto" Let the program choose (recommended).

   `trendma = n`: Specifies the length of the Henderson Trend Moving Average filter (odd numbers, eg, 9, 13, 23). 'auto' concession

   The order is based on model selection (recommended).

   `appendfcst = (yes/no)': Whether to append the regression prediction to the end of the series before adjustment to improve endpoint estimation (typically `yes`.).

   `appendbcst = (yes/no)': Whether to append a pushback value at the beginning of the series (usually 'yes').

   `save = (d10, d11, d12, d13)`: Save specific tables (eg, 'd11' final season factor, 'd12' final trend cycle, 'd13' final irregular components).

   Importance: Required if X-11 is selected. Define the specific parameters for X-11 tuning.

7. `seats` Function: Perform quarters using the SEATS method based on the ARIMA model section adjustments.

   Key Parameters:

   `mode = (add/mult/pseudoadd/logadd)`: Same as 'x11', usually determined by 'transform'.

   `trendcyclema = n`: Specifies the approximate length of the moving average of the trend loop components (affecting the output table).

   `save = (s10, s11, s12, s13, s16, s18)`: Save specific seat output tables (eg, 's11' final seasonal factor, 's12' final trend, 's13' final irregular component, 's16' final seasonal adjustment series).

   `out` / `print`: Controls the level of detail of the output of the seat module.

   Importance: Required if SEATS is selected. Define the SEATS adjustment behaviour.

8. `force` (Use with caution!) Feature: Enforce the assignment of an order, regression variable, or the seasonal adjustment factor for the ARIMA model, overriding the automatic selection results.

   Key parameters:

   `type = (arima/outlier/x11/seats)`: Specify the type of object you want to force.

   `data = . . .`: Specify the mandatory specifics (eg, 'data = (0 1 1) (0 1 1)' to force the use of aero models).

   Importance: Advanced/Use with caution. Override automatic results only for good reason. Misuse can result in an invalid analysis.

2. Output control

9. `estimate` Function: Controls the estimation process for ARIMA models and regression coefficients

   (usually using default values).

   Key parameter:

   'save = (model, coefs)' save the model and coefficient details.

   Importance: Usually default, sometimes you need to save model details.

10. `check` Function: Specifies the type of diagnostic check to be performed.

    Key Parameters:

    `maxlag = n`: Sets the maximum lag order (usually 24 or 36) for the residual autocorrelation test.

    `qstats = (lags)`: Specifies the lag order for calculating the Ljung–Box Q statistic (eg, 'qstats = (12, 24, 36)').

    `save = (acf, pacf, iacf)`: Save ACF, PACF, IACF plots of residuals.

    Importance: Commonly used. Evaluate the goodness-of-fit of the model.

11. `forecast` Function: Generate out-of-sample predictions (typically used to evaluate models or populate data).

    Key Parameter:

    'MaxLead = n' sets the number of forecast periods.

    Importance: Use as needed.

12. `save` Function: Specify the specific result items to be saved to the output database (.udb/ .sas7bdat / .csv). This is the key to getting the results of your analysis.

    Key Parameters:

    `database = . . .`: Specify the database format (usually set by the front-end or 'x13path').

    `format = (datevalue, date, aic, . . .)`: Specify date formats, etc.

    `results = (item1, item2, . . .)`: Core parameters! Specify the components you want to save:

    Final Result: 'd11' (X-11 Final Seasonal Factor), 'd12' (X-11 Final Trend Cycle), 'd13' (X-11 Final Irregular score), 'd16' (X-11 final seasonally adjusted series), 's11' (SEATS final seasonality factor), 's12' (SEATS most Final Trend), 's13' (SEATS Final Irregular Composition), 's16' (SEATS

    Final Seasonally Adjusted Series), 'fct' (prediction), 'bct' (backtest).

Diagnostic statistics: 'a1' (raw data), 'a19' (outlier-adjusted data), 'b1' (calendar-adjusted data), 'e1' (most Final model residuals), 'f3' (model fit), 'f2' (model fit including regression effects), 'aic' (model AIC), bic` (model BIC).

others: `c20` (Trading Day Factor), `c17` (The Easter factor), `out` (List of detected outliers and coefficients).

Importance: Required. Without 'save', it is not possible to obtain the main analysis result data.

13. `print` / `out` Function: Controls the level of detail output to the .out log file.

    Key Parameters: You can specify a specific table or diagnostic section to print (eg,' print = (acf, mdl, est)'). 'print = all' Print everything (usually too verbose).

    Importance: Used for debugging or to obtain detailed diagnostic information. Usually default or front-end control.

3. Other commonly used

14. `composite` Function: Combine multiple Spec files into a single composite Spec (for easy management of complex analyses).

    Key Parameter: 'spec = 'specfile.spec'' contains the child Spec file.

    Importance: Useful when organising large analyses.

15. `x11path` / `x13path`

    Function: Set the path of the X-13ARIMA-SEATS program and data files (usually set in the external environment or front-end).

4. Important reminder

- '.spc' File: Menu commands are written in a '.spc' (Spec) file.
- Order: The order of the menu items is important (eg,' series' must come first, 'regression' usually after 'transform', 'automdl'.).
- Dependencies: Some menu items depend on the results of other items (eg, 'automdl' depends on 'regression' and 'transform').
- Defaults: Most menu items have sensible defaults. Usually, only the core items need to be configured('series', 'transform', 'regression', 'outlier', 'automdl', 'x11'/'seats', 'save').
- Official Documentation: The most authoritative and complete menu reference is always the X-13ARIMA-SEATS Official Reference Manual (commonly named 'refman.pdf' or similar), which explains in detail all the parameters, usage, and examples of each menu item.
- Front-end software: Used JDemetra+, SAS, R 'seasonal'/'x13story', Python 'statsmodels' and so on. When the front-end software is used, we usually provide a graphical interface or more advanced encapsulation functions to generate '.spc' files; the user may not need to do it directly.

This reference covers the most critical and commonly used menu items in daily analysis. In the event of complications, refer to the official manual for detailed descriptions of the corresponding menu items.

Notes

1 For more detailed information on which diagnostics can be stored in the log file, please refer to the X-13ARIMA-SEATS Reference Manual.
2 For OxEdit's 'macro keys', 'X-13 TOOLS' can be used. See Appendix 5.1.
3 'x13a.tool' needs to be built by yourself; in x13a.tool, you need to set the directory of x13a.exe, x13a.mdl on your own computer. For the detailed information about x13a.tool, you can contact the 'Time Series Analysis Branch (TSAB)' in the UK Office for National Statistics.
4 For more detailed operating instructions, please refer to the official documentation or user manual of Win-X13.
5 For more detailed information, see U.S. Census Bureau. (2024). *X-13ARIMA-SEATS Reference Manual*, version 1.5, Build 64. U.S. Department of Commerce, Washington, DC. Retrieved from https://www2.census.gov/software/x-13arima-seats/x13as/refman.pdf.
6 See the sections on AIC, AICC, and BIC models for detailed information. Negative values are biased towards choosing a logarithmic transformation. The default value of -2 is not for statistical reasons but for convenience. Δ_{AICC}.
7 The Ljung-Box Q statistic is used to test whether the time series residuals are autocorrelated (ie, white noise).
8 That's because Box and Jenkins used it to simulate the number of airlines per month.
9 These parameters are detailed in the full X-13ARIMA-SEATS documentation. In general, these values should be left at their default values.
10 Trigonometric Regression Variables are a mathematical method for modelling seasonal effects, representing periodic variations through linear combinations of sine (sin) and cosine (cos) functions. The core principle leverages the orthogonality and periodicity of trigonometric functions to flexibly capture seasonal patterns with fewer parameters. For seasonal time series with periodicity *S* (eg, monthly data where S=12, quarterly data where S=4), the trigonometric regression variables consist of harmonics of the following form:

$$\sum_{k=1}^{K}\left[\alpha_k \cos\left(\frac{2\pi kt}{S}\right)+\beta_k \sin\left(\frac{2\pi kt}{S}\right)\right]$$

where *t* is the time index (eg, month *t* or quarter *t*); *k* is the harmonic order (k = 1, 2,. . ., K), typically with $K \leq$ floor(S/2); α_k, β_k are coefficients to be estimated. Trigonometric regression achieves parameter efficiency through (1) Comparison to seasonal dummy variables: (a) Traditional approach requires "S-1" dummy variables (eg, 11 for monthly data); (b) Trigonometric regression requires only 2K parameters (*K* sin/cos pairs), where K < S typically. *Example:* For monthly data (*S*=12), K=2 harmonics (4 parameters) often capture dominant seasonality. (2) Spectral flexibility: (a) Low-order harmonics (k=1) model the fundamental seasonal cycle; (2) Higher-order harmonics (k=2,3,. . .) capture complex variations (eg, bimodal seasonality, sub-cycle fluctuations);(3) Functional representation: (a) Dummy variables assign discrete seasonal effects (eg, fixed 'January effect'); (b) Trigonometric regression fits seasonality via continuous periodic functions: (c) More efficient for long periods (eg, daily data with *S*=365).
11 The (0,1,1) seasonal ARIMA model, ie, the Airline Model.
12 The default spec file settings can be different, and the default spec file described here is based on the default spec file designed by the UK Office for National Statistics (ONS).
13 You can compare the results in the two output files, "No Prior Pre-Tuning" and "With Prior Pre-Tuning".
14 The correct range of values for the M and Q statistics is discussed later in this chapter in the diagnosis.

6

X-13ARIMA-SEATS Advanced Topics and Practice

6.1 Introduction

Chapter 5 introduced the operational environment of the X-13ARIMA-SEATS program, particularly the application of adjustable spec file functions, parameter specification, the seasonal adjustment procedure, standard outputs, and post-adjustment tests and diagnostics. This chapter will continue to focus on other important topics related to seasonal adjustment.

As a seasonal adjustment tool, the design and various options of X-13ARIMA-SEATS are grounded in scientific principles, enabling it to tailor adjustment results to different data characteristics. However, achieving the desired seasonal adjustment objectives requires users to be thoroughly familiar with the capabilities and options of X-13ARIMA-SEATS, a proficiency demanding both solid theoretical knowledge and extensive practical experience.

The structure of this chapter is organised as follows: following the introduction to Section 6.1, Section 6.2 discusses series length selection; Section 6.3 discusses the consistency of annual totals before and after seasonal adjustment; Section 6.4 discusses direct and indirect seasonal adjustments; Section 6.5 discuss seasonal disruptions; Section 6.6 discusses the 'regression' spec user-defined variable; Section 6.7 discusses the adjustment of the Chinese New Year holidays (including the Dragon Boat Festival and the Mid-Autumn Festival); Section 6.8 discussion of Sliding Span; Section 6.9 discusses revisions and updates; Section 6.10 discusses forecasting; and finally, Section 6.11 gives a summary and conclusions.

6.2 Series Length Selection and Processing

6.2.1 General Principles for Series Length Selection

Any series longer than three years can be adjusted according to the X-13ARIMA-SEATS. However, if the series is short and the program does not have a enough data to process, it may be harder to find important evidence, such as the Easter impact or trading days. Very short time series may undergo substantial revisions, as new data can significantly alter the estimated seasonal factors. On the other hand, a very long time series may contain substantial historical data that no longer reflects the current pattern of the series. For example, the data from the 1970s does not accurately reflect current patterns, which affects the seasonal adjustment. For seasonal adjustment, at least three years of data are required for X-13ARIMA-SEATS modelling or for any type of seasonal adjustment. The X-13ARIMA-SEATS requires at least five years of data to support the automatic selection of

DOI: 10.1201/9781003716655-6

seasonal moving averages and the identification of dynamic seasonal patterns. Five years and three months (five years and one quarter) of data automatically fit the ARIMA model and correctly test all criteria of the model (especially the average forecast error over the past three years).

A short series refers to a time series with fewer than five complete calendar years of available data. When the sample contains less than five years of data, X-13ARIMA-SEATS may employ the entire series to estimate constant seasonality within the ARIMA modelling framework. The seasonal adjustment procedure in X-13ARIMA-SEATS operates under a strict assumption: that seasonal factors remain constant throughout the series. If fewer than three years of data are available, the seasonally adjusted version of the series should not be released until additional observations become accessible. With 3–5 years of data, X-13ARIMA-SEATS can implement ARIMA modelling or provide trading day and Easter adjustments; however, the resulting seasonal adjustments are typically highly unreliable and may require significant revisions as future observations are added. Consequently, seasonally adjusted data should not be released until more observations are available.

6.2.2 Special Treatment for Short Series

When seasonal adjustment of data spanning 3–5 years becomes necessary, three methodological options exist for handling short series.

The first approach involves utilising the ARIMA option within X-13ARIMA-SEATS to forecast and backcast the series, thereby generating sufficient data length for seasonal adjustment. Since automatic model selection requires over five years of data, practitioners should specify simple fixed models in the 'arima' spec, such as (0,1,1) (0,1,1), and apply a logarithmic transformation where appropriate. The 'forecast' spec should be configured with a 1–2 year 'maxlead' setting to extend the series to 5–6 years. Seasonal moving averages in the 'x11' spec may be automatically selected to optimise the treatment of seasonal components. While trading day and Easter regressors can be incorporated, it is critical to recognise that adjustment quality may prove highly unreliable and subject to substantial revisions as new observations accumulate.

The second option entails performing seasonal adjustments without ARIMA modelling. This approach circumvents certain limitations inherent to ARIMA extrapolation, which assumes stable seasonal patterns reflecting historical averages. Thus, the adjustment of extrapolated data yields results comparable to the direct adjustment of actual observations. The primary drawback lies in forfeiting ARIMA-based outlier and calendar effect detection; however, such identification remains statistically unreliable for short series regardless of methodology.

Thirdly, practitioners may link the short series with a longer indicator series exhibiting similar behavioural patterns. This technique extends the series backwards through temporal overlap, creating a composite series exceeding five years in length. This option is viable only when the indicator series covers periods preceding the availability of the short series. The effectiveness of this method depends fundamentally on the behavioural congruence between the two series.

The selection of these approaches involves multiple considerations. Options one and two offer implementation efficiency, making them preferable under time constraints. The third option, however, requires the existence of a compatible indicator series and demonstrates efficacy proportional to the similarity between the benchmark and target series.

Although any data series under five years is considered short-term, a minimum of five years is preferable to ensure acceptable revision stability. At least ten years of data are required to guarantee that adjustments in the initial years are unlikely to undergo substantial revisions. When dealing with series spanning five to ten years, they may be modelled and adjusted similarly to longer series; however, practitioners must recognise that revisions become highly probable as additional data accumulate, potentially including model modifications and identification of previously undetected seasonal breaks.

6.2.3 Challenges and Solutions for Long Series

At the opposite extreme, lengthy time series may exhibit structural discontinuities or significant shifts in seasonal patterns. When requiring a seasonally adjusted series across extended periods, it is generally better to adjust the entire dataset rather than segment it into discrete periods. The sole exception arises when fundamental changes occur over such extended durations that treating the series as a single invariant entity becomes methodologically unsound. Where historical breaks in trend or seasonality exist, relevant regressors should be incorporated within the 'regression' spec. A critical prerequisite for implementing such regressors is the availability of sufficient data both preceding and following the breakpoint to ensure reliable identification. This requirement further depends on the series' irregular component magnitude – greater volatility necessitates longer data segments for robust estimation.

Given these complex variables, establishing rigid thresholds for determining the minimum data years required for releasable adjustments remains impractical. In summary,

- Sub-5-year series warrant extreme caution; publication should occur only when the series' critical importance outweighs methodological risks, accompanied by prominent caveats.
- 5–10-year series demonstrate improved reliability yet retain significant revision risks. Release decisions should incorporate objective criteria (eg, M and Q diagnostic statistics), with supplemental warnings if substantial revisions emerge during updates.
- 10+ year series are generally publishable unless diagnostics indicate severe issues, such as instability revealed through sliding spans analysis, which would necessitate withholding the adjustment.

6.3 Annual Consistency Constraints and Adjustments

When seasonally adjusting time series, a discrepancy may arise between annual totals of seasonally adjusted estimates and their original counterparts. This occurs because monthly or quarterly data, when adjusted independently, account for trading day effects by scaling to 'average' trading day patterns, whereas original estimates reflect actual day-of-week distributions. To enhance consistency for end-users, annual constraints are occasionally applied to align fiscal year totals (or other annual aggregates) of seasonally adjusted estimates with the original totals. Where full consistency is required, X-13ARIMA-SEATS provides the force=total option within the 'X11' spec (invoked as X11{force=total}), ensuring seasonally adjusted annual sums precisely match original data aggregates.

6.3.1 Rationale and Conditions for Annual Constraints

The primary justification for annual constraints is to provide users with consistent annual totals, regardless of whether they reference original or seasonally adjusted values. The practice, however, entails a trade-off: constrained adjustments sacrifice statistical optimality, particularly when seasonality exhibits rapid shifts or trading day effects are pronounced. The decision to implement constraints requires careful evaluation against four criteria.

Firstly, consider the data application context. National Accounts datasets, for instance, often require constraints to meet accounting standards, especially critical for GDP series.

Secondly, assess the conceptual validity of annual constraints based on data collection methodologies. For example,

- The UK Labour Force Survey computes rolling three-month averages.
- Retail sales data follow 4–4–5 week collection cycles rather than calendar months.
- Point-in-time series (prices, incomes, claims) may lack meaningful annual averaging interpretations.

Thirdly, maintain methodological consistency. If constrained adjustments have been applied historically, altering this approach requires compelling justification, particularly avoiding intermittent switching between constrained and unconstrained methods across revision cycles.

Finally, examine Table E4 in X-13ARIMA-SEATS outputs. Where differences between constrained and unconstrained series (or multiplicative ratios) prove negligible, and other criteria are satisfied, annual constraints may be inadvisable despite meeting technical feasibility conditions.

6.3.2 Implementation and Revision of Annual Constraints

When implementing annually constrained adjustments, revision policies must explicitly account for this methodological choice. For instance, a revision policy requiring monthly updates to previous observations, such as revising December data when January figures are released, fundamentally disrupts annual additivity. The UK National Accounts dataset follows a revision policy that revises only observations within the current fiscal year, which constitutes the primary rationale for enforcing equality between seasonally adjusted and original annual totals through X-13ARIMA-SEATS. By default, X-13ARIMA-SEATS does not constrain annual totals to match. Users seeking enforced annual consistency may employ two constraint methods:

(1) Specifying 'force=totals' guarantees conceptual equality between adjusted and original aggregates, though printed totals may exhibit nominal discrepancies due to rounding effects.
(2) The force=both or force=round options enforce both conceptual equality and printing consistency, ensuring summed rounded monthly figures equal rounded annual totals, though without constraining to unadjusted aggregates.

Unless the precise digit requirements for final published figures are redetermined, 'force=totals' constitutes the recommended implementation. This approach preserves methodological integrity while acknowledging presentational rounding as a distinct consideration from statistical consistency.

```
series{title="Example of constraining the annual totals"
      start=1994.1
      period=4
      file="mydata.txt"}

transform{function=log}
arima{model=(0,1,1)(0,1,1)}
x11{mode=mult
force=totals}
```

Two essential considerations warrant particular attention when applying annual constraints. First, utilising the force parameter requires a minimum of five years of data. Extreme caution is required when combining this parameter with the 'composite' specification, as methodological interactions may compromise the integrity of the adjustment. Second, when annual constraints reference fiscal years, such as the UK fiscal year spanning April 1 to March 31 of the following year, specific configurations are necessary: for monthly series, include 'forcestart=april' or 'forcestart=apr' in the spec; for quarterly series, employ 'forcestart=Q2' to align with the fiscal quarter structure.

Invoking the force parameter generates supplementary output in Table D11A, which contains the annually constrained seasonally adjusted series. A comparative examination of Table A1 (original, unadjusted data) and Table D11A (constrained, adjusted series) will confirm identical annual aggregates, thereby validating the enforced consistency between raw and adjusted fiscal year totals. This verification mechanism provides transparency regarding the constraint's operational success.

A1 Time series data

	1st	2nd	3rd	4th	TOTAL
2000	95.54	101.48	115.21	106.12	418.34
2001	113.28	121.72	122.30	114.64	471.95
2002	113.04	131.79	119.39	124.S2	488.75

D 11.A Final seasonally adjusted series with forced yearly totals

	1st	2nd	3rd	4th	TOTAL
2000	99.96	99.97	110.77	107.64	418.34
2001	118.59	119.39	117.36	116.61	471.95
2002	118.06	128.97	114.80	126.92	488.75

6.3.3 Temporal Consistency Challenges

Seasonal adjustment introduces non-additivity across time horizons when monthly and quarterly series are processed independently. Substantial differences can occur between quarterly figures that are seasonally adjusted directly and those obtained by summing the adjusted monthly components, resulting in inconsistencies between the composite series and its disaggregated counterparts. This phenomenon parallels, yet differs fundamentally from, aggregate-component discrepancies: here, seasonal characteristics shift not between compositional elements but across months within quarters. For instance, if corporations collectively shift bonus payments from March to February in a given year, the monthly seasonal adjustment would fail to recognise this as seasonality, whereas the quarterly adjustment would detect the displacement. Similar patterns occur in energy consumption series where usage peaks migrate between months due to weather variations, or in agricultural statistics where harvest timing fluctuates with climatic conditions.

Three methodological approaches exist to reconcile monthly and quarterly adjustments:

(1) Indirect Quarterly Adjustment: Derive quarterly estimates by summing seasonally adjusted monthly components, potentially yielding suboptimal quarterly adjustments;
(2) Direct Independent Adjustment: Process both frequencies separately, accepting inevitable additivity loss;
(3) Constrained Direct Adjustment: Enforce additivity constraints during adjustment, but risk distorting monthly statistics. The selection among these methods necessitates careful consideration of the series' analytical purpose, coupled with rigorous evaluation of their technical properties. No universally optimal solution exists; the appropriate approach depends on whether preservation of high-frequency patterns or cross-horizon consistency holds greater relevance for end-use applications.

6.4 Direct versus Indirect Seasonal Adjustment Methods

Composite series (also termed aggregate series) consist of two or more component subseries combined through additive relationships such as A = B + C, where A represents the composite series while B and C denote its constituent elements. Component series may be aggregated through multiple hierarchical levels, where a composite series at one tier may itself serve as a component at higher aggregation tiers. Many seasonally adjusted time series exhibit such composite structures – for instance,

- UK total visitor numbers aggregate arrivals from (a) North America, (b) Western Europe, and (c) other regions;
- Overall unemployment rates sum gender-specific or age-cohort-specific sub-series;
- Constant-price national accounts series derive from current-price counterparts deflated by appropriate indices.

When seasonally adjusting composite series, two methodological approaches exist:

(1) Direct Adjustment: Independently adjusting series A, B, and C without reference to their compositional relationships.
(2) Indirect Adjustment: Constraining the composite series A's adjustment to the sum of its seasonally adjusted components (B + C), achieved by first adjusting constituent series individually before aggregation.

These approaches typically yield divergent results. Direct and indirect methods produce equivalent outcomes only under stringent conditions: Absent calendar/outlier adjustments, decomposition remains perfectly additive, making forecasting unnecessary. Such conditions are rarely met in practice, with discrepancies between the two methods potentially substantial depending on series characteristics. The choice fundamentally involves trade-offs between statistical optimality at component levels versus coherence within composite structures.

6.4.1 Additivity Challenges in Component Series

The seasonal adjustment of composite series centres on reconciling consistency between adjusted components and their aggregates, a fundamental tension termed the 'additivity problem'. While composite series inherently derive from component summation, enforcing additivity may conflict with the core objectives of seasonal adjustment. Unemployment statistics illustrate this: seasonally adjusted male and female estimates should aggregate to the unadjusted overall total. Yet, direct independent adjustment of gender-specific series and the aggregate series invariably yields discrepancies. This approach fails to guarantee component-aggregate consistency, particularly when seasonal patterns shift between demographic segments.

The indirect method, summing gender-adjusted components to derive total unemployment, preserves additivity at the cost of potentially suboptimal aggregate adjustment. This dilemma intensifies with multidimensional classifications (age, region, education, etc.). For example, summing the seasonally adjusted age-specific unemployment series yields totals that differ from those obtained by indirectly adjusting the gender-specific series, and both sets of results diverge from the directly adjusted aggregate. Three methodological pathways exist to enforce additivity within standard frameworks:

(1) Granular Indirect Adjustment: Apply seasonal adjustment at the lowest classification level (eg, gender-specific cohorts within age groups) before aggregation;
(2) Non-Additive Acceptance: Explicitly acknowledge and publish series with inherent non-additivity;
(3) Linearity Constraints: Restrict adjustment methods to purely linear operations preserving additivity.

Current institutional implementations reveal trade-offs:

- Eurostat's industrial production adjustment employs method (1), requiring thousands of component adjustments;
- UK Office for National Statistics frequently adopts (2) for the Labour Force Survey series;
- The Bank of England utilises (3) for monetary aggregates.

All approaches entail theoretical compromises and practical challenges, as constraints may distort seasonal signals across the dataset. Analysts thus face a trilemma:

- Indirect adjustment prioritises coherence but risks suboptimal component adjustments;
- Direct adjustment maximises individual series fidelity but sacrifices additivity;
- Constrained direct adjustment ensures additivity while potentially distorting component dynamics.

The choice ultimately depends on whether the analytical residual value lies in preserving composite integrity or in capturing component-level behavioural accuracy. This determination requires careful consideration of end-user needs and the underlying statistical properties.

6.4.2 Selection Criteria for Direct versus Indirect Methods

The choice of adjustment method requires evaluating characteristics of the composite series. Indirect adjustment becomes preferable when (1) granular component-level information is available; (2) heterogeneous data types within the dataset necessitate selective treatment; (3) cross-dataset consistency is paramount; (4) strict additivity between components and aggregates must be preserved; and (5) classified data hold greater analytical value than aggregates – as exemplified by demographic unemployment rates outweighing broader labour force metrics.

However, indirect adjustment carries significant limitations for aggregate quality:

- Post-adjustment processing may reintroduce seasonality.
- Interdependent component series generate distinct seasonal dynamics at aggregate levels.
- Excessive disaggregation amplifies sampling variability, complicating seasonal estimation.
- Component-level adjustments typically require longer moving averages (elevated I/S and I/C ratios), slowing responsiveness to short-term changes.
- Operational burden escalates exponentially with the series count, potentially limiting analytical oversight.
- Residual seasonality may persist in the composite series unless components exhibit divergent seasonal patterns.

Where a direct adjustment is mandated, yet additivity preservation is required, component series must be constrained to align with the directly adjusted composite. Minimising discrepancies involves the following:

- Applying identical models and prior adjustments to both composite and components.
- Selecting models optimised for the composite series, accepting potential suboptimal component adjustments.

Persistent additivity gaps necessitate discrepancy allocation (composite minus the sum of components) through three approaches:

(1) Attributing the entire discrepancy to the largest or least significant component.
(2) Distributing proportionally by component magnitude.
(3) Scale-based allocation using irregular component proportions.

Comprehensive implementation further requires considering:

- Dataset scale and temporal coverage constraints.
- Potential residual seasonality resolution through direct adjustment.
- Differential outlier treatment between composite and component adjustments.

6.4.3 Spec File Configuration for Indirect Adjustments

The X-13ARIMA-SEATS program can generate diagnostics to aid in decision-making between direct and indirect methods. To compare direct and indirect adjustments in the program, a 'composite' indirect adjustment Spec can be used to generate a diagnosis to evaluate the two methods. Take the three series of A, B, and C mentioned earlier as an example, A is the total amount, and B and C are the constituent components of A, that is, A = B + C, and the indirect adjustment spec is divided into three steps as follows.

```
(1) Create a 'series' spec file for each component Series (B, C), ie,
B.SPC and C.SPC:

series{title="B"
       file="B.txt"
       name="B"
       comptype=add
       }
x11{}
```

If the C series is a 'non-seasonal' Series and does not require seasonal adjustment, then C.SPC:

```
series{title="C"
       file="C.txt"
       name="C"
       comptype=add
       }
x11{type=summary}

(2) Create a 'composite' spec file for composite Series A: ie, A.SPC

Composite{title="A"
          name="A"}
x11{}
(3) Create a metafile that lists all the constituent Series (B, C) and
composite Series (A) together, and the composite Series (A) must be
listed at the end of the 'meta' spec file as follows:
       B
       C
       A
```

Save as a file of 'test.mta'. Run 'test.mta' to generate an output (of an aggregate Series) of indirect adjustment diagnostics. At the same time, direct adjustments can be made to A, B, and C separately, allowing comparison of direct versus indirect seasonal adjustment performance for the aggregate series. The choice of direct and indirect adjustment methods is mainly based on the following factors: (1) There should be no residual seasonality in the seasonally adjusted series. X-13ARIMA-SEATS provides a set of spectral diagnostics that can be used to quickly check for the presence of significant seasonal peaks, where the spectral amplitude at a seasonal frequency must be at least 'six stars' higher than its two adjacent frequencies. If there is a significant peak in the indirect adjustment, an alternative method should be chosen; (2) In general, a stable seasonal adjustment series is the preferred alternative. X-13ARIMA-SEATS generates a set of sliding span diagnostics as a method for determining whether seasonal adjustments are stable; (3) Interpretability of seasonally adjusted series: The M-value measures the regions in the seasonally adjusted mass, while the Q-value is the weighted average of all M-statistics. If the Q statistic fails, there may be a short-term variation problem in the seasonally adjusted series. In general, the method chosen should minimise the Q statistic; (4) Check the smoothness of the seasonal adjustment series; in general, the adjustment method that provides a smoother series should be used. Measurements of roughness are given at the end of the output (for ensemble series), or before sliding spans and historical diagnostics, and the measure of roughness describes the magnitude of the deviation from the smoothing trend of the adjustment series (where R1 and R2 use different trend estimation methods). Direct adjustments are smoother than indirect adjustments. However, smoothness is not necessarily the most desirable feature of a seasonally adjusted Series.

6.5 Identification and Handling of Seasonal Break

A seasonal break is a sudden and continuous change in the seasonal pattern of the original data. Potential causes of seasonal breaks include changes in data sources, methodological revisions, or shifts in management practices. If there is a seasonal break in the series, the estimates of the seasonal component will be distorted. Distorting the seasonal factor estimates (Table D8) could result in residual seasonality in the seasonally adjusted series. The primary diagnosis affecting seasonality (F-test in Table D8A) is that irregular values are artificially exaggerated and reduce the significance of F-scores. In both cases, the result is a more volatile seasonal adjustment series and a greater likelihood of a correction. Identifying seasonal disruptions can sometimes be difficult and requires some hands-on experience.

6.5.1 Methods for Identifying Seasonal Break

Seasonal break can be identified in two ways: (1) pattern recognition, and (2) SI ratio values. Draw a series of diagrams to identify examples of seasonal breaks:

As shown in Figure 6.1, before January 1999, a stable seasonal pattern characterised by a single annual peak and trough was observed. After 1999, this pattern permanently

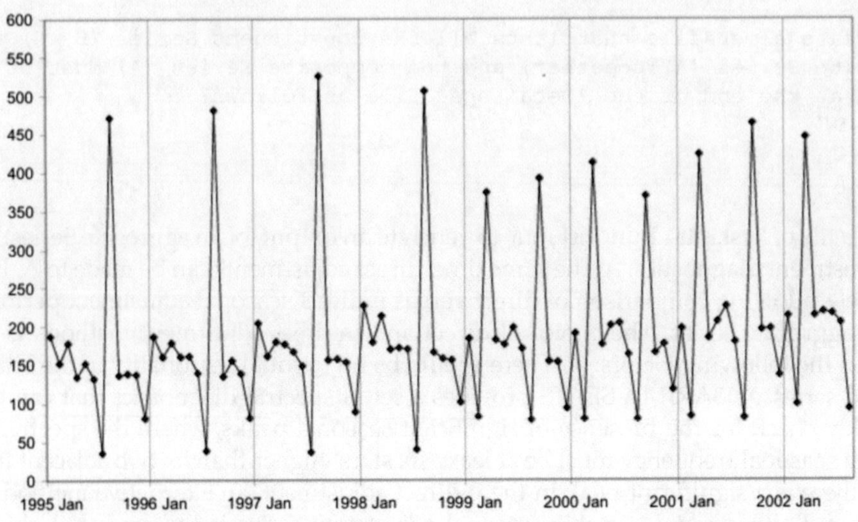

FIGURE 6.1
The Example of Seasonal Break.

Source: Using the simulated data for the example

changed to two troughs per year. 1 January 1999 marked the seasonal interruption in this series. (2) In Table D8(SI) ratios, a sudden increase or decrease in the value of a month/quarter may indicate a seasonal interruption:

D 8 Final unmodified SI ratios From 1991.1 to 2002.4 Observations 48

	1st	2nd	3rd	4th	AVGE
1991	83.7	100.6	104.1	111.9	100.1
1992	83.2	100.8	104.5	110.6	99.8
1993	84.7	100.0	104.0	112.1	100.2
1994	83.1	101.6	103.9	110.7	99.8
1995	80.9	101.1	104.9	112.9	99.9
1996	81.8	100.3	104.7	107.0	98.5
1997	89.1	100.1	104.9	103.2	99.3
1998	92.4	101.6	104.2	101.1	99.8
1999	94.1	102.8	103.6	100.1	100.1
2000	93.8	102.3	104.8	98.2	99.8
2001	94.1	104.5	102.4	98.6	99.9
2002	94.2	104.5	103.4	97.5	99.9

As Table D8 indicates, Q4 values exceed 110 until 1995 and decline to 107 by Q4 1996, 103 in 1997, and finally below 100. This may be due to changing seasonality or seasonal disruptions. However, this rapid shift, confined to three periods, with stability preceding and following, suggests a seasonal hiatus. Inspections of the first quarter support this suspicion; In the period before 1996, the SI ratio was between 80 and 85, jumped to 89 in 1997, and has been above 90 ever since. Notably, the simultaneous rapid evolution in Q1 and Q4 further corroborates the presence of seasonal disruptions. Finally, note that not all quarters need to be affected to argue for a seasonal break. In the example above, Q2 and Q3 were fairly stable, and the disruption primarily involved a shift in activity from Q4 to Q1.

6.5.2 Testing and Adjustment of Regression Models

The most important and reliable way to identify seasonal disruptions is to test and adjust using the 'regression' spec, which is a two-step process:

Step 1: Test for seasonal breaks

Assuming that the date of the break is January 1999, it is necessary to test the seasonal pattern before the break, firstly:

```
regression{
          variables= (seasonal/year.period//)
          save= (rmx)
          }
```

Or test seasonality after a seasonal break:

```
regression{
          variables= (seasonal//year.period/)
          save= (rmx)
          }
```

Verify the significance of the regression variables by testing the t-values and Chi-square values of the monthly/quarterly regression variables in the 'regression' spec regression output, and if the regression variables are not statistically significant, change the time point date of the seasonal break until the regression variable of the selected seasonal break achieves statistical significance. The Chi-square test is recommended only when the p-value for the seasonal break regression variable is below 0.05, as the test tends to overstate significance due to its high Type I error rate.

If you judge that the series is not seasonal, it is not recommended to use the seasonal break analysis. The outcomes of the seasonal break analysis can be used to derive permanent prior adjustments for the observations in the initial period. The regression variable and associated data can be saved in the 'regression' spec with the command 'save=(rmx)' in the next step for seasonal adjustment.

It is important to note the following:

- One possible drawback when using 'regression' to detect seasonal breaks 'variables=(seasonal/1999.jan//)' is that the permanent prior estimates are not always balanced, resulting in the problem that the level of the series does not remain constant. If the data chart shows a possible seasonal break, you also need to find an economic justification for the existence of the suspected seasonal disruption. If not, then don't adjust for it. The total length of the entire series takes at least five years to use the 'regression' of the seasonal break.
- At least one year of data is required on both sides of the break. If less than one year of data is available after the break, the use of seasonal break analysis should be considered.
- If data is available for 1–2 years (12–24 months or 4–8 quarters) after the break, consideration should be given to not publishing seasonally adjusted data for that series until more observational data is available, as while X-13ARIMA-SEATS can provide adjustments, they are often of poor quality and may be heavily revised as future observations become known. If data is available for 2–3 years (24–36 months or 8–12 quarters) after the break, consider using regARIMA to predict data for one year after the end of the series. If the seasonal pattern after the break looks very regular (both the model and the SI ratios are stable), you can resume seasonally adjusted publishing at this time.
- If more than three years of data are available after the break, seasonally adjusted releases can usually be resumed. However, if the model is inadequate, the series tends to exhibit heightened instability, or the SI ratio is particularly inconsistent between years, do not resume publishing at this time.

Step 2: Fit to adjust for seasonal breaks

Turn off variables=(seasonal/1999.jan//) in the 'regression' spec and use the 11 variables in the previously saved (save) at '.rmx' regression matrix as user-defined regression variables, (for monthly series, there should be 11 variables), ie,

```
regression{# variables=(seasonal/1999.jan//)
           # save=(rmx)
           user= (M1, M2, M3, M4, M5, M6, M7, M8, M9, M10, M11)
           usertype=(seasonal)
           file="name.rmx"
           format="x12save"
           save=(usr)
          }
```

Suppose this is the final permanent prior adjustment. In that case, all regression variables (Easter, trading day, level shift, and seasonal break) must be combined into one regression matrix, called '.ppp', and saved as a file. The order of the variables is usually the order in which the regression matrix is generated. This can be done by reading the saved '.rmx' file and all the variables used in the model – Easter, trading day, level transfer, etc. – into Excel, and then generating a '.ppp' save file (as a text file) in x12save format. Then turn off all 'regression', using 'transform' as a permanent adjustment, for example,

```
transform{
        Function=log
        File="prior.ppp"
      format="x12save"
        Type=permanent
         }
```

The length of the series before and after the seasonal break affects the choice of method. Using a seasonal break should be done with caution. If the data available after the break is less than three years old, especially if the parameters need to be re-estimated frequently until they become stable, this makes the process of updating a priori adjustments more complex and difficult. It is important to note that when a seasonal break is detected, it may be necessary to revise the historical data for the entire series, not just the most recent data point. If a seasonal disruption results in a significant revision to historical data, the user should be notified promptly, and the reason and impact of the revision should be explained. Seasonal breaks should typically be addressed during annual updates rather than at each data update. This reduces the frequency of revisions and improves the stability of the data. Seasonal breaks are an important issue in time series analysis and can significantly affect the quality of seasonal adjustments. The key to effectively identifying and addressing seasonal disruptions is ensuring the accuracy and consistency of adjustment results. At the same time, when dealing with seasonal breaks, care should be taken to revise policies to ensure data stability and user trust.

6.6 Application of User-Defined Regression Variables

Scientific economic evaluation or comparison requires assessing the same economic phenomenon at regular intervals; however, in practice, it is rare to obtain a series of measurements of the same phenomenon taken at equal intervals. For the effects of factors within a time series relationship, such as calendar effects, trading day variables, 'additive outliers', 'level shifts', 'temporary changes', 'slopes', permanent changes in seasonal patterns of trends or series, (seasonal breaks), and moving holidays, the regARIMA model of X-13ARIMA-SEATS provides built-in variables and built-in regressors. Time series can also be influenced by external deterministic factors, such as monthly bank holidays in the UK, the Chinese New Year, or Islamic Ramadan holidays, as well as by non-calendar-based data, ie, data collected on a basis different from the standard monthly calendar, which lack built-in regressors and must be manually specified by the user.

6.6.1 Constructing User-Defined Regression Matrices

The UK Retail Sales Index (RSI) exemplifies a critical methodological challenge: its data collection follows a 4–4–5 week pattern across rolling 13-week periods, operating on a non-calendar monthly basis. This structure causes December observations to span different calendar dates each year – for instance, December 2008 covered November 30 to January 3, while December 2009 encompassed November 29 to January 2. Such temporal misalignment induces 'December offsets' where year-over-year December data points lack consistent chronological positioning. To correct this structural bias, practitioners must define specialised regressors (eg, a December offset variable), quantify temporal displacement magnitudes, and construct a user-defined regression matrix for RSI offset calibration. This matrix explicitly models the phase-shifted seasonal effects arising from the survey's non-standard periodicity, enabling statistically coherent adjustment despite the underlying collection anomaly.

Within the X-13ARIMA-SEATS framework, this regression matrix functions as an external regressor set integrated into the time series model. Each column represents a monthly regression variable, while rows correspond to temporal observations (eg, specific year-month points), designed to quantify non-calendar effects such as the 4–4–5 week pattern displacement in the UK RSI. The matrix values reflect monthly offsets relative to a standardised 'zero' baseline, with December's offset serving as the primary illustration. Construction involves two critical phases:

TABLE 6.1
RSI Regression Matrix File (.rmx file)

Date	January	February	July	October	November	December
….	……	……	……	……	…….	…….
198601	1.125	0	0	0	0	0
198602	0	–1.175	0	0	0	0
198603	0	0	0	0	0	0
198604	0	0	0	0	0	0
198605	0	0	0	0	0	0
198606	0	0	0	0	0	0
198607	0	0	–2	0	0	0
198608	0	0	0	0	0	0
198609	0	0	0	0	0	0
198610	0	0	0	–2	0	0
198611	0	0	0	0	–2	0
198612	0	0	0	0	0	–2
198701	–1.375	0	0	0	0	0
198702	0	–0.75	0	0	0	0
198703	0	0	0	0	0	0
198704	0	0	0	0	0	0
198705	0	0	0	0	0	0
198706	0	0	0	0	0	0
198707	0	0	-1	0	0	0
198708	0	0	0	0	0	0

First, offset values are computed for each observation point, measuring chronological deviation from ideal calendar alignment. These raw offsets undergo centring (standardisation) to normalise their distribution, resulting in values scaled between −3 and 3. Negative values indicate backwards temporal shifts, positive values forward displacements, and zero denotes perfect alignment.

Second, the processed offset values populate a structured .rmx file (regression matrix file).[1] This machine-readable specification enables X-13ARIMA-SEATS to dynamically adjust seasonal estimates for phase-shifted months, effectively neutralising structural biases inherent in non-standard data collection regimes. The finalised matrix format ensures each temporal coordinate receives precise offset calibration during model estimation.

Using the month as a user-defined variable, the regression matrix is named 'regressor.rmx', testing the offset effect of the month, and running it with the 'regression' spec as follows:

```
regression{
        user=(Jan Feb July Oct Nov Dec)
        usertype=(USER USER USER USER USER USER)
        file = "regressors.rmx"
        format ="x12save"
        }
```

For the multiplicative RSI series, the output for the test month is as follows.

Regression Model

Variable	Parameter Estimate	Standard Error	t-value
User-defined			
Jan	0.0077	0.00418	1. 84
Feb	0.0041	0.00298	1.37
July	0.0028	0.00315	0. 90
Oct	0.0067	0.00329	2. 04
Nov	−0.0059	0.00334	−1.76
Dec	−0.0029	0.00331	−0.88

Chi-squared Tests for Groups of Regressors

Regression Effect	df	Chi-Square	P-Value
User-defined	6	16. 17	0.01

For months where the t-statistic is not significant, a significant regression is fitted (using the reduced regressors.rmx) and saved in the previous adjustment of 'User-Defined (USR)':

```
regression{
        user= (Oct)
        usertype = (USER)
        file = "regressors.rmx
        format="x12save"
        save=(usr)
        }
```

User-defined adjustments can be found in the X-13ARIMA-SEATS output in Table A9.

A 9 RegARIMA
user- Frora 1987.Jan defined 2009.Aug
to Observations 212 regression component

	Jan Jul	Feb Aug	Mar Sep	Apr Oct	May Nov	Jun Dec	AVGE
1987	100.00	100.00	100.00	100.00	100.00	100.00	
	100.00	100.00	100.00	99.24	100.00	100.00	99. 94
1988	100.00	100.00	100.00	100.00	100.00	100.00	
	100.00	100.00	100.00	100.77	100.00	100.00	100.06
1989	100.00	100.00	100.00	100.00	100.00	100.00	
	100.00	100.00	100.00	101.54	100.00	100.00	100.13
1990	100.00	100.00	100.00	100.00	100.00	100.00	
	100.00	100.00	100.00	102.32	100.00	100.00	100.19
1991	100.00	100.00	100.00	100.00	100.00	100.00	
	100.00	100.00	100.00	97.73	100.00	100.00	99.81
1992	100.00	100.00	100.00	100.00	100.00	100.00	

In this way, a priori adjustment is made for the calendar effect of the RSI, using a 445 survey method. This approach treats many of the user-defined regression variables similarly, eg, a priori adjustments for UK bank holidays, Easter, etc.

If we take the UK moving bank holiday effect into account, MBH and ABH move between May/June and August/September, respectively. When a bank holiday falls in the 'correct' reporting period, 1 if MBH is in May and ABH is in August, –0.8 if MBH is in June/-0.8 if ABH is in September, then it is calculated in the middle.

Easter effect modulation is further taken into account. It should be noted that while X-13ARIMA-SEATS provides a default Easter regressor, certain datasets, such as those collected using a 4–4–5 weekly pattern, like the RSI, do not align strictly with calendar months. In such cases, the division of the year differs from the model's assumed monthly calendar periods. In this case, there may be an Easter effect. But it should not be estimated using the default regressor provided by the X-13ARIMA-SEATS. The user needs to use the 'user' and 'usertype' parameters in the 'regression' spec to define regression variables and estimate the Easter effect. The X-13ARIMA-SEATS Easter regressor allows the effect to span 1–25 days before Easter. This length can be specified in the regression spec using the automatic selection option aictest=(Easter) or with fixed selections such as Easter[1], Easter[8], and Easter[15]. If RSI data cannot use the default Easter regressor, the user must customise the Easter variables (Eas1, Eas2, and Eas3) using the same method applied to compute the RSI calendar effects and moving-holiday effects, and then calculate the corresponding regression matrix (.rmx), as illustrated in the following table:

TABLE 6.2

The Corresponding Regression Matrix

Date	Jan	Feb	July	Oct	Nov	Dec	MBH	ABH	Eas1	Eas2	Eas3
....	....	....	....	....	....	....	....	....	....	....	....
198601	1.125	0	0	0	0	0	0	0	0	0	0
198602	0	–1.175	0	0	0	0	0	0	0	0	0
198603	0	0	0	0	0	0	0	0	0.588235	–0.16471	–0.14118
198604	0	0	0	0	0	0	0	0	–0.73529	0.205882	0.176471
198605	0	0	0	0	0	0	0.571429	0	0	0	0
198606	0	0	0	0	0	0	–0.45714	0	0	0	0
198607	0	0	–2	0	0	0	0	0	0	0	0
198608	0	0	0	0	0	0	0	0.714286	0	0	0
198609	0	0	0	0	0	0	0	–0.57143	0	0	0
198610	0	0	0	–2	0	0	0	0	0	0	0
198611	0	0	0	0	–2	0	0	0	0	0	0
198612	0	0	0	0	0	–2	0	0	0	0	0
198701	–1.375	0	0	0	0	0	0	0	0	0	0
198702	0	–0.75	0	0	0	0	0	0	0	0	0
198703	0	0	0	0	0	0	0	0	–0.21177	–0.16471	–0.14118
198704	0	0	0	0	0	0	0	0	0.264706	0.205882	0.175471

All user-defined variables (Jan, Feb, Jul, Oct, Nov, Dec, MBH, ABH, Easl, EAS2, EAS3) are listed in the 'user' parameter in the 'regression' spec. Unless all user variables have the same type, their type must be listed in the 'usertype' parameter and match the order of the names in the 'user' parameter. The values of the variables for all periods (including the forecast period) and for all user-defined variables must be listed in a table (.rmx), stored in a data text named 'regression.rmx', and referenced in the 'regression' spec via the 'file' parameter:

```
regression{
      user=(Jan Feb Jul Oct Nov Dec MBH ABH Easl EAS2 EAS3)
      usertype=(user user user user user user Holiday Holiday Holiday
      Holiday Holiday)
      file="regressor.rmx"
      format="x12save"
      save=(usr)
      }
```

6.6.2 Test Methods for Regression Models

Once the model has been estimated, it needs to be checked to see if it has been fitted correctly. Not all Specifier and User-Defined Variables will be the last optional regression variables. Specifier Variables and User-Defined Variables are tested and are removed if the statistical test is not significant. Then you need to select another 'speculative variable' and run another run to re-estimate the model. For example,

Regression Model

Variable	Parameter Estimate	Standard Error	t-value
User-defined			
Jan	0.0076	0.00411	1.86
Feb	0.0040	0.00293	1.35
Jul	0.0028	0.00309	0.90
Oct	0.0068	0.00323	2.12
Nov	–0.00S7	0.00328	–1.7S
Dec	–0.0029	0.00326	–0.8888
MBH	0.0071	0.00902	0.79
ABH	0.0103	0.00997	1.03
Easl	0.0234	0.01090	2.IS
Eas2	0.0067	0.01265	0.53
Eas3	0.0020	0.01215	0.17

Chi-squared Tests for Groups of Regressors

Regression Effect	df	Chi-Square	P-Value
User-defined	11	23. 86	0. 01
User-defined Holiday Regressors S	5	7.04	0. 22

For the user-defined Jan, Feb, Jul, Oct, Nov, Dec, MBH, ABH, Eas1 Eas2, and Eas3 variables, only the Oct and Eas1 variables have t values greater than 2. The statistical detection of the other variables is not significant, so only the significant regression variables (Oct) and (Eas1) are fitted, and the previous adjustments of 'user-defined' and 'holiday' are saved:

```
regression{
     user=(Oct Eas1 Eas2 Eas3)
     usertype=(user Holiday Holiday Holiday)
     file="finalregressor.rmx"
     format="x12save"
     save=(usr hol)
     }
```

Regression Model

Variable	Parameter Estimate	Standard Error	t-value
User-defined			
Oct	0.0090	0.00412	2.19
Eas 1	0.0044	0.00300	1.45
Eas2	0.0033	0.00316	1.05
Eas3	0.0083	0.00319	2.59

Chi-squared Tests for Groups of Regressors

Regression Effect	df	Chi-Square	P-Value
Use r-defined	4	12. 69	0.01
User-defined Holiday Regressors	3	9.34	0.03

For testing models with regression sections, the 't-statistic', 'Chi-square', and 'AIC' tests are commonly used.

(1) Automated inspections. Typically, all required inspections are generated automatically. Specifically, the program reports a t-statistic for every regression coefficient. The program also provides Chi-square statistics for groups of variables, particularly when they are more meaningful when evaluated collectively. For example, it is more informative to test whether the six trading day variables jointly have a significant effect than to assess the significance of a single variable (eg, Tuesday) in isolation. When a variable is tested individually, it should be considered statistically significant if its t-statistic exceeds the cut-off for the desired level of significance. Typically, a t-statistic with an absolute value greater than 2 is deemed significant at the 5% level. On the other hand, if a set of variables, such as a trading day regression variable, is tested jointly, then the Chi-square is the appropriate statistic. For Chi-square, the program automatically gives a probability (P) value, and any value less than 5% (0.05) is usually considered significant. In some cases, explicit testing is not required, and this applies to level shifts and other outlier terms selected by the 'outlier'. In other cases, individual or group items should be properly validated and retained or deleted as needed. For example, it is a good practice to verify that the p-value or t-value of a trading day or Easter item selected by 'aictest' is statistically significant.

(2) Non-automated inspections. Although automated inspection is sufficient in most cases, the user must manually inspect cases where automation is not feasible, typically by manipulating or combining elements of the output.

- 'Alternative models' may need to run X-13ARIMA-SEATS separately, as it is not possible to designate one model as a special version of another (these are referred to as non-nested models). In this case, the model with the minimum value can be selected based on model statistics, such as AIC, AICC, and BIC. However, there are some limitations to the use of these statistics, in particular, that the dependent variable in the regression must contain the same outliers in the series span, differencing order, transformation, and model.
- The autoregressive and moving average parameters do not have accompanying t-statistics. However, they come with a standard error value, and the t-value is the ratio of the two. In general, automatic model selection ensures that only significant variables are retained; however, such a check is necessary when the model is manually refined.
- If you test whether the parameters of two variables are equal, you can define a new variable, include it in the model, replace one of the other variables, rearrange the model, and test the significance of the difference. Test whether a set of explanatory variables is jointly significant. If it is not possible to pass them as a group to generate Chi-square values, you can run models with and without these variables separately and compare the results. The most straightforward comparison is to use AICC or BIC, one of the model statistics. The other is to compare log-likelihood values to give a numerical significance level. The difference in log-likelihood obeys a Chi-square distribution, with degrees of freedom equal to the number of additional parameters.
- The 'regression' spec should contain all the salient variables from the last run, as well as any variables that are excluded but on the edge.

- Run X-13ARIMA-SEATS to generate model selection and regression statistics. Check the significance of the regression variables as described earlier. If any previously included variables are now strongly insignificant (t-value below 1), they can be removed. If any previously included regression variable has a t-value between 1 and 2, it is best to keep that variable to avoid unacceptable revisions to previously published seasonally adjusted data. This variable should be tested again at the next annual reanalysis.

6.6.3 Spec Configuration for Final Prior Adjustments

If all identified regression variables are statistically significant, the last prior-adjusted spec file can close all 'regression' spec files and finally convert to the 'transform' spec file. As in the example above, 'Oct' and 'Eas1', 'Eas2', 'Eas3', and the specified regression variables (AO, LS, RP, etc.) have been identified, and before importing the 'transform' spec into X-13ARIMA-SEATS, it can be merged into a permanent prior 'priors.ppp' file stored as Table 6.2 in Excel, and all 'regression' can be turned off, and finally 'transform' can be used. The spec is as follows:

```
transform{
      function=log
      file="priors. ppp"
      type=permanent
      }
```

Complete all a priori adjustments.

6.7 Adjusting for Chinese Lunar Holidays

6.7.1 Overview and Adjustment Requirements

Seasonal adjustment enhances the comparability of economic data by enabling the evaluation of phenomena across equivalent periods. China's reliance on year-on-year growth metrics, such as annual GDP growth or inflation rates, is subject to distortion due to the shifting timing of the Lunar New Year, which can fall anywhere between January 21 and February 20 on the Gregorian calendar. Using unadjusted series directly for growth calculations can introduce significant inaccuracies, similar to the distortions caused by the shifting date of Western Easter between March and April.

The public holiday effect of the Lunar New Year, alongside other lunar festivals such as Dragon Boat and Mid-Autumn, causes substantial economic time series distortions as it migrates across Gregorian months. Consequently, Chinese economic data require specialised prior adjustments similar to those implemented for the UK RSI's non-calendar effects and Easter.

As X-13ARIMA-SEATS lacks built-in regressors for lunar holidays, practitioners must do the following:

(1) Utilise the regression spec within regARIMA modelling.

(2) Define user-defined regressors for Lunar New Year, Dragon Boat, and Mid-Autumn festivals.

(3) Construct dedicated regression matrices (.rmx files) mirroring the UK RSI methodology.

Core Implementation Workflow

(1) Offset Calculation: Quantify each festival's annual deviation from ideal dates.
(2) Centering: Subtract mean offset value → centred offset = (raw offset – mean offset).
(3) Standardisation: Divide centred offsets by their standard deviation.
(4) Matrix Population: Populate .rmx files with standardised values.

This process generates normalised regressors that objectively capture lunar holiday displacements, enabling statistically robust prior adjustments in economic time series.

6.7.2 Application of the Genhol Tool

Although X-13ARIMA-SEATS lacks built-in regressors for the Chinese New Year, the U.S. Census Bureau (USCB) developed the Genhol utility. This stand-alone program generates user-defined regressors for user-defined holidays (including pre-configured templates for the Lunar New Year). Genhol is specifically engineered to produce regression matrices for lunar-based Spring Festival and diverse trading holidays (or stock market Easter variations), serving as a critical component within the X-13ARIMA-SEATS seasonal adjustment software for Chinese lunar calendar adjustments and other moving holiday effects.

6.7.2.1 Implementation of Chinese New Year Adjustment

The standard command syntax for adjusting the Chinese Spring Festival is
Genhol lunar [a b] center=weighted
Where:

- a = Days *before* New Year (negative integer, eg, -15)
- b = Days *after* New Year (positive integer, eg, 15)
- center=weighted = Weighted centring method for holiday effect normalisation

Computational Workflow

(1) Define Impact Window
  - Baseline: New Year's Day (t=0)
  - Standard range: 15 days pre-festival (t=-15) to 15 days post-festival (t=15)
  - Total days: 31
  - Command: genhol lunar [-15 15]

(2) Weight Allocation & Centring
Weight Assignment (w):
  - New Year's Day (t=0): weight = 1
  - Other days: Linearly decreasing weights from centre to endpoints

Weighted Centring (center=weighted):

- Calculate total weight(W): W = Σw(t)
- Compute weighted mean (c): C = Σ(t × w(t))/W;
- Adjust weights to anchor C at an integer date (Adjust the weights so that the weighted centre falls on the integer date)

The holiday effect was calculated using the weighted centre method. The specific calculation formula is as follows:

$$W(t)\begin{cases}1-\dfrac{|t-D|}{D-a+1} & \text{if } (a \le t < D)\\ 1-\dfrac{|t-D|}{b-D+1} & \text{if } (D \le t \le b)\\ 0 & \text{if others}\end{cases}$$

(3) The final regressor X(t) is the weight w(t) of each date divided by the total weight (W):

$$X(t)=\frac{W(t)}{W}$$

After the treatment, the sum of the weights is 1 for the entire influence period. It can be explained as the distribution of the holiday effect in time: (1) on the day of the Spring Festival (t =0), the weight (D) is 1; (2) The weights of the 15th day before the Spring Festival (t=-15) and the 15th day after the Spring Festival (t=15) were 0; (3) the weight of the intermediate date changes linearly; (4) The sum of the weights of the final regressor is 1.

When using Genhol's separate utility, the following should be noted:

(1) It is necessary to determine a reasonable range of [a, b] according to the actual economic behaviour, that is, the choice of the influence period, which is commonly used for the Chinese New Year[-7,7], considering the influence period from 7 days before the Spring Festival to 7 days after the Spring Festival (15 days in total), or [-15,15], considering the influence period from 15 days before the Spring Festival to 15 days after the Spring Festival (31 days in total), etc.

(2) 'centre=weighted' is a centralised method, which is usually more reasonable than a simple average, and can better reflect the actual distribution of the holiday effect.

(3) After the application of Genhol, the model diagnosis, model residuals, and fitting effects should be checked, and the range of the influence period [a, b] should be adjusted if necessary.

(4) Sometimes it is necessary to combine multiple Genhol devices to capture complex holiday effect patterns.

(5) The Genhol device is usually used in combination with other regressors (such as outliers, trends, etc.) adjustments.

Genhol offers a flexible approach for moving holidays such as the Chinese New Year, effectively modelling their economic impact and enhancing seasonal adjustment through strategic weight allocation and centralised processing. In practice, it is necessary to determine the optimal parameter settings in combination with specific data and domain knowledge.

The Genhol tool can be accessed by using Win-Genhol. Win-Genhol is a Windows interface for utilities provided by the U.S. Census Bureau (2015). Genhol can create regressors for intermediate effects before, during, or after the holidays. The U.S. Census Bureau (2020) further incorporated these regressors into the regARIMA component of the X-13ARIMA-SEATS model. A separate statement on how to use Win-Genhol is available in Appendix 6.2.

6.7.3 Configuration of the 'Regression' Spec

Once the Moving Holiday Regression 'cny.rmx' (Regression Matrix) is generated by Win-Genhol, it can be used as an input file for X-13ARIMA-SEATS, running the 'regression' spec to pre-adjust for Chinese New Year or any moving holiday. The following is an example of a 'regression' spec:

```
Regression{variable=(cny1 cny2 cny3)
           user=(beforecny betweencny aftercny)
           File=(cny.rmx)
           Formate=("datavalue"
           Start=2000.1
           Usertype=(hol hol hol)
           chi2test=yes
           savelog=chi2test
          }
```

Win-Genhol is an excellent tool for creating moving holiday returners.[2]

6.7.4 Adjustment of Dragon Boat Festival and Mid-Autumn Festival

Adjustments for the Chinese Dragon Boat Festival and the Mid-Autumn Festival can be made using the same seasonal adjustment methods applied to the Chinese New Year. Using the same method to calculate the offset regression for the Chinese New Year, the offsets for the Chinese Dragon Boat Festival and the Mid-Autumn Festival were calculated separately and standardised to match the regression matrix for the Chinese New Year. Alternatively, the regression matrix generated by Genhol for Dragon Boat and Mid-Autumn Festival adjustments was integrated into the regARIMA model alongside user-defined Chinese New Year regressors, with final estimation performed by X-13ARIMA-SEATS. Earlier versions of the X-12-ARIMA and X-13ARIMA-SEATS treated this integrated return as a holiday. Synthetic regression can be tested as a whole, but individual moving holidays cannot be specified individually. The current version of X-13ARIMA-SEATS allows the user to specify up to five user-defined sets of holiday regressors, which can be listed separately in the program output, and to generate diagnostics for individual groups, such as Chi-square statistics, for each of them, so that the user can determine whether a particular effect can be removed.

6.7.5 Significance Test and Diagnosis

Once you have specified this user-defined type of regression variable, you can use the 'chi2test' parameter of the 'regression' spec to check if the user-defined set of holiday regression variables (cny1 cny2 cny3) should be kept in the regARIMA model. When 'chi2test = yes', chi2test metrics will be generated for all user-defined sets of holiday regression variables, while those that are not significant (given by the coefficient of variation (CV) of the parameter chi2test) will be removed from the regARIMA model. The Chi-square statistic formula is as follows:

$$\hat{\chi}^2 = \hat{\beta}'\left[Var(\hat{\beta})^{-1}\right]$$

where $\hat{\beta}$ represents the coefficient estimation vector for the group of regression variables, all of which will remain in the regARIMA model. When testing multiple sets of moving holidays in a set of series, the user can use the Chi2 test, CV to set the significance level of this test.

The X-13ARIMA-SEATS has a long history of automated testing based on Akaike's information criterion corrected (AICC). The formula for AICC is as follows:

$$AICC = -2In(L) + 2k + \frac{2k(k+1)}{n-k-1}$$

where L is the maximum likelihood estimate of the model; k is the number of free parameters in the model (including the intercept term); n is the sample size (number of observations); $2k$ is the penalty term of AIC. $\frac{2k(k+1)}{n-k-1}$ is a small-sample modifier (which tends to 0 when n is large). Choose the model with the lowest AICC. The regression effect was tested by comparing models that 'contained' and 'did not contain' target regression quantities.

If the regression volume and user-defined regression variable effects of the trading day, Easter, month length, quarter length, or leap year have been added to the regression quantities that can be specified by the 'aictest' parameter, the order in which the regression quantities are tested is as follows: (1) trading day regression variables; (2) the length of the month, the length of the quarter or the regression of leap years; (3) Easter regression variables; and (4) user-defined regressors. It should be mentioned that the regressor specified by the 'aictest' parameter must not conflict with the variables and other regressors specified in the 'aictest' parameter.

6.8 Sliding Span Analysis and Stability Assessment

6.8.1 Fundamentals of Sliding Span

The 'slidingspans' spec is to check the stability of the seasonal adjustments. A series is considered stable if adding or removing endpoints has a negligible impact on seasonal adjustments. The stability of the seasonal adjustment is a good indicator of its quality. The 'slidingspan' spec is 2, 3, or 4 overlapping spans seasonally adjusted for the NSA series, respectively. When two or more sliding spans overlap, compare different seasonal adjustments. The X-13ARIMA-SEATS are available in 6-, 7-, 8-, or 11-year spans. If the length of the data series is not an integer year, the span length is partially longer. For example, if a series from January 1990 to March 2001 runs and uses a 3 × 3 seasonal filter, the span (7 years + 3 months) = 87 months, ie, the first span will run between January 1991 and March 1998. The second span would run between January 1992 and March 1999. The third span will run between January 1993 and March 2000. The final span will run between January 1994 and March 2001 (the most recent data point).

The X-13ARIMA-SEATS adjust each span separately according to the season. Run the program with overlapping spans, comparing the seasonality factor and monthly and annual percentage changes for each data point in each span. As shown in the following diagram of 4 slide spans with a length of seven years:

	1st span	2nd span	3rd span	4th span
Q1 1991	100			
Q2 1991	106			
Q3 1991	103			
Q4 1991	110			
Q1 1992	105	105		
Q2 1992	110	no		
Q3 1992	108	108		
Q4 1992	115	115		
Q1 1993	110	110	110	
Q2 1993	114	114	114	
Q3 1993	111	111	111	
Q4 1993	118	118	118	
Q1 1994	107	107	107	107
Q2 1994	116	116	116	116
Q3 1994	114	114	114	114
Q4 1994	121	121	121	121
Q1 1995	115	115	115	115
Q2 1995	124	124	124	124
Q3 1995	120	120	120	120
Q4 1995	125	125	125	125
Q1 1996	120	120	120	120
Q2 1996	128	128	128	128
Q3 1996	123	123	123	123
Q4 1996	129	129	129	129
Q1 1997	124	124	124	124
Q2 1997	133	133	133	133
Q3 1997	127	127	127	127
Q4 1997	135	135	135	135
Q1 1998		128	128	128
Q21998		139	139	139
Q3 1998		132	132	132
Q4 1998		138	138	138
Q1 1999			131	131
Q2 1999			143	143
Q3 1999			138	138
Q4 1999			144	144
Q1 2000				134
Q2 2000				149
Q3 2000				138
Q4 2000				150

As you can see, the value of 124 for Q1 1997 is the same across all four spans. However, since each span is seasonally adjusted individually, seasonally adjusted values for the first quarter of 1997 may vary. This difference is the basis for the sliding span analysis. Usually, 'slidingspans' are contained at the bottom of the spec file:

```
Series {title="Example of sliding spans spec"
                start=1996.1
                period=4
                file="mydata.txt"}
  arima{model=(0,1,1)(0,1,1)}
  x11{mode=mult}
  slidingspans{}
```

6.8.2 Output and Interpretation of Sliding Spans

TABLE 6.3

Sliding Span Output

Table	Description	use
S0	Summary of options	
S1	The period refers to seasonal factors	
S2	Percentage of unstable cycles	The primary output of the sliding span
S3	Collapse during periods of instability	Particularly volatile months and years
S7	Full slip span analysis	An idea of the distribution of unstable months is given, and the SA span is also included for analysis

Tables S2, S3, and S7 are generated multiple times, labelled A through E:

- Table A shows seasonal factors.
- Table B covers the trading day and is only printed when 'fixmdl' is set to no or clear.
- Table C is the final seasonally adjusted series, which is printed only if the model includes one or more regression variables (eg, trading day or Easter effect) and is not fixed.
- Table D is a monthly or quarterly change.
- Table E shows the year-on-year change.

In Table S2, the seasonally adjusted series (S) is used as the basis for calculating month-on-month (MM) or quarter-on-quarter (QQ) changes. These short-term growth rates (MM/QQ), *alongside* year-on-year (YoY) changes, are then analysed to generate guidance percentages, accounting for seasonal patterns and trend direction. Adjustment quality is assessed through S and MM diagnostic statistics.[3]

Interpretation of the S-Statistic for Assessing Seasonal Adjustment Quality:

The S-statistic serves as a critical diagnostic metric in X-13ARIMA-SEATS to evaluate seasonal stability:

- S < 15%: Indicates a *stable seasonal pattern.*
- Adjustment is statistically reliable.
- Minimal revisions expected with new data

- 15% ≤ S ≤ 25%: Warrants *caution and scrutiny.*
- Requires supplementary diagnostics (eg, sliding spans analysis)
- May represent borderline acceptable adjustment
- Example: Acceptable for preliminary estimates with clear caveats
- S > 25%: Signifies *unreliable seasonal adjustment.*
- Strongly suggests fundamental issues in decomposition
- Should not be published without methodological review
- Typically requires:
- Model re-specification
- Outlier re-examination
- Series transformation
- S-Statistic (Seasonal Strength):
- If S < 15%, it indicates stable seasonal adjustment
- If S > 25%, it rarely yields a reliable seasonal adjustment
- If 15% < S ≤ 25%, it requires a critical scrutiny; may represent borderline acceptable adjustment
- MM/QQ Statistics (Moving Seasonality Ratio):
- If MM > 40% (multiplicative model) or QQ > 40% (additive model), it should not be seasonally adjusted
- YY-Statistic (Year-on-Year Variation):
- If YY < 2%, it is common in many series
- If YY > 10%, it considered excessively high (though no strict upper threshold)
- Note: High YY values typically correlate with poor S/MM results, diminishing YY's diagnostic significance

Table S3 output provides critical diagnostics for evaluating the quality of the adjustment when tracking unstable regions. With the 'print = brief' option, it does not print automatically, so you should use 'default' or 'print = all' when analysing a single series in more detail. The 'slidingspans' diagnosis is one of the useful statistics in the X-13ARIMA-SEATS output and can be viewed in the 'first wave'. Sliding span diagnostics measure the stability of seasonal adjustments and are related to quality, but they are not necessarily the same thing. Series that fail to pass 'slidingspans' may have periods of seasonal interruption or instability. Sliding spans are especially useful in these cases, as M and Q statistics can misleadingly indicate that seasonal adjustments are good. Sliding spans can also be used to compare different seasonal adjustment methods and indicate which option results in a more stable adjustment Series.

6.9 Revision Policy and Update Management

When new data points are available from the original data, this additional information may lead to revisions to published seasonally adjusted and trend estimates. Methods widely used to update seasonal adjustments are as follows: (1) annual updates, which involve an annual review and evaluation of each immediate seasonal adjustment Series by existing prior adjustments, where possible, with the need to correct certain prior adjustments; (2) X-13ARIMA-SEATS diagnostics (eg, M and Q statistics, slip span, and revision history) are used to analyse and evaluate the appropriateness of seasonal adjustment parameters. The current refresh runs the latest time series through X-13ARIMA-SEATS to produce updated seasonal adjustments. When performing these seasonal adjustment runs, the seasoning options used are not the defaults but are determined during the most recent annual update.

6.9.1 Basic Principles of Updates

The general principles for whether seasonally adjusted parameters should be fixed or should be re-estimated on each run are as follows:

- The model that has been identified for the use of the 'ARIMA' spec should be fixed.
- The parameters in the appropriate 'transform' should be fixed.
- The 'Easter' and 'trading days' that were originally used in the 'regression' spec identification should continue to be included in the 'regression' spec as regression variables, rather than using the AIC test for automatic selection.
- Additive Outliers and Date of Horizontal Offset, as defined by automatic outlier detection, should be converted to fixed regression variables in Regression. In this way, the impact of the regression variables will be fixed in the model, while the adjusted parameters will be re-estimated. Another approach is to fix a parameter estimate of known influence at the same time, so that it is not re-evaluated in each period.
- Seasonal breaks should be corrected using a user-defined seasonal regression amount in the 'regression' spec. These regression variables should be converted to permanent priors only if the time series has multiple seasonal breaks.
- For important time series, the trend moving average and seasonal moving average should be fixed in the 'x11' spec.

6.9.2 Impact and Management of Amendments

Each new raw data point that becomes available will affect the estimation of the seasonal and trending components from previous periods. Each seasonally adjusted update may result in a revision of the seasonally adjusted estimates. Major revisions are typically applied after new data is added to the previous period and to the corresponding period one year ago. For example, when new raw data becomes available in December 2023, the seasonally adjusted trend estimates will be revised retrospectively, beginning with November 2023. Larger revisions are likely to occur in November 2023 and December 2022.

However, if the annual totals are limited and the entire year cannot be revised without revising, it is generally not necessary to make revisions. Seasonal adjustments are inherently revision-sensitive, often requiring updates at multiple lags. For instance, if adding new data triggers significant revisions at lags 2 or 3, those specific periods should be reviewed. Typically, the seasonal adjustment must be corrected back to any period in which the original data must be modified. Substantial revisions to source data may necessitate seasonal adjustment updates for neighbouring observations. Always compare pre- and post-revision adjustment charts to verify this pattern. If seasonal adjustment issues are found between annual updates, they should be corrected as soon as possible. This means that seasonally adjusted estimates and trend estimates may be revised. If the seasonally adjusted series is particularly smooth, the annual update may allow modification only of the previous two years, or even only the previous year. Conversely, if seasonal adjustments are driven by irregular components, earlier recalibration may be necessary, potentially up to four years in advance.

When formulating a revision strategy, both user requirements and changes in the original data must be considered. The revision history, diagnosis, and revision triangle approach is important to determine which revision strategy is best, when, and how much data to revise from a seasonal adjustment perspective. Revising a policy can have a significant impact on user confidence and the quality of published data. Revision policies play a crucial role in producing high-quality seasonally adjusted series, making it essential to consider all the factors mentioned earlier. Generally, trend estimates and seasonally adjusted revision strategies depend on the nature of the time series and the customer responsible for publishing the data.

6.9.3 Application of History Spec

The 'history' specification is a diagnostic tool used to assess the impact of new data on the time series, and is one of the stability diagnostics available in the X-13ARIMA-SEATS. The 'history' spec can be used with different sets of parameters, and the choice of parameters depends on the scope of the revision analysis.

```
series{title="Example of slidingspans spec"
             start=1996.1
             period=4
             file="mydata.txt"
      }
arima{model=(0,1,1)(0,1,1)}
x11{mode=mult}
history{start=1999.1
       sadjlags=(1,2)
      }
```

There are two parameters in the 'history' spec: (1) 'start' determines the start date of the series, such as 'start=1999.1', the user-specified start parameter. If the 'start' parameter is not used, the X-13ARIMA-SEATS will select the default start date. Because the user is most interested in the execution of the seasonal adjustment on the current data (and not on the data from several years ago), it is recommended to limit the length of the historical diagnosis to the last few years when using the 'start' parameter; (2) 'sadjlags' is a lag parameter specified by the user in the revision history, which will allow the user to specify up to 5 lags.

The 'history' spec run is only used to generate Table R1, which gives the revision between the first and the final estimates of the seasonally adjusted series, which is to compare the stability of two competing adjustments after the introduction of new data, and is the simplest use of revision history diagnosis, as shown in Table R1:

R 1 Percent revisions of the concurrent seasonal adjustments

From	**1999.1 to 2000.2**	
Observations	6	
1999		
1st	0.89 -2.08	−2.27
2nd	−9.11 -2.88	−1.77
3rd	4.00 3.13	2.61
4th	**1.08** -0.25	−0.52
2000		
1st	−1.00 -1.03 *********	
2nd	−2.55 ********* *********	

The first column in Table R1 is the year. The second column is the revision history of the initial adjustment, which is the percentage difference between the first and final years. The third column shows the revision with a lag of 1, which is the percentage difference between the values for the fourth quarter of 1999 and the first quarter of 2000 (the second year). The fourth column is the revision with a lag of 2, which is the percentage difference between the final year and the third year. In general, for columns representing lagged *n* revisions, the reported value indicates the percentage difference between the final year's data and that of year $(1 + n)$.

Revisions select different lag positions because these reveal information about the magnitude of each lag's revision; for example, the relative size of revisions at each lag indicates how quickly the corrected data converges to its final value. This information can provide evidence for revision policy decisions for the series. Diagnostics also prints out a summary table, as shown below.

R l.S Summary statistics: average absolute percent revisions of the seasonaladjustments

Quarters:			
1st	0.95	1.55	2.27
2nd	5.83	2.88	1.77
3rd	4.00	3.13	2.61
4th	1.08	0.25	0.52
Years	3.77	2.08	1.79
1999			
2000	1.78	1.03	*********
Total:	3.11	1.87	1.79
hinge values:			
Min	0.89	0.25	0.52
25%	1.00	1.03	1.14
Med	1.82	2.08	2.02
75%	4.00	2.88	2.44
Max	9.11	3.13	2.61

Of these statistics, the most useful value is the 'Total' revision. This number represents the first (or second, third, etc., depending on the 'sadjlags' option) seasonal adjustment, as well as adjustments using all available data, and can be used to compare the quality of competing adjustments.

Other values are also useful. Users are typically more concerned with recent data, so the size of revisions resulting from incorporating the most recent observations may carry greater importance in choosing an adjustment method than revisions in earlier periods. If you want to compare the results of the revision, you can simply change the ARIMA model without changing the other options and repeat the series to compare the results:

TABLE 6.4

Seasonally Adjusted Models and Revised Average Absolute Percentages

Model	Revision Average Absolute Percentage		
	1st estimate	Lag 1	Lag 2
Log (0 1 1) (0 1 1)	3.11	1.87	1.79
Log (0 1 2) (0 1 1)	3.11	1.92	1.90
Log (2 1 0) (0 1 1)	3.17	1.91	1.89
Log (0 2 2) (0 1 1)	3.11	1.86	1.79
Log (2 1 2) (0 1 1)	3.20	1.83	1.81

The table shows that changing the ARIMA model used does not have a large impact on the size of the revision. The first and fourth models appear to perform best, as the average absolute revision percentage shows a minimal change. In practice, the (0 1 1) (0 1 1) model might be chosen because it uses the fewest parameters.

6.9.3.1 Other Options

Tables R2, R4, and R5 are all similar to Table R1, but reference slightly different target variables. The following table summarises the target variables that are available and the commands required to generate each.

The variables to be found must be included in the estimated parameter, such as lags 1, 2, 3, and 12, which are of interest in examining the level and monthly changes in the final trend component. Then the history part of the spec file would be as follows:

TABLE 6.5

Other Options

Table	Directives	Details
R1	Sadj	Last Season Adjustment Sequence (default)
R2	Sadjchng	Period changes in the final season adjustment sequence
R3	Trend	The ultimate Henderson trend component
R4	Trendchng	Inter-period variation of the Henderson trend components

```
history{start=1998.1
        estimates=(trend trendchng)
        trendlags=(1,2,3,12)
        }
```

A complete explanation of the variables, such as the more complex growth rate, ie, the change in the last three months versus the previous three months (April–May–June, vs. January–February–March), requires a different approach.[4] Usually, the 'history' spec is contained near the end of the spec file:

```
x11{mode=mult
    seasonalma=s3x3
    trendma=5
    }
history{start=1999.1
        sadjlags=(1,2)
     }
slidingspans{ }
```

6.10 Forecasting Methods and Implementation

The X-13ARIMA-SEATS program can be used for forecasting seasonally adjusted data both forward and backward, and it can also generate forecasts for purposes beyond seasonal adjustment.

6.10.1 Forecasts in Seasonal Adjustments

The regARIMA portion of the X-13ARIMA-SEATS program is used for forecasting (forward and backward) in seasonal adjustments, modelling time series to estimate calendar effects, outliers, or breaks, and selecting a fitting model before the seasonal adjustment itself. If no forecast is specified, the default range for the forecast is one year. The 'standard' approach for the X-13ARIMA-SEATS is to use the 'automdl' selected model and a one-year forecast.

6.10.2 Realisation of Independent Economic Forecasting

Forecasting techniques are widely used in national accounts, and X-3ARIMA-SEATS can also extend the series to projections beyond the most recent period for which data are available, such as forecasts for economic data in the coming years. The method of pure economic forecasting is no different from the forecasting method in seasonal adjustment, with the exception that some factors must be accounted for in determining the forecast range and series length. For monthly data, the default Spec files available are as follows:

```
# FORECASTING default spec file for MONTHLY series using ADTOMDL
# 2 things may change - the file and the maxlead
series{
      title="DefaultTitle"
      file="Filename.dat"
```

```
      format="datevalue"
      period=12
      decimals=2
      }
transform{function=auto}
regression{aictest=(easter, td)
         }
check{print=all}
automdl{maxdiff=(2 1)
      maxorder=(2 2)
      }
outlier{types=(ao Is)}
forecast{maxlead=12
      save=fct
      }
estimate{
      maxiter=5000
       }
```

For quarterly data, the default Spec files available are slightly different:

```
#      FORECASTING default spec file for QUARTERLY series using AUTOMDL
#      2 things may change - the file and the
maxlead series{
              title="DefaultTitle"
              file="Filename.dat"
              format="datevalue"
              period=4
              decimals=2
              }
transform{function=auto}
regression{aictest=(easter)
         }
check{print=all}
automdl{maxdiff=(1 1)
      maxorder=(2 2)
      }
outlier{types=(ao Is)}
forecast{maxlead=4
      save=fct
      }
estimate{
      maxiter=5000
       }
```

It should be noted that all predictive models used here employ automatic model selection. However, if the available data is less than three years, other forecasting methods (eg, simple extrapolation) should be used. The X-3ARIMA-SEATS require at least three years of data to operate. If there is 3–5 years of data, the X-3ARIMA-SEATS can be modelled using ARIMA or provide trading day and Easter adjustments, but they are often of low quality and may be significantly revised in the future as new data is added.[5] If more than five years of data are available, X-3ARIMA-SEATS can provide a good estimate for predictions using automatic model selection.

6.11 Special Handling of Problem Series

6.11.1 Seasonally Adjusted Negative Data

In the original series, some negative values are acceptable in the seasonally adjusted series if they are reasonable and easy to interpret. When there are values in the original series that are 0 or very close to 0, using the additive decomposition model, those values become negative in the seasonally adjusted series. In some cases, these negative values are difficult to interpret or even unreasonable, for example, in a series where the number of people is negative. The most logical solution to avoid the risk of negative values in the seasonally adjusted series is to employ multiplicative decomposition rather than additive decomposition. If only one 0 exists, the series scalar can be moved before the seasonal adjustment is performed; If there are several zeros in the series, the best solution needs to be determined on a case-by-case basis.

6.11.2 Handling of Unusual Feature Series

The problem sequence refers to the difficulty of achieving satisfactory seasonal adjustment effects, and most of them have very unusual characteristics, such as (1) large irregular components, and it is difficult to distinguish between normal seasonal effects and irregular effects; (2) unstable seasonality, (3) seasonal heteroskedasticity, mostly caused by some irregular events, such as the duration and intensity of frost and snow periods, which may affect the production of some industries; (4) specific seasonal patterns occur only at a few times of the year; and (5) the time-varying calendar effect, often generated by strong sociological or legal evolution. These features produce high nonlinearities or a large number of outliers.

Although the X11 algorithm allows for different seasonal filters to be applied across periods to account for non-uniform seasonality, the automatic detection of extreme values can partially address the issue of numerous outliers. Shortening the series is sometimes a way to address the time evolution feature. However, in some cases, conventional algorithms may yield inadequate seasonally adjusted results, necessitating individualised adjustments such as using state space models with seasonally specific components, non-Gaussian generalisations with time-varying regression coefficients, ARIMA models, or moving window-based solutions. The quality of seasonally adjusted data depends on the suitability of the solution employed. Perform a seasonal adjustment on the problematic sequence. A case-by-case seasonal adjustment approach is preferable to an auto-adjustment approach.

6.12 Summary and Conclusions

This chapter explores advanced topics in X-13ARIMA-SEATS seasonal adjustment, including time series length selection, annual consistency, direct and indirect adjustments, handling of seasonal break, user-defined regression variables, adjustments for Chinese lunar holidays, sliding span analysis, revision strategies, and forecasting methods. These contents provide technical guidance and practical methods for seasonal adjustment in complex data scenarios, help users optimise adjustment results and ensure data quality, and emphasise that scientific adjustment needs to combine theoretical knowledge and practical experience.

1. The quality of seasonal adjustment is generally poor for short series (less than five years) and should be interpreted with caution. A minimum three years of data is recommended, though a series length of at least five years is considered more reliable. A long series may contain structural changes (eg, seasonal break) that need to be corrected using regression variables. Special processing can be used to predict extended data with ARIMA, without modelling adjustments or linking related sequences.
2. Although seasonal adjustment typically preserves annual totals, the adjusted annual sum may not always match the original data exactly. Usually, the solution 'force=totals' parameter enforces consistency, and the optimality needs to be adjusted at a trade-off. National accounts, etc., data need to be strictly consistent. The consistency of annual totals before and after seasonal adjustment requires the application of annual constraints, depending on the following: (1) how the data is used, (2) how the data is collected and compiled and whether it makes conceptual sense for the time series, (3) maintaining consistency considering the practice of the past few years, and (4) looking at the E4 table output by X-13ARIMA-SEATS to see if the difference between the constrained and unconstrained data (or the ratio of multiplication) is significant.
3. Direct and indirect seasonal adjustments, with direct adjustments being independent of each series and potentially losing aggregate consistency. Indirect adjustment adjusts the aggregation sequence via the component sequence, ensuring consistency in the sum, but may leave seasonality unchanged. Selection criteria include data usage, stability of seasonal patterns, and diagnostics such as the sliding span analysis.
4. Identification methods of seasonal break: graphical analysis (eg, abrupt pattern changes), SI ratio abrupt changes. Adjustment steps are as follows: (1) Use the 'regression' spec to detect the breakpoint (eg, 'variables=(seasonal/1999.jan//)'); (2) Fitting permanent adjustments via user-defined regression variables (eg, '.rmx' files). It should be noted that the data should be at least one year before and after the interruption to avoid excessive revision. This chapter provides detailed examples of testing, adjustments, and permanent prior adjustments within the 'regression' specification. The key to effectively identifying and managing seasonal disruptions lies in ensuring the accuracy and consistency of the adjustment results.
5. The 'regression' spec is defined by non-built-in effects (eg, 4–4–5 week pattern of UK RSI, moving holidays). Methods: (1) The offset was calculated and standardised. (2) Construct a regression matrix ('.rmx' file) and import the 'regression' spec. (3) T-value, Chi-square, and AICC were used to screen significant variables. Adjusting for the Chinese New Year, along with the Dragon Boat Festival and the Mid-Autumn Festival, is particularly challenging, as these lunar holidays shift within the Gregorian calendar and require treatment similar to that used for Easter.
6. Establishing the correct regression matrix (.rmx) file is key to achieving the Chinese New Year adjustment. You can use the Genhol tool to generate a holiday

returner (eg, 'genhol lunar [-15 15]'); The Win-Genhol graphical interface simplifies operation. User-defined regression variables were merged into the regARIMA model, and significance was verified by 'chi2test'.

7. Slidingspans ('slidingspans' spec) is used to evaluate the stability of the adjustment (eg, differences in results between span overlapping intervals). Diagnostic criteria include stability of seasonal factors (S-score < 15% preferred) and monthly/annual change thresholds (eg, MM>40% are unreliable adjustments). Stability is a good indicator of the quality of seasonal adjustments. The X-13ARIMA-SEATS adjusts each span separately based on the season, running the program with overlapping spans, comparing the seasonal factors and monthly and annual percentage changes for each data point in each span.
8. Revisions and updates, using fixed parameters: ARIMA models, seasonal moving averages, etc., should be fixed to ensure consistency. Revision strategies include using new data that can alter recent and lagged adjustments. Analysing the magnitude of revisions in conjunction with the 'history' spec (eg, 'R1' table).
9. Forecast: The default one-year forecast in seasonal adjustment relies on the regARIMA model; independent forecasts can be extended to multi-year economic forecasts, requiring at least three years of data (five years is preferred).
10. The special treatment of the problem sequence can avoid negative values in the seasonal adjustment sequence by using multiplicative decomposition instead of additive decomposition. Unusual feature sequences can be individualised, eg, using state space models, non-Gaussian generalisations with time-varying regression coefficients, ARIMA models, or moving window-based solutions.

Appendix 6.1: Calculation of RSI Regression Matrix

X-13ARIMA-SEATS includes built-in regressors for performing regression tests on deterministic factors such as calendar effects, trading day variables, additive outliers, level shifts, temporary changes, slopes, trends, and seasonal disruptions. However, the X-13ARIMA-SEATS needs to provide the user with some user-defined variables, such as the offset regressor (regression matrix) required to adjust the UK retail sales index. The regression matrix is the key to adjusting the user-defined variables of the X-13ARIMA-SEATS. This appendix will illustrate the calculation and creation of a regression matrix using the UK Retail Sales Index as an example.

Background to the RSI Data Survey

The UK RSI data is collected based on retail sales information over a rolling 13-week period, divided into the first 4 weeks, the second 4 weeks, and the last 5 weeks (ie, 445 cycles). This means that the 'months' vary from year to year, and the data for each 'month' actually spans different calendar dates, resulting in different dates overlaid by the December data for each year. For example,

- December 2008: 30 November to 3 January
- December 2009: 29 November to 2 January

This change in date coverage can lead to a 'shift' in December data, ie, the December data for different years are not perfectly aligned in time. To adjust for this shift, you need to define a regression variable (eg, December), an offset, calculate the offset of the data for this regression variable, and build a regression matrix. The columns of the regression matrix are the corresponding months (regression variables), and the rows of the regression matrix are the offsets of the corresponding time points (a certain month of a certain year). The value of this shift (regression) scale from –3 to 3 indicates the degree to which the December data is offset from the standard '0' value. X-13ARIMA-SEATS uses these regression variables (numerical values) in the regression matrix to test whether their impact on the retail sales index is significant. For example December 2008 = November 30 – January 3, December 2009 = November 29 to January 2, To adjust for the 'offset' of December, it is necessary to calculate the offset of the month 'between years', define a regression variable with a value from -3 to 3 based on the change in the standard '0' value for December (ie, November 28 to January 1) – and then centre, This shift of the monthly regression variables was checked by fixing the 'regression' spec of X-13ARIMA-SEATS, adjusting back to the equivalent time interval.

Calculation of Regression Matrices

The calculation of the regression matrix involves three steps: (1) offset calculation, (2) centring, and (3) standardisation. The specific steps are as follows:

Step 1: Calculate the offset. the core of the offset calculation is to compare the difference between the actual date range and the standard date range. (1) Determine the standard date range: The standard date range is fixed, for example, for January, the standard date range might be January 1 to January 31; for December, the standard date range might be November 28 to January 1. (2) Determine the actual date range: The actual date range is determined based on the data collection method for the 445 periods. For example, the actual date range for January 2008 is likely to be from 28 December to 31 January; the actual date range for January 2009 is likely to be from 27 December to 30 January. (3) Calculate the offset: The offset is the difference between the actual date range and the standard date range, including the calculation of (a) the offset of the start date, that is, the difference in days between the actual start date and the standard start date; (b) the offset of the end date, which is the difference in days between the actual end date and the standard end date, and (c) the total offset, which is usually taken as the total offset.

The offset is calculated as follows: Offset = Actual Start Date – Standard Start Date. For example, if the standard January start date is January 1, the actual start date of January 2008 is December 28, offset = (December 28 – January 1) = -4 days (ie, the actual start date is 4 days earlier than the standard start date). If the data is in weeks or a finer unit of time (such as hours of the day), the calculation offset can be decimal (such as +1.125). For example, if the actual start date is 1 day and 3 hours later than the standard start date, the offset can be expressed as +1.125 days (3 hours = 0.125 days).

Step 2: Centring Processing: The purpose of centring processing is to set the regression mean to 0, thereby eliminating bias. It includes (1) calculating the mean of the offsets of all time points, such as $\mu = \frac{\sum_{i=1}^{n} \text{offset}}{n}$, where μ is the mean of the offsets, and n is the total number of time points; (2) centring treatment, subtracting the mean value from each offset to obtain the offset after centring, that is, the offset $= (\text{offset} - \mu)$.

Step 3: Normalisation (optional): Normalisation is typically performed by subtracting the mean and dividing by the standard deviation. If further normalisation is required, the centred offset can be divided by its standard deviation. For example, let's say we calculate the offset for January for the following years:

APP TABLE 6.1

The Offset Calculation

Year	Actual Start Date	Standard Start Date	Offset (days)
2008	12/28	1/1	–4
2009	12/27	1/1	–5
2010	12/29	1/1	–3
2011	12/30	1/1	–2
2012	12/31	1/1	–1

Then, the mean of the offsets is calculated.

$$\mu = \frac{(-4)+(-5)+(-3)+(-2)+(-1)}{5} = \frac{-15}{5} = -3$$

Then, centring: subtract the mean (-3) from each offset.

Finally, the normalisation process (optional) calculates the standard deviation of the offset after centring:

$$\sigma = \sqrt{\frac{(-1)^2+(-2)^2+0^2+1^2+2^2}{5}} = \sqrt{\frac{1+4+0+1+4}{5}} = \sqrt{2} \approx 1.414$$

Normalised offsets:

$$\text{Normalised offset} = \frac{\text{The offset after centering}}{1.414}$$

The normalised offset is the difference between the actual date range and the standard date range in the regression matrix. In summary, the offset is calculated by comparing the difference between the actual date range (start date and end date) and the standard date range; Then, centring is processed to calculate the mean of the offset. Subtract the mean from each offset so that the mean of the regression is 0. Finally, the standard deviation of the centring offset is calculated (optional), and the centring offset is divided by its standard deviation so that the regression variable has a uniform scale. Use this method to develop a regression matrix (.rmx) as a user-defined regression variable in x-13ARIMA-SEATS using a regression file to pre-adjust the RSI. The same method applies to UK moving holidays, Easter or Chinese New Year, Dragon Boat Festival, Mid-Autumn Festival, and other moving holiday adjustments.

APP TABLE 6.2

Offset After Centring

Year	Offset (number of days)	Offset after centring)
2008	−4	-4-(-3)=-1
2009	−5	-5-(-3)=-2
2010	−3	-3-(-3)=0
2011	−2	-2-(-3)=1
2012	−1	-1-(-3)=2

Appendix 6.2: Use the Win-Genhol Tool to Generate a Regression Matrix for Chinese New Year Adjustments

Overview

The key to the Chinese New Year adjustment (including the Dragon Boat Festival and the Mid-Autumn Festival) using the X-13ARIMA-SEATS is to create a regression matrix for adjusting the Chinese New Year. Genhol is the X-13ARIMA-SEATS seasonally adjusted software developed by the U.S. Census Bureau, specifically designed to generate regressors (regression matrices) for different stock holidays (or stock Easters) based on the Lunar New Year. Genhol is a stand-alone utility and a vital tool for handling Chinese New Year adjustments and other moving holiday effects. The Genhol tool offers a flexible approach to handling moving holidays, effectively capturing their impact on economic time series and enhancing the quality of seasonal adjustment through appropriate weight allocation and centralised processing. In practice, it is necessary to determine the optimal parameter settings in combination with specific data and domain knowledge.

Win-Genhol

Win-Genhol is the Windows interface for the Genhol utility (U.S. Census Bureau, 2015). Genhol created moving holiday regressors that could be incorporated into the regARIMA model of X-13ARIMA-SEATS (U.S. Census Bureau, 2020). Genhol can create regressors for intermediate effects before, during, or after the holidays.[6] The current Win-Genhol is version 1.0, Build 3. The program will create the input files needed to create a moving holiday returner using Genhol, run Genhol, and display Genhol input and output files. To install Win-Genhol, (1) download the WinGenhol_V1.0_B3.zip to a local drive (eg, C:\WinGenhol_V1.0_B3.zip); (2) extract the WinGenhol_V1.0_B3.zip to the same directory (ie, Extract to here); (3) open the wingenhol folder, which contains all the files required by Win-Genhol; (4) double-click.

Win-Genhol Main Screen

Use the Win-Genhol home screen to open and run the Genhol input file as well as the Genhol output file. The Genhol input file has a '.inp' extension and is syntactically very similar to the X-13ARIMA SEATS spec file. An example of Genhol input file is as follows:

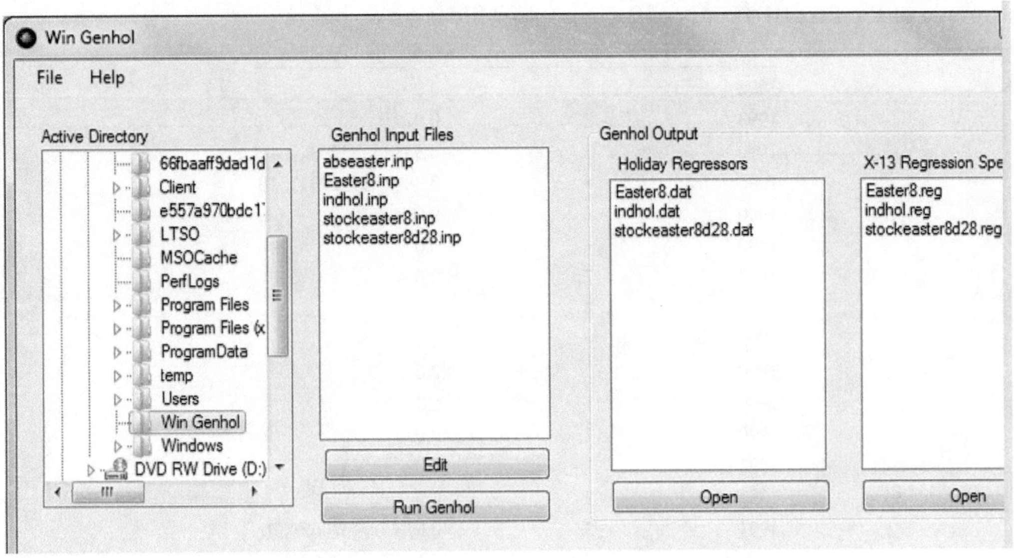

APP FIGURE 6.1
Win-Genhol.

```
global{outfile = "newholiday.dat"
       outspec = "newholiday.reg"
       period = 12
       type = ratio
       first = 1990
       }
holiday1{name = aNewHoliday
         infile = "holidayDates.txt"
         center=calendar
         begbefore = -8
         endbefore = -1
         }
```

(1) Using the folder browser on the left, select the folder where the Genhol input files are stored;

(2) Open the Genhol Input File, select it in the Genhol Input Files list, and press the Edit button;

(3) Run Genhol, select an input file in the Genhol Input Files list, and press the Run Genhol button or double-click the file name;

(4) There are two types of output files generated by Genhol: (a) regressor files and (b) regression spec files. Each successful Genhol run creates a regressor file, a regression matrix in date-value format for user-defined holiday regression quantities generated by Genhol. This file has a .dat (.rmx) extension.

A sample of the first two years of the regressor file is given below.

1990	1	0.0
1993	2	0.0
1990	3	–0.38323353293413lB
1990	4	0.3832335329341318
1990	5	0.0
1990	6	0.0
1990	7	0.0
1990	8	0.0
1990	9	0.0
1990	10	0.0
1990	11	0.0
1990	12	0.0
1991	1	0.0
1991	2	0.0
1991	3	0.61676646706586B2
1991	4	–0.6167664670658682
1991	5	0.0
1991	6	0.0
1991	7	0.0
1991	B	0.0

The 'regression' spec file is a text file containing the regression spec of the user-defined regression variables generated using Genhol. This file has a .reg extension and is only created when requested. An example of a 'regression' spec file is as follows:

```
regression{user= (aNewHoliday)
                    file="newholiday.dat"
                    format="datevalue"
                    # start=1990.1
                    usertype=holiday
                   }
```

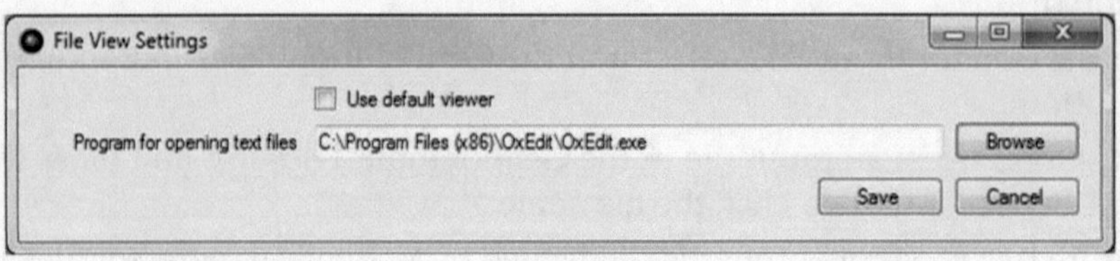

APP FIGURE 6.2
File View Settings.

Create a New Input File

The Genhol input file contains two types of specs: (1) global spec and (2) holiday spec. The global spec specifies global options and is first displayed in the input file. The holiday spec provides information about each holiday returner, including the name of the file containing the holiday date and the length of the holiday window.

To create a new Genhol input file, select File -> Create Input File from the menu, or press Ctrl + I. This will open the Global Spec window.

- Save spec as: full path to the location where you want to save the input file and file name (with extension).
- Save Regressor As: The filename (with extension) of the holiday regression matrix will be stored.
- The number of holidays to create. The default value is 1.
- Period: The period for the holiday regression variable. Select 12 to create regression for the monthly time series, 6 to create regression for the bi-monthly series, and 4 to create regression for the quarterly series. The default value is 12.
- Type: The type of holiday regression volume that will be generated. For count, the output is the number of days found in the holiday window for a given observation.

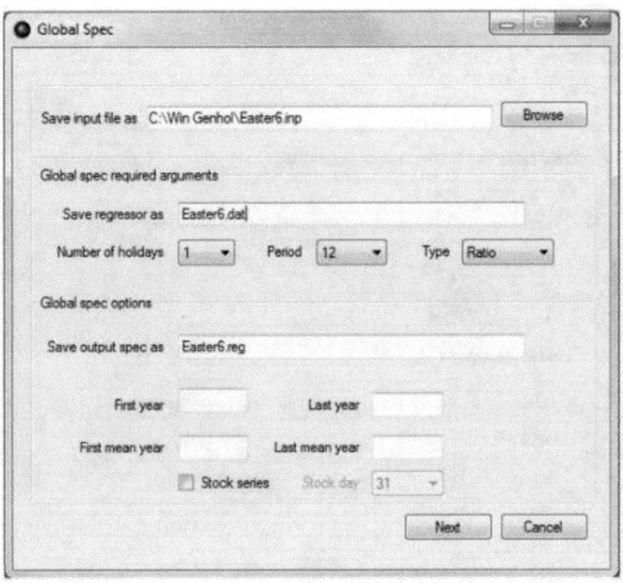

APP FIGURE 6.3
Global Spec (1).

For 'ratio', output the proportion of holiday windows that occur in a given observation. The default value is ratio.

- Save Output Spec As: The filename (with extension) of the regression spec will be stored.
- First year: The four-digit year will be the first year saved in the output file. The default value is the first year that is common to all data files provided by the infile 'parameter of each holiday spec'. This parameter is optional.
- Last year: The four-digit year will be the last year saved in the output file. The default value is the last year that is common to all data files provided by the infile parameter of each holiday spec. This parameter is optional.
- First Mean Year: A four-digit year that represents the first year used to calculate the mean of the holiday regression. This parameter is optional.
- Last Mean Year: A four-digit year that represents the last year used to calculate the mean of the holiday regression. This parameter is optional.
- Inventory Date: The inventory date of the inventory series, that is, the month and day of the count. This option should only be set when creating a holiday regression volume for a stock series.

With the global Spec option selected, press the Next button to move to the Holiday Spec window. The Holiday spec window has a tab for creating each holiday as specified in the global spec.

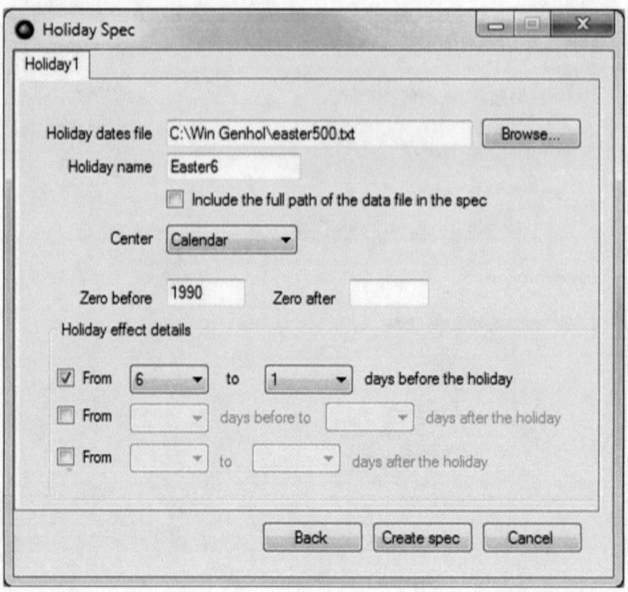

APP FIGURE 6.4
Holidays Spec (1).

'Holiday' spec parameter

Holiday Date File: The full path and file name (with extension) of the text file containing the holiday date. Holiday dates must be stored in month, day, and year format.[7]

- Holiday Name: The name of the holiday. Holiday names can't contain spaces and must be less than 17 characters long. The name you enter must be unique for each holiday.
- Centre: Specifies that the sample mean or seasonal mean is removed from the user-defined regression variables. For centre = mean, subtract the overall mean (sample mean) of the user-defined regression volume from the regression volume. For centre = calendar, subtract the average of each calendar month, two months, or quarter (depending on the selected period) from the user-defined regression amount. When centre = none, the regressor is not modified.
- Before zero: The user-defined regression is equal to the four-digit year before zero.
- After zero: A four-digit year after which the user-defined regression is equal to zero.
- Holiday Effect Details: Use the drop-down box in this section to select when the holiday effect will occur. Use the checkboxes to activate drop-down boxes for the three possible holiday effect windows before, during, or after the holiday. You can select multiple holiday effect windows for each holiday. All holiday windows are inclusive.

Press the Back button to return to the Global Spec window. Press the Create spec button to create a Genhol input file with the selected parameters. This will display the newly created Genhol input file and return it to the Win-Genhol home screen.

Template Examples

Here are two samples: (1) generate a stock resurrection holiday regression matrix; (2) generate a regression matrix for Chinese New Year holidays.

Example 1: Generate a Stock Easter Holiday Regression Matrix. To generate an end-of-month stock Easter Regression Matrix, you need an input file like the following:

```
global{numhol = 1
      outfile = "stockeaster8.dat"
      outspec = "stockeaster8.reg"
      first = 1990
      last = 2015
      period = 12
      stockday = 31
      }
```

```
holiday1{name = stockeaster8
         begbefore = -8
         endbefore = -1
         infile = "easter500.txt"
         center = calendar
         }
```

To create this input file with Win-Genhol, determine the name of the input and output file and the first and last year, but now check the Stock series box and set the Stock day box to 31 as follows. See APP Figure 6.5 (2).

In 'holiday1' spec, let 'holiday name' be 'stockeaster8' as follows. See APP Figure 6.6 (2).

To generate a stock Easter regressor for another sample day, such as the 28th day of the month, select 28 from the Stock Day drop-down box in the Global spec window. See APP Figure 6.7 (2).

Example 2: Generate a regression matrix for the Chinese New Year holiday and 'Idul Fitri' combinations. Create a file with user-defined regression variables for the Lunar New Year and 'Idul Fitri' holidays. In this example, both the Lunar New Year holiday effect and the 'Idul Fitri' holiday effect will consist of three parts, for a total of 6 regression variables. For the Global spec, name the input file 'Indhol' and change the number of holidays to 2. See APP Figure 6.8 (2).

In this way, the Holiday Spec window now has two options: Holiday1 and Holiday2. In the Holiday1 option, set the Lunar New Year regression variable. Here, three windows of Lunar New Year impact will be defined:

- 10 days before the holiday to 1 day before the holiday;
- From the day of the holiday to 4 days after the holiday;
- 5 days after the holiday to 14 days after the holiday; See APP Figure 6.9;

The Centre is set to Calendar.

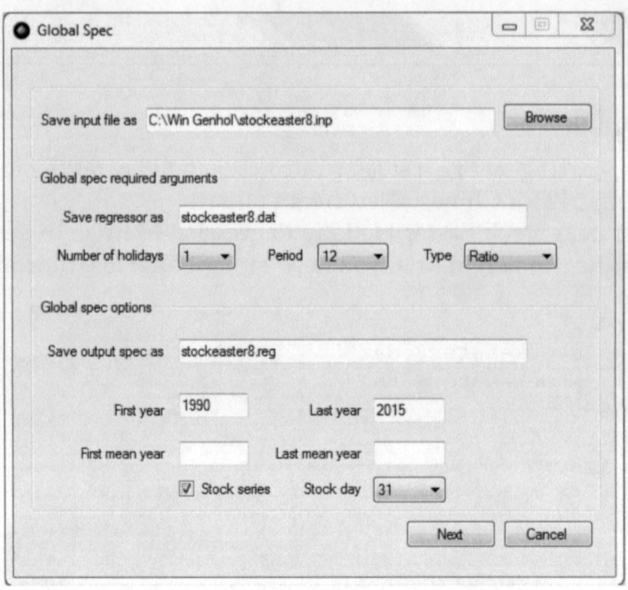

APP FIGURE 6.5
Global Spec (2).

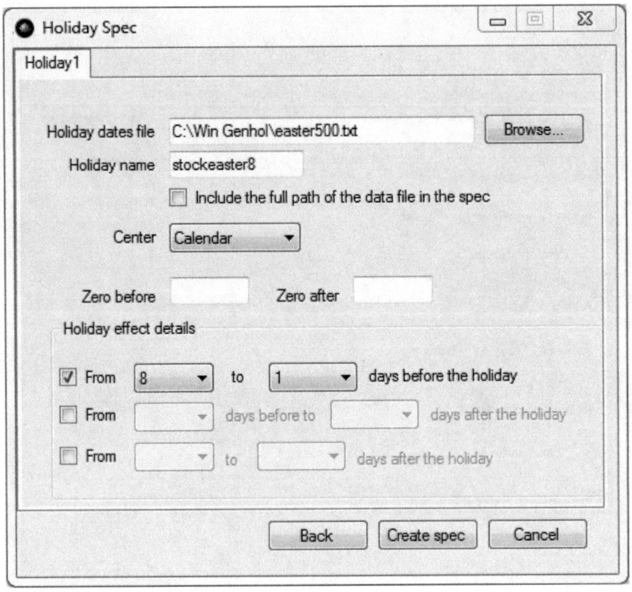

APP FIGURE 6.6
Holidays Spec (2).

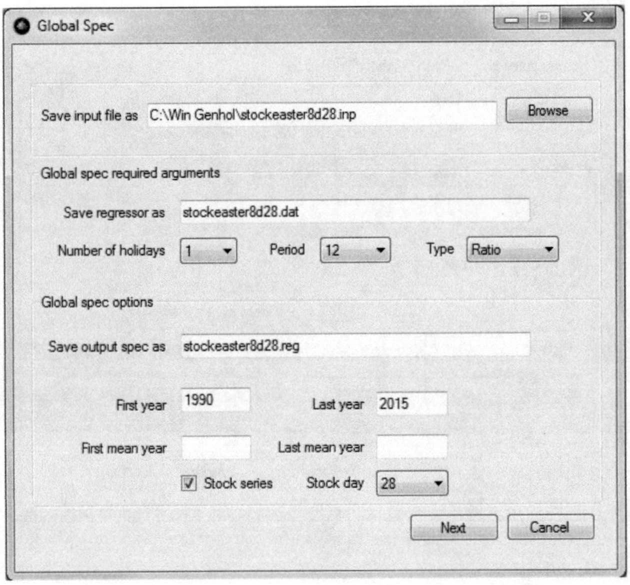

APP FIGURE 6.7
Global Spec (3).

APP FIGURE 6.8
Global Spec (4).

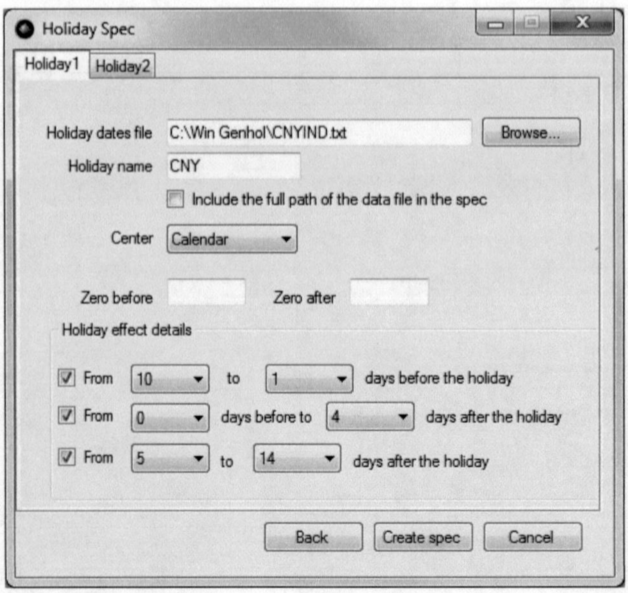

APP FIGURE 6.9
Holidays Spec (3).

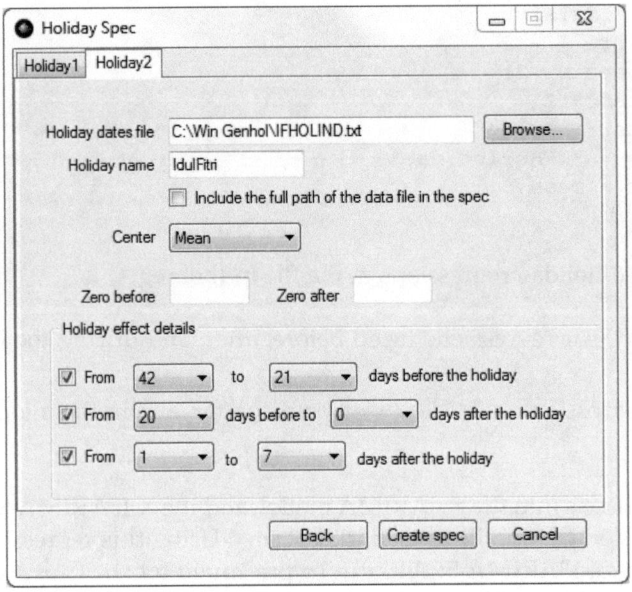

APP FIGURE 6.10
Holidays Spec (4).

Set up the Idul Fitri Regressor in the Holiday2 options:

- 42 days before the holiday to 21 days before the holiday;
- 20 days before the holiday to the day of the holiday;
- 1 day after the holiday to 7 days after the holiday;

The Centre is set to 'Mean' because it is assumed that Idul Fitri is moving in all months. See APP Figure 6.10.

These options will create an input file (indhol.inp) as follows:

```
global{numhol = 2
       outfile = "indhol.dat"
       outspec = "indhol.reg"
       period = 12
       type = ratio
          }
holiday1{name = CNY
          begbefore = -10
          endbefore = -1
          begafter = 5
          endafter = 14
          infile = "cnyInd.dat"
          center = calendar
        }
```

```
holiday2{name = IdulFitri
         begbefore = -42
         endbefore = -21
         begafter = 1
         endafter = 7
         infile = "IFHolInd.dat"
         center = mean
         }
```

This will generate 6-holiday regressions in the file indhol.reg:

- 3 Lunar New Year regressions (used before, after, and during the holiday effect) and
- 3 Idul Fitri regression quantities (used before, after, and in the middle of the holiday effect).

Merge these regressions into the regARIMA model, and the X-13ARIMA-SEATS command will store a regression matrix file named indhol.rmx. Using this regression matrix generated by Genhol, the X-13ARIMA-SEATS can be pre-tuned for the Chinese New Year with the 'regression' spec along with any other user-defined regression variables (Dragon Boat Festival, Mid-Autumn Festival, etc.), such as

```
regression{user=(BeforeCNY BetweenCNY AfterCNY BeforeIdulFitri
BetweenIdulFitri AfterIdulFitri)
           file="indhol.rmx"
           format="datevalue"
           # start=1970.1
           usertype=holiday
         }
```

After running, the following should be checked:

(1) coefficient estimation and significance of the regression variables;

(2) outlier detection results;

(3) model fit statistics;

(4) autocorrelation test of residuals.

The following should be noted:

- The selection of the influence period: the Spring Festival influence period [-15,15] is a common setting, but it should be adjusted according to the actual data.
- Outlier cut-off value: 'critical 3.5' is more conservative than the default value of 4.0 and reduces false positives.

When dealing with multiple holidays at the same time, pay attention to potential collinearity issues. Regression settings can be further customised based on specific data and adjustment needs.

Notes

1 For the calculation of the offset and the establishment of the .rmx regression matrix, see Appendix 6.1.

2 For a detailed description of the use of Win-Genhol, see Appendix 6.2.

3 Findley, D. F., Monsell, B. F., Shulman, H. B., & Pugh, M. G. (1990). "Sliding Spans Diagnostics for Seasonal and Related Adjustments". *Journal of the American Statistical Association*, 85, 345–355.

4 Please refer to the user manual for the X-13ARIMA-SEATS.

5 If only 3–5 years of data are available, another X-3ARIMA-SEATS method can be used: the Holt–Winters method. The basic idea of Holt–Winters involves a process called exponential smoothing. In this process, the predicted values are updated each time new information becomes available at the end of the sequence. In the Holt-Winters method, the horizontal, slope, and seasonal forecasts of the series are updated using exponential smoothing. These predictions are then combined to give a prediction of the next observation in the series. Holt–Winters has additive and multiplicative forms that determine whether the seasonal component is added to the elevation component or multiplied by it, along with the slope component. If simple predictions are required for a large number of sequences, and 3–5 years of data are available, then the most sensible approach is to use 'Holt–Winters' across the board. This will provide high-quality predictions for most sequences without the need for lengthy manual modelling of the 'ARIMA' method.

6 For more information on Genhol utilities, see the Genhol documentation.

7 See Genhol documentation.

7

TRAMO-SEATS: A Model-Based Time Series Decomposition Method

7.1 Introduction

TRAMO-SEATS, abbreviated as the AMB (AutoModel-Based) method, is an ARIMA model-based time series decomposition procedure developed by Víctor Gómez and Agustín Maravall[1] of the Bank of Spain. TRAMO evolved from original methodologies by Hillmer, Bell and Tiao (1983); Tsay (1986);[2] Chen and Liu (1993);[3] and Gómez and Maravall (1994).[4] SEATS originated from foundational work by Box, Hillmer and Tiao (1978);[5] Burman (1980);[6] Hillmer and Tiao (1982);[7] Maravall and Pierce (1987);[8] and Maravall (1995).[9] TRAMO and SEATS can operate independently or collaboratively. A modified version of SEATS was integrated with X12ARIMA to form X-13ARIMA-SEATS (2016), providing an alternative to the X11 method within this seasonal adjustment framework.

This chapter focuses on the TRAMO-SEATS methodology, structured as follows: Following Section 7.1 Introduction, Section 7.2 discusses TRAMO's ARIMA-based pre-adjustment; Section 7.3 covers SEATS' spectral decomposition using ARIMA models, which relies on Wiener–Kolmogorov (WK) filters; Section 7.4 compares the advantages and limitations of TRAMO-SEATS and X-13ARIMA-SEATS; Section 7.5 lists software implementing TRAMO-SEATS; and finally, Section 7.6 provides conclusions.

7.2 TRAMO: Time Series Pre-Adjustment Method

TRAMO (Time Series Regression with ARIMA Noise, Missing Observations, and Outliers) is a preprocessing method used to detect outliers, process missing values, and fit ARIMA models. Similar to ARIMA (regARIMA) in X-13ARIMA-SEATS, TRAMO is a pre-adjustment program for estimating, predicting, and interpolating regression models with missing observations and ARIMA errors, particularly in the presence of various multiple types of outliers. The program is suitable for monthly or low-frequency data, with a maximum of 600 observations and a minimum periodicity (eg, at least 16 quarterly and 36 monthly data points).

The first official release of TRAMO – SEATS around 2001 was called 'Time Series Workbench' (TSW)[10]. In TSW, the TRAMO method provides a robust automatic model recognition (AMI) procedure that, by default, includes ARIMA automatic identification of models[11] and automatic detection and correction of outliers (additional, temporary, or level offset values) and calendar effects (eg, trading days and Easter).[12] SEATS is used to decompose time series: the time series is split using the ARIMA model's Canonical Decomposition

DOI: 10.1201/9781003716655-7

into trend, seasonal, irregular, and optional calendar effects. SEATS decomposes a time series using a Wiener–Kolmogorov (WK) filter. While suitable for linear Gaussian models, the WK filter's theoretical foundation differs from that of state space models.

7.2.1 The Basic Formula of the TRAMO Model

If the time series data is $Y_t = (y_{t_1}, \ldots, y_{t_n})$, where $0 < t_1 < \ldots < t_n$ and fit the regression model,

$$Y_t = X_t'\beta + Z_t \tag{7.1}$$

where $\beta = (\beta_1, \ldots, \beta_m)$ is the vector of the regression coefficient, $X_t' = (X_{1t}, \ldots, X_{mt})$ representing m regression variables, Z_t is a random component (ARIMA process) and follows the general ARIMA (Seasonal ARIMA) process,

$$\varnothing_p(B)\Phi_P(B^s)\nabla^d\nabla_s^D Z_t = \theta_q(B)\Theta_Q(B^s)\varepsilon_t \tag{7.2}$$

The model may contain a constant term (μ) equal to the mean of the difference series, $\nabla^d\nabla_s^D = (1-B)^d(1-B^s)^D$. *In practice, the μ parameter is one of the regression parameters in equation (7.1). The initial estimate of the μ parameter can be entered by the user, set as a default, or calculated by the program. Regression variables (X_i) are typically generated from intervention variables such as additive outliers, level drift, temporary changes, and slope effects, and can be input from relevant economic variables or programmed. Dummy variables are binary indicators (typically 0 or 1) that represent the presence or absence of a specific event or effect at a given time point or over a defined period.*

Additive outliers (AO): $y_t = y_t^* + \omega I(t = t_0)$, at the time $t = t_0$ the value of the dummy variable is 1, and at other time points it is 0:

$$D_t^{AO} = \begin{cases} 1, t = t_0 \\ 0, \text{others} \end{cases}$$

$I(t = t_0)$ make sure that the outliers are only found in $t = t_0$ ie $y_t = y_t^* + \omega$ Other times, $y_t = y_t^*$, ω is the magnitude of the outlier (parameter to be estimated).

Level shift $y_t = y_t^* + \omega I(t \geq t_0)$

$$I = \begin{cases} 1, t \geq t_0 \\ 0, \text{others} \end{cases}$$

$I(t \geq t_0)$ is always 1 at $(t \geq t_0)$, causing the overall level of the time series to rise or fall permanently from t_0.

Temporary change (TC): a temporary change indicates that an event has a temporary effect on the time series, and the effect gradually disappears, $y_t = y_t^* + \omega \cdot \delta^{t-t_0} I(t - t_0)$.

$$TC_t = \begin{cases} \delta^{t-t_0}, t \geq t_1 \\ 0, t < t_0 \end{cases}$$

where δ is a decay parameter $(0 < \delta < 1)$ that represents the decay rate of the influence.

Ramp Effect (ramp) is an event that causes a trend in a time series to change linearly:

$$y_t = y_t^* + \omega \cdot R_t$$

$$R_t = \begin{cases} 0, t < t_0 \\ t - t_0 + 1\, t \geq t_0 \end{cases}$$

$t - t_0 + 1$ means that the effect accumulates linearly with each additional time unit, starting from t_0. At $t < t_0$, $R_t = 0$, no effect; and at $t \geq t_0$, R_t increases by 1 in each period, and the effect accumulates to $w \times (t - t_o + 1)$. The trend rises linearly and $\omega < 0$, and the trend decreases linearly from t_0.

7.2.2 TRAMO Parameter Estimation Method

TRAMO first obtained initial values for the ARMA parameters using the Hannan–Rissanen method and then optimised them using exact maximum likelihood (ML). The Hannan–Rissanen method is a classic two-stage algorithm for parameter initialisation and order selection of ARMA (p, q) models. Proposed by E.J. Hannan and J. Rissanen in 1982, it is one of the core tools.[13] For pre-adjustment programs such as TRAMO and X-12-ARIMA, it efficiently estimates ARMA parameters using regression frameworks and information criteria.

1. Parameter Estimation Process:

Step 1: Higher-order AR fitting

(1) Choose a larger AR order *k*, eg, *k*= [log(*n*)]+{seasonally adjusted}, where *n* is the sample size;

(2) Fitting the AR(*k*) model:

$$y_t = \sum_{i=1}^{k} \alpha_i y_{t-i} + \eta_t \tag{7.3}$$

The decomposition of SEATS relies on the Wiener–Kolmogorov (WK) filter and is suitable for linear Gaussian models.

(3) Calculate the residual estimate:

$$\hat{\eta}_t = y_t - \sum_{i=1}^{k} \hat{\alpha}_i y_{t-i} \tag{7.4}$$

These residuals, $\hat{\eta}_t$ were used as ε_t initial estimates.

Step 2: ARMA Parameter Regression:

(1) Construct a regression equation:

$$y_t = \sum_{i=1}^{p} \varnothing_i y_{t-i} + \sum_{j=1}^{q} \theta_j \hat{\eta}_{t-j} + \varepsilon_t \tag{7.5}$$

Where the AR lagged part, y_{t-i} and the MA lagged residual part, $\hat{\eta}_t$ is the explanatory variable. Original series (y_t) is a response variable.

(2) Estimating the parameters $\hat{\varnothing}_i$ and $\hat{\theta}_j$ by OLS.
(3) Calculate the final residual:

$$\hat{\varepsilon}_t = y_t - \sum_{i=1}^{p} \hat{\varnothing}_i y_{t-i} - \sum_{j=1}^{q} \hat{\theta}_j \hat{\eta}_{t-j} \tag{7.6}$$

Use the initial ARMA model to calculate residuals and detect outliers (eg, AO, LS).
Step 3: Order selection (determination of p, q):

The Hannan–Rissanen method often selects the optimal (p, q) in combination with the information criterion:

- AIC (Akaike Information Criterion): AIC=2k-2ln($\hat{L}$), where $k = p + q$, and $\hat{L}$ is the model likelihood value.
- BIC (Bayesian Information Criterion): BIC=kin(n)-2ln($\hat{L}$) Select the (p, q) combination that minimises AIC/BIC.

In a nutshell, TRAMO initialises the ARIMA parameters by (1) obtaining the initial values of the ARMA parameters using the Hannan–Rissanen method. Then, exact ML optimisation was employed for the preliminary fitting of the ARIMA model. (2) Based on the preliminary fitting of the ARIMA model (the airline model (0,1,1) (0,1,1) is used by default), the standardised residuals are calculated (the residuals are calculated using the initial ARMA model). (3) The t-statistic of 4 types of outliers was calculated for each time point, and the critical value of |t|>C, C=3.5 (the default value was selected as the significant outlier); detect outliers (eg, AO, LS). (4) Re-evaluate the model containing outliers for iterative operations, ie, repeatedly apply Hannan–Rissanen and ML until convergence.

The program's auto-detection is based on the original method of Tsay (1986) and Chen and Liu (1993).[14] Detect outliers one by one and use multiple regression to detect spurious outliers. The process for merging or rejecting outliers is similar to the stepwise regression used to select the 'best' regression equation.

Estimation of regression parameters (including intervention parameters and outliers, as well as missing observations in the initial values of the series) and ARIMA model parameters in the fitted regression model (7.5) can be performed by exact maximum likelihood or in conjunction with unconditional and conditional least squares methods. By default, the exact likelihood method is used. Unconditional and conditional least squares are available as options. Estimation of the regression parameters begins by transforming the regression equation using the Cholesky decomposition of the inverse error covariance matrix.[15] The resulting least squares problem is then solved using the QR algorithm. The Householder Orthogonal Transform is used. This process provides an efficient and numerically stable method for computing the generalised least squares estimators of the regression parameters, thereby avoiding explicit matrix inversion. The regression parameters are initially estimated using OLS, while the ARMA model parameters are obtained through two regression steps. Subsequently, updated estimates of the regression parameters and residuals are computed using the Kalman filter and the QR algorithm. For each observation, t-test values for four types of outliers (additive outliers, innovative outliers, level shifts, and transient changes) will be calculated. If the absolute value of t is greater than the outlier of the pre-selected critical level C, the outlier with the maximum absolute value of t is selected. Otherwise, there are no outliers in the series, and the algorithm stops. If certain outliers are detected, the series is corrected by its effects, the ARMA model parameters are re-estimated, and multiple regression is performed using the Kalman filter and QR algorithm. If there are some outliers whose absolute t-values are below the critical level C, the outliers with the lowest absolute values are removed from the regression and multiple regression is re-estimated. In the next step, using the regression residuals provided by the last multiple regression, the four types of outliers and the t-test for each observation are calculated. If an outlier is detected with an absolute t-value exceeding the critical threshold C, the outlier with the largest absolute t-value is selected. The algorithm then proceeds to re-estimate the ARMA model parameters in the next iteration. Otherwise, the algorithm will stop.

7.2.3 Handling of Missing Values

TRAMO provides a variety of ways or handling missing values. (a) Skipping approach: Missing data is handled by adjusting the dimensions of the State Space model, ie, the missing values are ignored directly. Instead of interpolating or estimating it, the structure of the model is adjusted to accommodate the missing data; (b) Deal with it by introducing dummy variables (additive outlier method): treat missing values as additive outliers (AO) and model these outliers by introducing additional parameters.[16] The X-13ARIMA-SEATS also supports missing value interpolation, but the implementation may differ from TRAMO. These two methods can be seen as the equivalent treatment of missing observations; (c) TRAMO can also use the State Space Model + Kalman Filter for imputation of missing values, but its equation of state depends on the fitted ARIMA structure, the update process is not exactly equivalent to a simple Random Walk which is modelled and imputed based on a more general ARIMA process.

7.2.3.1 TRAMO Models Time Series as State Space Models

The time series regression model is assumed to be

$$Y_t = X_t'\beta + \varepsilon_t \tag{7.7}$$

where X_t' are deterministic regression variables (trends, seasons, regression variables, outliers, etc.); ε_t is the ARIMA noise process.

The Observation Equation for a time series state space model can be written as

$$Y_t = Z\alpha_t + \varepsilon_t \quad \varepsilon_t \sim N(0,H) \tag{7.8}$$

The state equation for a time series state space model can be written as

$$\alpha_t = T\alpha_{t-1} + R\eta_t \quad \eta_t \sim N(0,Q) \tag{7.9}$$

Where α_t is the state vector (eg, trend, seasonal, ARIMA component); T is the state transition matrix (depending on the ARIMA model); Z is the observation matrix, usually Z is [1,0, . . .,0]; R and Q are the noise structure matrices.

7.2.3.2 Koman Filter

The Koman filter method is used for predicting and smoothing. For missing values Y_t TRAMO uses the Kalman filter for optimal imputation through two steps:

a. Prediction Step:

$$\alpha_{t|t-1} = T\alpha_{t-1|t-1}$$
$$P_{t|t-1} = TP_{t-1|t-1}T^T + RQR^T \tag{7.10}$$

b. Update Step:
  - If not missing y_t

$$v_t = y_t - Z\alpha_{t|t-1} \ (\textit{New information})$$
$$K_t = P_{t|t-1}\left(ZP_{t|t-1}Z^T + H\right)^{-1} \ (\textit{Kalman gain})$$
$$\alpha_{t|t} = \alpha_{t|t-1} + K_t v_t$$
$$P_{t|t} = (I - K_t Z)P_{t|t-1} \tag{7.11}$$

  - If missing y_t:

$$\alpha_{t|t} = \alpha_{t|t-1} \ \ P_{t|t} = P_{t|t-1} \tag{7.11a}$$

That is, no updates are made to the state, relying only on the predicted values. It is important to note that the state transition matrix T of the TRAMO depends on the ARIMA model fitted:

- If the ARIMA model is (0,1,1) (ie, IMA (1,1)), then the state equation is similar to Random Walk + MA(1) noise:

$$\alpha_t = \alpha_{t-1} + \eta_t + \theta \ \eta_{t-1} \tag{7.12}$$

- If the ARIMA model is (1,1,0), ie, the AR (1) difference, then the state equation is similar to AR (1) with a trend:

$$\alpha_t = (1 + \varnothing)\alpha_{t-1} - \varnothing\alpha_{t-2} + \eta_t \tag{7.13}$$

- If the ARIMA model is (0,0,0), ie, white noise, the state equation degenerates to a constant mean.

Thus, the update mechanism of TRAMO depends on the ARIMA structure, not just the random walk.

The advantage of TRAMO is that it can handle ARIMA noise (eg, MA, AR structures) and can be combined with regression variables, outliers, seasonal adjustments, and parameter optimisation using MLE.

TABLE 7.1

Differences Between TRAMO's Update Mechanism and Simple Random Walk Imputation

Method	Model assumptions	Imputation	Applicable scenarios
Random walk imputation	$y_t = y_{t-1} + \varepsilon_t$	Use the previous observation directly	Ideal for simple data with no trends and no seasons
TRAMO Imputation	A state space model based on ARIMA	Kalman filter optimal prediction	Data for complex trends, seasons, and outliers

7.3 SEATS: Signal Extraction and Component Decomposition

SEATS is a decomposition method based on the ARIMA model for extracting trend, seasonal, and irregular components. SEATS are typically used in conjunction with TRAMO, which is responsible for data preprocessing, and SEATS is responsible for seasonal adjustment and trend extraction. SEATS can also be used on their own, especially if you don't need to deal with outliers or missing values. The X-13ARIMA-SEATS software introduces the SEATS method, which stands alongside the traditional X-11 method as a user-selectable seasonal adjustment tool.

7.3.1 SEATS Decomposition Method

The SEATS decomposition method is based on the ARIMA (Autoregressive Integrated Moving Average) model. The ARIMA model describes a time series using the following three components. (1) Autoregressive (AR): the relationship between the current value and several lagging values; (2) differential (I): differential the series to make it smooth; and (3) the relationship between the current value of the moving average (MA) and the lag error term.

SEATS decomposes the time series into four main components: (1) the trend component reflects the long-term trend in the time series; (2) the cyclical component: usually related to the economic cycle, which is manifested as fluctuations longer than one year; (3) seasonal component: reflects the seasonal fluctuations that occur regularly in the data; and (4) irregular component: random fluctuations that cannot be attributed to other components.

Before applying SEATS, it is first necessary to identify and estimate the parameters of the ARIMA model on the time series data. The model identification phase determines the appropriate ARIMA model order, while the parameters are estimated by methods such as MLE.

7.3.2 SEATS Decomposition Process

Once the ARIMA model has been identified and fitted, SEATS uses it to extract the individual components of the time series. Specifically, (1) signal: usually a combination of trend and cycle components; (2) noise: including seasonal and irregular components.

Signal extraction is performed through a series of mathematical transformations and filtering techniques, based on the structure of the ARIMA model, to separate the individual components. The core idea of SEATS is to decompose a time series spectrum into components associated with different series elements via frequency-domain analysis, thereby achieving component separation.

7.3.3 Frequency Domain Analysis and Spectral Decomposition

The decomposition process of SEATS relies on frequency domain analysis, assuming that the time series y_t can be expressed as follows:

$$y_t = T_t + C_t + S_t + I_t \tag{7.14}$$

where T_t is the trend component; C_t is the cyclical component; S_t is the seasonal component; I_t is an irregular component. Each component follows an ARIMA model, and the components are orthogonal (ie, independent of each other). SEATS decomposes the spectrum of a time series into fractions associated with different components through frequency domain analysis of the ARIMA model. To put it simply, the decomposition process of SEATS can be divided into three stages.

Phase 1: Building the foundation of the ARIMA model

The starting point for the SEATS decomposition is the established ARIMA model, which fully describes the dynamic structure of the time series:

$$\varnothing(B)\nabla^d y_t = \theta(B)\epsilon_t \tag{7.15}$$

where $\varnothing(B)$ and $\theta(B)$ are autoregressive (AR) and moving average (MA) polynomials, respectively, ∇^d denote the difference operation of the d-order, and ϵ_t is the white noise with a variance of σ^2 This model provides a mathematical basis for subsequent decomposition, in particular, the calculation of the total spectrum by its frequency domain representation.

The innovation of SEATS lies in translating the problem to a frequency domain solution. The time series total spectrum $f(\omega)$ is derived according to the ARIMA model. The total spectrum $f(w)$ is the total PSD of the time series, which represents the energy distribution of the time series at different frequencies ω. Calculation of total and component spectra is as follows:

$$f(\omega) = \frac{\left|\theta\left(e^{-i\omega}\right)\right|^2}{\left|\varnothing\left(e^{-i\omega}\right)\right|^2}.\sigma^2 \tag{7.16}$$

The formula maps the energy distribution of the time series to the frequency domain, where ω denotes the angular frequency, and $e^{-i\omega}$ is a complex exponential function. $\varnothing\left(e^{-i\omega}\right)$ and $\varnothing\left(e^{-i\omega}\right)$ are frequency domain representations of AR and MA polynomials, respectively. σ^2 is the variance of white noise ϵ_t.

Phase 2: Frequency domain decomposition and filter design

Based on the assumption of component orthogonality, the total spectrum $f(\omega)$ can be decomposed into four sub-spectra: the sum of the component spectra of trend, seasonal, cycle, and irregular.

$$f(\omega) = fT(\omega) + fS(\omega) + fC(\omega) + fI(\omega) \tag{7.17}$$

where $fT(\omega)$ is the spectrum of the trend component; $fS(\omega)$ is the spectrum of the seasonal component; $fC(\omega)$ is the spectrum of the periodic component; $fI(\omega)$ is the spectrum of irregular components.

The spectrum of each component can be derived from its ARIMA model. For example, the ARIMA model of the trend component, typically exhibits low-frequency characteristics. The ARIMA model of the seasonal component, $fS(\omega)$, exhibits peaks at seasonal frequencies. The ARIMA model of the irregular component, $fI(\omega)$ exhibits a white noise process. The calculation of the component spectrum is similar to the calculation of the total spectrum (7.16) as follows.

7.3.3.1 Calculation of the Trend Spectrum fT(ω)

$fT(\omega)$ denotes the power spectral density of the trend component T_t. Trend components typically capture low-frequency changes in a time series. The ARIMA model of the trend component typically exhibits a low-pass filter. $fT(\omega)$ can be calculated by the following formula:

$$fT(\omega) = \frac{\left|\theta_T\left(e^{-i\omega}\right)\right|^2}{\left|\varnothing_T\left(e^{-i\omega}\right)\right|^2} \cdot \sigma_T^2 \tag{7.19}$$

Where $\varnothing_T\left(e^{-i\omega}\right)$ and $\theta_T\left(e^{-i\omega}\right)$ are a frequency domain representation of the AR and MA polynomials of the trend components; σ_T^2 is the variance of the trend component.

7.3.3.2 Calculation of the Periodic Spectrum fC(ω)

$fC(\omega)$ represents the power spectral density of the periodic component C_t. Cyclical components typically capture fluctuations in a time series that are associated with the economic cycle. The ARIMA model of the periodic component is typically a bandpass filter. $fC(\omega)$can be calculated by the following formula:

$$fC(\omega) = \frac{\left|\theta_C\left(e^{-i\omega}\right)\right|^2}{\left|\varnothing_C\left(e^{-i\omega}\right)\right|^2} .\sigma_C^2 \tag{7.20}$$

where $\varnothing_C\left(e^{-i\omega}\right)$ and $\theta_C\left(e^{-i\omega}\right)$ are the frequency domain representations of the AR and MA polynomials of the periodic components; σ_C^2 is the variance of the periodic components.

7.3.3.3 Calculation of the Seasonal Spectrum fS(ω)

$fS(\omega)$ denotes the power spectral density of the seasonal component S_t. Seasonal components typically capture cyclical fluctuations in time series. The ARIMA model of the seasonal component is typically a bandpass filter, $fS(\omega)$ can be calculated by the following formula:

$$fS(\omega) = \frac{\left|\theta_S\left(e^{-i\omega}\right)\right|^2}{\left|\varnothing_S\left(e^{-i\omega}\right)\right|^2} \cdot \sigma_S^2 \tag{7.21}$$

where $\varnothing_S\left(e^{-i\omega}\right)$ and $\theta_S\left(e^{-i\omega}\right)$ is the frequency domain representation of the AR and MA polynomials of the seasonal components; σ_S^2 is the variance of the seasonal components.

7.3.3.4 Calculation of Irregular Spectrum fI(ω)

$fI(\omega)$ represents the power spectral density of the irregular component. Irregular components usually capture random fluctuations in a time series. The irregular component is usually a white noise process with a constant spectral density $fI(\omega)$:

$$fI(\omega)=\sigma_I^2 \tag{7.22}$$

where σ_I^2 is the variance of the irregular components.

Each sub-spectrum is determined by the ARIMA model of its components (eg, the trend spectrum $fT(\omega)$ is concentrated in the low-frequency band). Taking advantage of this feature, SEATS constructs a Wiener–Kolmogorov (WK) filter to extract the individual components from the time series. The WK filter is the theoretical optimal solution under the minimum mean square error criterion, which ensures the maximum fidelity of the extracted signal, and the frequency domain of the WK filter is expressed as follows:

$$H_T(\omega)=\frac{fT(\omega)}{f(\omega)},\ H_S(\omega)=\frac{fS(\omega)}{f(\omega)},\ H_C(\omega)=\frac{fC(\omega)}{f(\omega)},\ H_I(\omega)=\frac{fI(\omega)}{f(\omega)} \tag{7.23}$$

where $H_T(\omega)$ are the filters of the trend components; $H_S(\omega)$ is the filter of the seasonal component; $H_C(\omega)$ is the filter of the periodic component; $H_I(\omega)$ is a filter with irregular components.

The computation of these frequency domain filters reveals that at a specific frequency ω, the value of $H_T(\omega)$ (or $H_S(\omega), H_C(\omega)$, $H_I(\omega)$ represents the proportion of energy attributed to the trend (or seasonal, cyclical, irregular) component at that frequency point, that is, the pass rate of the filter at that frequency.

Phase 3: Time domain signal extraction and filter conversion

Frequency domain filters $H_T(\omega)$ need to be converted to the time domain form $H_T(B)$to process the actual data. When a frequency domain filter $[H_T(\omega), H_S(\omega), H_C(\omega), H_I(\omega)]$ is converted to time domain form $H_T(B)$, $H_T(B), H_S(B), H_C(B), H_I(B)]$, the individual components can be expressed as follows:

$$T_t=H_T(B)y_t, S_t=H_S(B)y_t, C_t=H_C(B)y_t, I_t=H_I(B)y_t \tag{7.24}$$

$H_T(\omega)$ to $T_tH_T(B)$ is achieved by the inverse Fourier transform:

(1) Calculate the impulse response function, $H_T(\omega)$. Perform the inverse Fourier transform to obtain the time domain impulse response series, $h_T(k)$

$$h_T(k)=\frac{1}{2\pi}\int_{-\pi}^{\pi}H_T(\omega)e^{i\omega k}d\omega,\ k\in(-\infty,\infty) \tag{7.25}$$

$h_T(k)$ refers to the output series produced when the input is a unit impulse signal ($\delta_t=1$ at $t=0$, and 0 elsewhere)

(2) Construct hysteresis operator polynomials:

The impulse response series is expressed as the lag operator B (defined $B^k y_t = y_{t-k}$) of infinite-order polynomials:

$$H_t(B) = \sum_{k=-\infty}^{\infty} h_T(k) B^k \tag{7.26}$$

In practice, $h_T(k)$ with $|k|$ increases and attenuates, and can be approximated by intercepting finite terms such as $|k| \le 50$.

(3) Apply the time domain filter to extract the trend component through a convolution operation:

$$T_t = H_T(B) y_t = \sum_{k=-M}^{M} h_T(k) y_{t-k} \tag{7.27}$$

where B is the lag operator.

7.3.4 Examples of Frequency Domain Analysis and Spectrum Decomposition

7.3.4.1 Extraction of the Trend Component (T_t)

Step 1: Establish an ARIMA model of the trend component

The ARIMA model that the trend component follows is assumed to be (1,1,0), ie,

$$(1 - \varnothing_{T1} B) \nabla T_t = \epsilon_{Tt} \tag{7.28}$$

Step 2: Calculate the trend spectrum $fT(\omega)$: According to Equation (7.19), ie, $fT(\omega) = \frac{|\theta_T(e^{-i\omega})|^2}{|\varnothing_T(e^{-i\omega})|^2} \cdot \sigma_T^2$, then the trend spectrum is

$$fT(\omega) = \frac{1}{|1 - \varnothing_{T1} e^{-i\omega}|^2} \cdot \sigma_T^2 \tag{7.29}$$

If $\varnothing_{T1} = 0.8$ and $\sigma_T^2 = 0.8$, substituting the model parameters, the trend spectrum is

$$fT(\omega) = \frac{1}{|1 - 0.8 e^{-i\omega}|^2} \cdot 0.8$$

Step 3: Construct a frequency domain filter $H_T(\omega)$:

If the total spectrum of the original series is known to be $f(\omega)$, the frequency domain trend filter is

$$H_T(\omega) = \frac{0.8 / |1 - 0.8 e^{-i\omega}|^2}{f(\omega)}$$

Step 4: Time domain transformation, $H_T(\omega) \rightarrow H_T(B)$, computation of the inverse Fourier transform by numerical integration, ie,

$$h_T(k) = \frac{1}{2\pi}\int_{-\pi}^{\pi} H_T(\omega)e^{i\omega k}d\omega,\ k \in (-\infty, \infty) \tag{7.30}$$

The impulse response is obtained $h_T(k)$. Suppose the result has significant values at $k = 0, \pm 1, \pm 2$.

$$h_T(-1) = 0.15,\ h_T(0) = 0.70,\ h_T(1) = 0.15,$$

then the time domain filter can be approximated as

$$H_T(B) \approx 0.15B^{-1} + 0.70 + 0.15B$$

Step 5: Extract the trend component T_t. The final trend components are derived from the sliding-weighted average:

$$T_t = 0.15y_{t+1} + 0.70y_t + 0.15y_{t-1}$$

7.3.4.2 Extraction of Seasonal Components (S_t)

Step 1: Establish an ARIMA model of the seasonal components:

Assuming that the seasonal component of the seasonal ARIMA model formulas is

$$\nabla_{12}S_t = \left(1 + \theta_{S1}B^{12}\right)\epsilon S_t \tag{7.31}$$

This model describes annual seasonal fluctuations (subscript 12 indicates a 12-month cycle), and its structure suggests that the seasonal adjustment passes through a 12-step differential implementation.

Step 2: Calculate the seasonal spectrum $fS(\omega)$:

According to (7.21), the seasonal spectrum formula is

$$fS(\omega) = \frac{\left|1 + \theta_{S1}e^{-i\omega}\right|^2}{\left|1 - \varnothing_{S1}e^{-i\omega}\right|^2} \cdot \sigma_S^2 \tag{7.32}$$

If, $\theta_{S1} = -0.7\quad \sigma_S^2 = 0.9$, substitute the model parameters

$$fS(\omega) = \frac{\left|1 -_{0.7} e^{-i12\omega}\right|^2}{\left|1 - e^{-i12\omega}\right|^2} \cdot 0.9$$

This indicates that the molecule produces a peak at seasonal frequencies ($\omega = \pi\ 6\ \frac{\pi}{3}$), reflecting monthly/quarterly periodicity.

Step 3: Construct the frequency domain filter $H_S(\omega)$:

Let the total spectrum of the original series $f(\omega)$ be known (calculated by Equation 7.16), and the seasonal filter is

$$H_S(\omega) = \frac{f_S(\omega)}{f(\omega)} = \frac{0.9 \cdot \left|1 - 0.7e^{-i12\omega}\right|^2 / \left|1 - e^{-i12\omega}\right|^2}{f(\omega)}$$

The filter gains close to 1 at seasonal frequencies (allowing seasonal signals to pass through) and attenuates at other frequencies.

Step 4: Time Domain Translation $H_S(\omega) \to H_S(B)$:

(1) Inverse Fourier transform to find impulse response:

$$h_S(k) = \frac{1}{2\pi} \int_{-\pi}^{\pi} H_S(\omega) e^{i\omega k} d\omega \tag{7.33}$$

Since $e^{i\omega k}$. There are significant values at $h_S(k)$, $k = 0 \pm 12, \pm 24$ … …,

(2) Example of numerical results (hypothetical integral calculation):

$$h_S(0) = 0.50,\ h_S(12) = h_S(-12) = -0.25,\ h_S(24) = h_S(-24) = 0.10$$

Other lag order response values are close to 0 ($|k| \notin \{0, 12, 24\}$).

(3) Construct a time domain filter:

$$H_S(B) = \sum_k h_s(k) B^k \approx 0.50 - 0.25B^{12} - 0.25B^{-12} + 0.10B^{24} + 0.10B^{-24}$$

Step 5: Extract the seasonal components (S_t) and apply the time domain filter:

$$S_t = H_s(B) y_t = 0.50y_t - 0.25y_{t-12} - 0.25y_{t+12} + 0.10y_{t-24} + 0.10y_{t+24}$$

This weighted combination strengthens the harmonic component of the series with 12 months and suppresses non-seasonal fluctuations.

7.3.4.3 Extraction of Irregular Components (I_t)

Step 1: Model the irregular components:

According to the irregular components of the white noise process:

$$I_t = \epsilon_{It} \tag{7.34}$$

Suppose $E(\epsilon_{It}^2) = \sigma_I^2 = 0.3$, which is characterised by no autocorrelation (zero covariance for all lags).

Step 2: Calculate the irregular spectrum $fI(\omega)$, the power spectral density of the white noise is constant:

$$fI(\omega) = \sigma_I^2 = 0.3$$

This indicates that the energy is evenly distributed across all frequencies.

Step 3: Construct a Frequency Domain Filter $H_I(\omega)$:

$$H_I(\omega) = \frac{f_I(\omega)}{f(\omega)} = \frac{0.3}{f(\omega)}$$

Characteristics: In the frequency band where the energy of the original series is low ($f(\omega)$ is small), the filter gain is larger, ie, the preferential extraction is not converged residual fluctuations of potential/seasonal explanations.

Step 4: Time Domain Translation $H_I(\omega) \to H_I(B)$:

(1) Inverse Fourier transform:

$$h_I(k) = \frac{1}{2\pi}\int_{-\pi}^{\pi} \frac{0.3}{f(\omega)} e^{i\omega k} d\omega \tag{7.35}$$

The results depend on the specific form of $f(\omega)$ (determined by the original ARIMA model).

(2) Assuming that $f(\omega)$ corresponds to the ARIMA(1,1,1) model, let the numerical integration yield

$$h_I(0) = 0.85,\ h_I(1) = h_I(-1) = -0.20,\ h_I(2) = h_I(-2) = 0.05$$

The higher-order hysteresis response is negligible when $|k| > 2,\ h_I|k| < 0.01$.

(3) Construct a time domain filter:

$$H_I(B) \approx 0.85 - 0.20B - 0.20B^{-1} + 0.05B^2 + 0.05B^{-2}$$

Step 5: Extract the irregular components (I_t)

$$I_t = H_I(B)y_t = 0.85y_t - 0.20y_{t-1} - 0.20y_{t+1} + 0.05y_{t-2} + 0.05y_{t+2}$$

Statistical significance: This operation is equivalent to removing the autocorrelation of the time series, and the output result is pure random noise.

The essence of the SEATS method lies in three core characteristics:

(1) Mathematical guarantee of orthogonal decomposition. The additivity of the component spectra, $f(\omega) = fT(\omega) + fS(\omega) + fC(\omega) + fI(\omega)$ ensures a mathematically rigorous decomposition of the time series, $y_t = T_t + C_t + S_t + I_t$, with each component being mutually non-overlapping.

(2) Optimal filtering properties. The Wiener–Kolmogorov filter provides the theoretically optimal solution under the minimum mean square error criterion, guaranteeing maximum fidelity in signal extraction.

(3) Bridging role of time-frequency conversion: By transforming complex time domain convolution problems into intuitive frequency domain arithmetic operations through the Fourier transform, it overcomes the challenge of separating non-stationary signals that traditional methods struggle to handle.

Consequently, within the TRAMO-SEATS system, the complete technical implementation of the SEATS module – owing to its mathematical rigour and computational efficiency – has established it as the preferred seasonal adjustment method for central banks and economic research institutions worldwide.

7.3.5 Diagnostics and Validation

After the SEATS decomposition is completed, diagnosis and verification are required to ensure the goodness-of-fit and rationality of the decomposition. Diagnosis and validation include goodness-of-fit of the model, validity assessment, and residual analysis of components.

(1) Goodness-of-fit of the model: SEATS mainly evaluates the goodness-of-fit of the model through statistical indicators such as AIC and BIC.

- AIC(Akaike Information Criterion): $\text{AIC} = -2\log \text{L} + 2\text{K}$
- BIC(Bayesian Information Criterion): $\text{BIC} = -2\log \text{L} + 2\text{K}\log \text{n}$.

where L is the likelihood function of the model, k is the number of model parameters, n is the length of the time series. AIC and BIC's smaller values indicate better goodness-of-fit of the model. By comparing the AIC and BIC of different models, the optimal one can be selected as the ARIMA model.

(2) Rationality check of components: The extracted components need to be economically or statistically reasonable. Specifically,

- Trend component: It should reflect long-term changes in the time series, usually with low-frequency characteristics.
- Cyclical component: It should reflect fluctuations associated with the economic cycle, usually longer than one year.
- Seasonal component: It should reflect seasonal fluctuations that occur regularly in the data, usually with spectral peaks at seasonal frequencies.
- Irregular components: It should be random, usually white noise. By examining the spectrum and behaviour of the individual components, the rationality of the decomposition can be verified.

(3) Residual analysis: After the SEATS decomposition is completed, the residuals need to be analysed to ensure that the residuals are random and not structured. Residuals are defined as follows:

$$\epsilon_t = y_t - (T_t + S_t + C + I) \tag{7.36}$$

If the residuals are white noise, the decomposition is reasonable. Residual analysis includes the following:

- Autocorrelation Function (ACF): Checks the autocorrelation of residuals to ensure that there is no significant autocorrelation.
- Ljung–Box test: Tests whether the residuals are white noise in nature.

If the residuals exhibit significant autocorrelation or structure, it may be necessary to readjust the model or check for outliers.

7.4 TRAMO-SEATS versus X-13ARIMA-SEATS

Although TRAMO can be used on its own as a predictor program, it can also be viewed as serving a data-cleaning role for ARIMA series with missing values or outliers. Specifically, for a given time series, it interpolates missing data, identifies and adjusts for outliers, estimates trading day and Easter effects, and ultimately yields a linear process driven purely by random disturbances, namely the ARIMA model. Consequently, TRAMO can be used as a pre-adjustment process for SEATS. TRAMO-SEATS and X-13ARIMA-SEATS (hereinafter referred to as X-13) are mainstream seasonal adjustment tools, both based on the ARIMA model, although they differ in methodology and implementation.

Common:

- Both pre-adjust time series use the ARIMA (regARIMA) model to deal with outliers, missing values, calendar effects (such as trading days and Easter effects), and other factors that affect the quality of seasonal adjustments, with the goal of pre-adjusting to provide a 'clean' time series for subsequent seasonal adjustments (such as X-11 or SEATS).
- Both have the function of automatically detecting and correcting outliers (such as additive outliers, level shifts, transient changes, etc.); the detection of outliers is usually based on statistical tests (such as the t-test) and corrected by regression models.
- Both can deal with calendar effects, including the trading day effect and the Easter effect. These effects were modelled by introducing dummy variables or other regression variables.
- Both support interpolation of missing values to ensure the integrity of the time series; both support automatic model recognition (eg, automatic selection of the order of the ARIMA model) and parameter estimation (eg, maximum likelihood estimation or least squares estimation).

Differences:

- TRAMO was developed by the Bank of Spain, primarily based on the research of Víctor Gómez and Agustín Maravall. X-13ARIMA-SEATS, developed by the U.S. Census Bureau, is an enhanced version of X-12-ARIMA that integrates the X-11 seasonal adjustment approach with ARIMA modelling. Due to their different development backgrounds, these methods may differ in implementation details, such as the specific techniques for outlier detection or the initialisation procedures for model parameters. TRAMO uses a Tsay (1986) and Chen & Liu (1993)-based approach for outlier detection and selects outliers through a stepwise regression process; the X-13ARIMA-SEATS outlier detection method is based on similar statistical tests, but may differ in specific implementations, such as the selection of thresholds or the priority of outlier types.
- TRAMO uses the exact maximum likelihood method for parameter estimation by default, and supports unconditional/conditional least squares as an option; X-13ARIMA-SEATS also supports maximum likelihood estimation, but may differ from TRAMO in the specific steps of model identification and parameter estimation.

- TRAMO provides two ways to deal with missing values: one is direct interpolation, and the other is handled by introducing dummy variables (additive outliers); X-13ARIMA-SEATS also supports missing value interpolation, but the implementation may be different from TRAMO.
- TRAMO's default ARIMA model is the so-called airline model, ie, (0,1,1) (0,1,1), popularised by Box and Jenkins (1970); The X-13 also supports airline models, but may differ in default settings and model selection.
- TRAMO is often used in conjunction with SEATS to form a complete seasonal adjustment process (TRAMO-SEATS); X-13ARIMA-SEATS is a stand-alone software that combines X-11 and ARIMA modelling capabilities, with a potentially more user-friendly interface and more customisation options.

The SEATS method is based on the ARIMA model, which has a rigorous statistical theoretical basis and is capable of handling complex seasonal patterns and irregular data. SEATS can be used either in combination with TRAMO or independently, depending on the data preprocessing requirements. SEATS provides automatic model recognition and outlier detection, reducing the need for manual intervention. SEATS offers a statistically rigorous approach to decomposing time series. Capable of handling complex seasonal patterns and irregular data, it can be seamlessly integrated with X-13ARIMA-SEATS, a widely used tool for seasonal adjustment. This provides analysts with a powerful way to gain insight into the structure of time series data, especially by removing seasonal influences to reveal true trends and fluctuations. However, SEATS has high data requirements, particularly regarding the length and frequency of the data. For example, the minimum number of observations is 36 for monthly data and 16 for quarterly data. Frequency domain analysis and filter design in SEATS involve complex mathematical computations that may require substantial computational resources. The decomposition results from SEATS depend on the accuracy of the ARIMA model, which can lead to inaccurate decomposition results if the model is not properly selected.

In summary, TRAMO-SEATS is similar to X-13ARIMA-SEATS' core approach but differs in implementation details (eg, default model, outlier handling flow) and user experience. Thus, the choice of method (X-13ARIMA-SEATS or TRAMO-SEATS) and the selection of the tool usually depend on the user's specific needs, the characteristics of the data, and familiarity with the software. This discipline is both a science and an art, relying not only on theoretical knowledge but also on the practitioner's accumulated experience.

7.5 Software Implementation of TRAMO-SEATS

The core algorithm of TRAMO-SEATS is independently encapsulated, and the signal extraction is realised through the ARIMA model and the Wiener–Kolmogorov (WK) filter. Originally implemented in 2001 as part of the TSW program, a Windows-based extension of TRAMO and SEATS, the TRAMO-SEATS method has since undergone extensive refinements, significantly affecting key input parameters and resulting outputs. These changes have led to TSW+,[17] which also offers a Linux version of TRAMO-SEATS.

The use of TRAMO-SEATS is also available in (1) JDemetra+ (developed and maintained by the European Union Statistical Agency and the European Central Bank); (2) EViews (supported via plugins); (3) MATLAB (partial toolbox or custom implementation); (4) R (via the 'RJDemetra' or 'seasonal' package).

JDemetra+ is a seasonally adjusted software developed by the ECB in collaboration with Eurostat. JDemetra (Java-based version) was used in the early 2010s as a tool for implementing the TRAMO/SEATS+ methodology. Later, as an upgraded version of JDemetra+, it integrated the X-13ARIMA-SEATS and TRAMO-SEATS methods and was officially launched in 2015, becoming the European Union's official recommended seasonal adjustment tool. It is also the current mainstream tool. TRAMO – SEATS is part of the JDemetra+ (2016) package and is another option used by JDemetra+, distinct from the X13-ARIMA-SEATS approach. We will discuss the JDemetra+ program in detail in Chapter 8.

7.6 Summary and Conclusions

TRAMO-SEATS' ARIMA model-based approach is used for seasonal adjustments in its default automatic mode. Focusing on the TRAMO-SEATS program, this chapter examines its theoretical foundations, functional design, and applied methodology.

1. TRAMO is a pre-adjustment program for time series. It preprocesses data to estimate, forecast, and interpolate regression models with ARIMA errors and missing observations, while detecting outliers and handling multiple outlier types.
2. The TRAMO parameter estimation is based on the Hannan–Rissanen method to obtain the initial values of the ARMA parameters and then optimise them with exact ML. Combined with the information criterion, the optimal ARMA (p, q) model was selected.
3. TRAMO provides a variety of ways to deal with missing values: (a) Skipping approach: that is, the missing values are ignored directly, without interpolating or estimating, but by adjusting the structure of the model to accommodate the missing data; (b) Treating missing values as additive outliers (AO) by introducing dummy variables (additive outlier method) and modelling these outliers by introducing additional parameters.
4. TRAMO optionally implements missing value imputation via State Space Models and Kalman Filtering. However, the state equation is contingent on the estimated ARIMA specification, and the update mechanism, distinct from a basic Random Walk, operates under a broader ARIMA process for modelling and imputation.
5. SEATS decomposes time series using an ARIMA model and Wiener–Kolmogorov (WK) filter. Through frequency domain analysis, it separates components by decomposing the series' spectrum into constituent parts. The WK filter – optimally derived from the fitted ARIMA model's spectrum – calculates both the total spectrum and the component spectra. Then, the frequency → and time domain transformation extracts each component through the Fourier transform.
6. The extracted components need to be diagnosed and verified, including (1) the goodness-of-fit of the model; (2) the rationality check of the component; and (3) residual analysis to ensure the goodness-of-fit and rationality of the decomposition of the model.

7. TRAMO-SEATS and X-13ARIMA-SEATS are leading seasonal adjustment programs. While both share core seasonal adjustment functionality, their pre-adjustment modules, TRAMO (in TRAMO-SEATS) versus regARIMA (in X-13ARIMA-SEATS), exhibit methodological and implementation differences. Further divergence exists in their seasonal adjustment engines: SEATS employs signal extraction via ARIMA models, whereas X-11 relies on moving averages. Software selection depends critically on data characteristics and user expertise, requiring both theoretical knowledge and practical experience.

In summary, the TRAMO-SEATS model-based signal extraction method provides a more accurate and complete picture of reality than a fixed filter. Ultimately, as Hawking and Mlodinow assert, all reality-testing is model-dependent.[18]

Appendix 7.1: QR Decomposition and Cholesky Decomposition

1. **QR decomposition**

For an arbitrary $n \times m$ matrix X ($n \geq m$), the QR decomposes it into

$$X = QR$$

where Q is the orthogonal matrix (n × n) ($Q^T Q = I$); R is the upper triangular matrix of ($n \times m$). A regression model is assumed

$$y = X\beta + \epsilon$$

The goal is to minimise the sum of squares of the residuals: $\min_{\beta} \|y - X\beta\|^2$, where y is the vector of the observed value of ($n \times 1$), X is the design matrix of ($n \times p$), including explanatory variables, and β is the vector of the parameter to be estimated for ($p \times 1$). If the problem is weighted (eg, generalised least squares), the objective function becomes

$$\min_{\beta} \|y - X\beta\|_w^2 = (y - X\beta)^T W((y - X\beta)$$

Where W is the weight matrix (eg, the inverse of the error covariance matrix).

Since Q is orthogonal, the least squares problem translates to

$$\|y - QR\beta\|^2 = \|Q^T y - R\beta\|^2$$

Let $Q^T y = \begin{pmatrix} C_1 \\ C_2 \end{pmatrix}$, where C_1 M line before the time.

Solving:

- Upper triangular system $R\beta = C_1$, which can be solved efficiently by the substitution method.
- The sum of squares of the residuals is $\|C_2\|^2$.

Algorithm implementation (Householder transform):

(1) For each column k of X, ($k = 1, \ldots, m$):

- Compute the Householder vector (e_1) direction.
- Apply transformations: $X[k:n, k:m] = (I\text{-}2v_k v_k^T) X[k:n,\ k:m]$
- Storage v_k used for subsequent calculations of Q.

(2) The final R is stored in the upper triangle of (X), Q via $Q = H_1 H_2 \ldots \ldots H_m$ to be implicitly expressed.

The benefits of this algorithm are as follows:

- Avoid direct calculations $X^T X$.
- Applies to rank loss matrices (via column principal QR).

2. **Cholesky Decomposition**

Definition: For the symmetric positive definite matrix A, Cholesky decomposes it into

$$A = LL^T$$

where L is the lower triangular matrix.

Application in TRAMO:

Cholesky decomposition is used for covariance matrix updates for GLS and Kalman filters.

Step 1. GLS Questions:

When the covariance matrix of the error term (ε) is Σ, the objective function is

$$\min_{\beta} (y - X\beta)^T \Sigma^{-1} (y - X\beta)$$

Step 2. Decomposition & Transform:

- compute $\Sigma = LL^T$.
- defining $y^* = L^{-1} y$, $X^* = L^{-1} X$, the problem translates into ordinary least squares:

$$\min_{\beta} \left\| y^* - X^* \beta \right\|^2$$

Solving:

- perform a QR decomposition of X^* to further simplify the calculation

Algorithm:

(1) For $i = 1, \ldots, n$:

$$L_{i,i} = \sqrt{A_{i,i} - \sum_{k=1}^{i-1} L_{i,k}^2}$$

$$L_{i,i} = \frac{1}{L_{i,i}} \left(A_{i,i} - \sum_{k=1}^{i-1} L_{j,k} L_{i,k} \right) \quad (j = i+1, \ldots \ldots, n)$$

(2) if $A_{i,i} - \sum_{k=1}^{i-1} L_{i,k}^2 \leq 0$, the matrix is non-positive definite, and the algorithm terminates.

Advantages:

- Computational complexity of $0(n^3/6)$, which is about twice as fast as LU decomposition.
- Numerically stable (provided the matrix is positively definite).

(3) Parameter estimation process:

- Initialise: Initialise the ARMA parameters using the Hannan–Rissanen method.
- Generalised least squares: Whitening of the residual covariance matrix by Cholesky decomposition $\Sigma = LL^T$.
- Regression solving: Perform QR decomposition on the transformed design matrix, X^*, and solve the β.
- Iterative optimisation: Update ARIMA parameters and outlier effects until convergence.

Notes

1 See Gómez, V., & Maravall, A. (1996). "Programs TRAMO and SEATS: Instructions for the User". Beta Version. Working Paper No. 9628. Banco de España (Bank of Spain), Madrid, June.

2 Tsay, R. S. (1986). "Time Series Model Specification in the Presence of Outliers". *Journal of the American Statistical Association*, 81, 132–141.

3 Chen, C., & Liu, L. M. (1993). "Joint Estimation of Model Parameters and Outlier Effects in Time Series". *Journal of the American Statistical Association*, 88, 284–297.

4 Gómez, V., & Maravall, A. (1994). "Estimation, Prediction and Interpolation for Nonstationary Series with the Kalman Filter". *Journal of the American Statistical Association*, 89, 611–624.

5 Box, G. E. P., Hillmer, S. C., & Tiao, G. C. (1978). "Analysis and Modelling of Seasonal Time Series". In Zellner, A. (ed.), *Seasonal Analysis of Economic Time Series*, pp. 309–334. U.S. Department of Commerce- Bureau of the Census, Washington, DC.

6 Burman, J. P. (1980). "Seasonal Adjustment by Signal Extraction". *Journal of the Royal Statistical Society A*, 143, 321–337.

7 Hillmer, S. C., & Tiao, G. C. (1982). "An ARIMA Model-Based Approach to Seasonal Adjustment". *Journal of the American Statistical Association*, 77, 63–70.

8 Maravall, A., & Pierce, D. A. (1987). "A Prototypical Seasonal Adjustment Model". *Journal of Time Series Analysis*, 8, 177–193. Reprinted in Hylleberg, S. (ed.) (1992). *Modelling Seasonality*. Oxford University Press.

9 Maravall, A. (1995). "Unobserved Components in Economic Time Series". In Pesaran, H., & Wickens, M. (eds.), *The Handbook of Applied Econometrics*, chapter 1, pp. 12–72. Oxford University Press, Oxford.

10 See Gómez, V., & Maravall, A. (2001). "Programs TRAMO and SEATS: Instructions for the User (with Some Updates)". *Journal of Business & Economic Statistics*, 19(1), 115–127. https://doi.org/10.1198/07350010152472661.

11 Maravall, López, & Pérez (2015) and Maravall, López, & Pérez (2016), discussed the reliability of the AMI program for the monthly time series, and on a set of 50,000 simulation sequences from 50 different ARIMA models, they found about 3 out of 4 series, obtaining an accurate model order; For 99% of the series, the model obtained is a reasonable approximation. See Maravall, A., López, R., & Pérez, D. (2015). "Outlier Detection in Seasonal Adjustment: An Integrated Approach". Series: Banco de España Working Paper No. 1510. Banco de España (Bank of Spain), Madrid. https://doi.org/10.2139/ssrn.2621543; Maravall, A., López, R., & Pérez, D. (2016). "Outlier Detection in Seasonal Adjustment: A Further Improvement". Series: Banco de España Working Paper No. 1601. Banco de España (Bank of Spain), Madrid. https://doi.org/10.2139/ssrn.2745243.

12 TRAMO's default run (ie, with automatic model recognition, including detection and correction of outliers and calendar effects) produces acceptable models 90.3% of the time.

13 The core idea of the Hannan–Rissanen method: let the ARMA(p, q) model, $y_t = \sum^{p} \varnothing_i y_{t-i} + \sum^{q} \theta_j \epsilon_{t-j} + \varepsilon_t$. The Hannan–Rissanen method simplifies the Hannan–Rissanen method by two steps. (1) higher-order autoregressive (AR) approximation: fitting the data with a long-order AR model to obtain residual estimates T_t; (2) linear regression estimation: converting the ARMA problem into linear regression, using C_t as a regression variable.

14 Tsay (1986) and Chen & Liu (1993) are important methods for outlier detection and intervention analysis in the field of time series analysis. Tsay (1986) proposed an outlier detection method based on recursive residuals, which is mainly used to detect additive outliers (AO) and level shifts (LS) in time series. The core idea: calculate the residuals for each step by recursively fitting the time series model. If the residuals at a point in time deviate significantly from expectations, it may be an outlier (recursive residuals). A statistical test is performed on the recursive residuals to determine if there are outliers (hypothesis testing). Chen & Liu (1993) proposed a more comprehensive approach to outlier detection and intervention analysis, capable of detecting multiple types of outliers, including additive outliers (AO), horizontal drift (LS), temporary change (TC), and innovational outliers (IO). Core idea: Improve the accuracy of detection by federating outliers and estimating their impact (federated detection and estimation). Progressively identify all outliers by iteratively detecting and modifying the model (iterative process). See Tsay, R. S. (1986). "Time Series Model Specification in the Presence of Outliers". *Journal of the American Statistical Association (JASA)*, 81(393), 132–141; Chen, Chung, & Liu, Lon-Mu. (1993). "Joint Estimation of Model Parameters and Outlier Effects in Time Series". *Journal of the American Statistical Association (JASA)*, 88(421), 284–297.

15 The Cholesky decomposition method is a statistical technique for efficiently calculating regression parameters, especially in time series analysis or generalised least squares (GLS). The core idea of this approach is to transform the problem into a more manageable form by Cholesky's decomposition of the error covariance matrix, thereby improving computational efficiency and ensuring numerical stability. The QR algorithm is an efficient and numerically stable method for solving least-squares estimation in linear regression problems. The QR algorithm simplifies the process of solving least squares problems by decomposing the design matrix X into an orthogonal matrix Q and an upper triangular matrix R. For the specific algorithm, see Appendix 7.1.

16 See Gómez, V., & Maravall, A. (1994). "Estimation, Prediction, and Interpolation for Nonstationary Series with the Kalman Filter". *Journal of the American Statistical Association*, 89(426), 611–624; Gómez, V., Maravall, A., & Peña, D. (1999). "Missing Observations in ARIMA Models: Skipping Approach Versus Additive Outlier Approach". *Journal of Econometrics*, 88(2), 341–363. These two methods are mathematically equivalent, ie, they will end up with the same estimates. The skip method deals with missing values more directly, while the additive outlier rule deals with missing values indirectly by introducing additional parameters. Which method you choose depends on the specific application scenario and computational convenience.

17 See Maravall, A., Caporello, G., Pérez, D., & López, R. (2014). "Program TSW: Revised Reference Manual". Series: Banco de España Occasional Paper No. 1404, p. 120. Banco de España (Bank of Spain). https://doi.org/10.2139/ssrn.2524063.

18 Hawking, Stephen, & Mlodinow, Leonard. (2010). "*The Grand Design*", p. 46. Bantam Books.

8

JDemetra+: Advanced Seasonal Adjustment Tools

8.1 Introduction

Seasonal adjustment involves statistical (econometric) modelling or smoothing of the data using filters (moving averages) to isolate seasonal effects and reveal potential developments. The X-13ARIMA-SEATS method, developed by the U.S. Census Bureau, is a semi-parametric method based on moving averages. The main drawback of this approach is its reliance on similar filters for different types of time series. TRAMO-SEATS is a parametric estimation method based on econometric models developed by the Central Bank of Spain. The advantage of this approach is the use of data-specific filters. Both the TRAMO-SEATS and the X-13ARIMA-SEATS are under continuous development. Each method has its advantages and disadvantages, and there is no international consensus on which method is preferred. Both approaches can be implemented on the JDemetra+ (JD+ for short) platform, which was officially launched in 2015. Users can use both software programs to make seasonal adjustments and compare the adjusted quality of the two methods.

This chapter focuses on the four main steps of TRAMO-SEATS' seasonal adjustment at JD+:[1] data preparation, seasonal adjustment, results analysis, and user communication. The structure of this chapter is as follows: following the introduction in Section 8.1, Section 8.2 provides a brief background to the creation of JD+; Section 8.3 includes the quality and length of the original time series, and suitable data formats for use by JD+; Section 8.4 describes seasonal adjustments, including the steps involved in seasonal adjustments, the definition of specific calendar effects in JD+, in particular, the user-defined calendar regressor and various seasonal adjustment options; Section 8.5 discusses the analysis of the results, describing several quality assurance tools and options for obtaining outputs from the seasonally adjusted (SA) process (ie, the seasonal adjustment series); Section 8.6 discusses issues related to data recording and revisions to seasonal adjustment series; and finally, Section 8.7 provides a summary and conclusions.

8.2 Background

The TRAMO-SEATS method was first implemented in 2001 within the Time Series Workbench (TSW), a Windows-based extension of the TRAMO and SEATS programs. Since then, numerous modifications and enhancements have been made, affecting both key input parameters and the resulting outputs. These changes have led to TSW+,[2] which also offers a Linux version of TRAMO-SEATS. For the X-13ARIMA-SEATS, the U.S. Census

DOI: 10.1201/9781003716655-8

Bureau provides a Windows interface called Win X-13.[3] Both of these seasonal adjustment programs were originally written in FORTRAN, which is currently considered a declining language. The limitations of FORTRAN, particularly in creating reusable components and handling complex tasks, have made maintaining the associated IT code increasingly burdensome. These early seasonal adjustment programs are often regarded as challenging to operate, particularly for non-specialists. As a result, efforts have focused on enhancing the accessibility of these methods to a broader user base. Eurostat has launched a new software called Demetra. It provides a user-friendly interface for two seasonal adjustment algorithms, TRAMO-SEATS and X-12-ARIMA, to facilitate the comparison of the outputs of the two algorithms. Even so, Demetra's use of the FORTRAN library, combined with product development and inadequate handling of 'errors', was a factor in the rapid decline in software usage.

In 2009, the European Statistical System (ESS) launched the 'Seasonally Adjusted Guidelines',[4] which were endorsed by the European Statistical Program Committee and aim to harmonise European practices, improve the comparability of annual national statistics, and improve the overall quality of EU and Eurozone aggregates. The ESS Seasonal Adjustment Guidelines (2009) and their 2015 revision cover all key steps of the seasonal and calendar adjustment process. They discussed both the theoretical aspects and the practical implementation of seasonal adjustment. As Demetra was unable to adapt to the new requirements outlined in the guidelines, Eurostat, in partnership with the National Bank of Belgium (NBB), launched a project to develop an improved software, Demetra+. It was released in 2012. The tool offers a unified approach to seasonal adjustment through the TRAMO-SEATS and X-13ARIMA-SEATS methods, both of which are more closely aligned with established guidelines. It includes a unified graphical interface and input/output diagnostics for both methods. The Demetra+ source code is written in C++ and uses two original FORTRAN modules as well as a .NET library. As a result, Demetra+ software is not scalable and cannot be used in IT environments other than Windows. For these reasons, it does not appear capable of meeting users' expectations over the long term.

As a result, Eurostat took the initiative to develop a new software that builds on the Demetra+ experience while being platform-independent and scalable. The resulting program is called JD+. NBB has been developing it since 2012. From a typical user's perspective, JD+ introduces several enhancements over Demetra+ in both interface design and functionality. The most critical innovation was to rewrite the original FORTRAN code for X-12-ARIMA/X-13ARIMA-SEATS and TRAMO-SEATS in Java, following a true object-oriented approach.

The rewriting of the latest version of JD+, the leading seasonally aligned algorithm in Java, was a key factor in enabling long-term maintenance of the tool and integrating the library into the IT environments of many agencies. JD+ is not only a user-friendly graphical interface but also an open set of Java libraries that can be used to deal with problems related to time series, including SA processing of large-scale datasets, the use of non-standard SA methods, the development of advanced research modules, time decomposition, benchmarking, and business cycle analysis.

JD+ is written using an object-oriented programming (OOP) approach. It enables developers to design software in a modular fashion, that is, by dividing the application's functionality into independent, interchangeable modules. Such units serve specific functions and can be separated from the overall framework. OOP approaches are particularly effective in complex programs or situations where reusability is important. In addition to statistical algorithms, JD+ also provides a wide range of auxiliary services. The most important features are as follows. (1) Dynamic access to various 'time series programs': JD+ provides modules to work with time series from different sources: Excel, databases (via JDBC), web services, files (TXT, TSW, USCB, XML, SDMX, . . .); Because the time series is automatically refreshed, this consults the provider to download the new information. The model allows for asynchronous processing, in the sense that access is dynamic. (2) Common XML format: The seasonal adjustment process can be saved as XML, which can also be used to generate web services around the seasonal adjustment. The graphical interface of JD+ is based on the NetBeans framework. With this technology, external IT teams can develop their modules to extend the original software without altering the core application. This includes allowing users to add modules for downloading data from other databases, performing seasonal adjustment diagnostics, and generating outputs from the seasonal adjustment process. As a result, APIs can be used to build self-contained applications or to create extensions of current applications more easily. Since one of JD+'s goals is to develop a software program that can compare the results of TRAMO-SEATS and X-13ARIMA-SEATS. As a result, most analysis tools are common to both algorithms, such as revision history and sliding span, even if they were originally developed in only one of them. On the other hand, not all features developed in the original program were implemented in JD+; for example, in contrast with TRAMO-SEATS, JD+ does not separate long-term trend from the cyclical component.

JD+ runs on operating systems that support Java VM (virtual machine), eg, Microsoft Windows Vista SP1/Windows 7/8; Ubuntu 9.10; Solaris OS Release 11 Express (SPARC and x86/x64 Platform Editions); Macintosh OS X 10.6 Intel. JD+ runs on Java SE Runtime Environment (JRE) version 8 Update 21 or later.

JD+ consists of several software components that can be used independently in other applications. In addition to the graphical interface, a command-line tool called 'Cruncher' is available for batch seasonal processing. Since 2019, JD+ has had an R interface called RJDemetra1, thus a dedicated 'R' package that can be used as an alternative to the graphical interface.

JD+ is built around the concepts and algorithms used in two leading seasonal adjustment methods, namely TRAMO-SEATS and X-12-ARIMA/X-13ARIMA-SEATS. These methods have been redesigned using an OOP approach, making them easier to handle, extend, and modify. While TRAMO-SEATS and X-13ARIMA-SEATS differ in the methods used to identify seasonal changes, they generally give very similar results in practice. In both methods, the pre-adjusted portion, which can be defined as identifying the part of the non-seasonal effect, is almost identical.

JD+ offers two seasonal adjustment methods, TRAMO-SEATS and X-13ARIMA-SEATS, along with a unified analysis tool that facilitates easy comparison of results across these methods. Most of the analysis implemented in JD+ can be used in the core engine, such as (1) regARIMA modelling (using concepts developed in TRAMO and X-12-ARIMA); (2) residual analysis (mainly similar to TRAMO); (3) seasonal testing (similar to TRAMO and X-12-ARIMA); (4) spectral analysis (X-12-ARIMA definition); (5) sliding span (X-12-ARIMA use); (6) Wiener–Kolmogorov analysis (similar to SEATS for unobserved ARIMA component models). However, the results generated by JD+ may vary slightly for several reasons (different statistical/algorithm choices). In any case, the global messages from the seasons are always (almost) similar.

JD+ provides several predefined seasonally adjusted routines (RSA) based on TRAMO-SEATS or X-13ARIMA-SEATS, and statisticians must decide which to use. JD+ uses an easy

TABLE 8.1

JD+ Seasonal Adjustment Procedures

Main Steps	Intermediate Steps
Preparation of data	Open JD+
	Prepare source data
	Import data
	Check the original series
	Prepare Calendar
Seasonal adjustment	Select an approach (updating as a calendar)
	Define specification and regressors
	Seasonally adjust
	Benchmarking
Analysis of the results	Visual check
	Quality diagnostics
	Refine and readjust
	Export data
User communication	Reports
	Documentation
	Supporting users

comparison of the two methods along with a unified visual tool for both. While different RSAs often produce very similar results, the quality diagnostics provided by JD+ are critical to confirming and refining process results. JD+ facilitates batch analysis of one-time SA (single processing) and mass production (multiple processing), and is therefore suitable for the routine production of SA time series by statistical offices. This chapter concentrates on the application of TRAMO-SEATS, both for illustrative purposes and to maintain simplicity. This is because the X-13-ARIMA-SEATS have been described in detail in previous chapters.

With JD+'s SA program, there are typically four steps: (1) data preparation; (2) seasonal adjustment; (3) results analysis; and (4) user communication.

8.3 Data Preparation

8.3.1 Data Preprocessing

Prepare the source data in an appropriate format and import it into JD+. Subsequently, examine the original (unadjusted) time series and prepare it for seasonal adjustment. Examine the raw data in terms of accuracy, time series length, time consistency, and any other issues that may affect the SA process and the quality of the SA series. Visual analysis tools help in identifying irregular effects, missing values, volatility, trends, seasonal patterns, and other irregular disturbances in the series.

Many economic time series are affected by calendar effects. JD+ also allows the inclusion of a national calendar of working days and holidays, which can be integrated into the seasonal adjustment process.

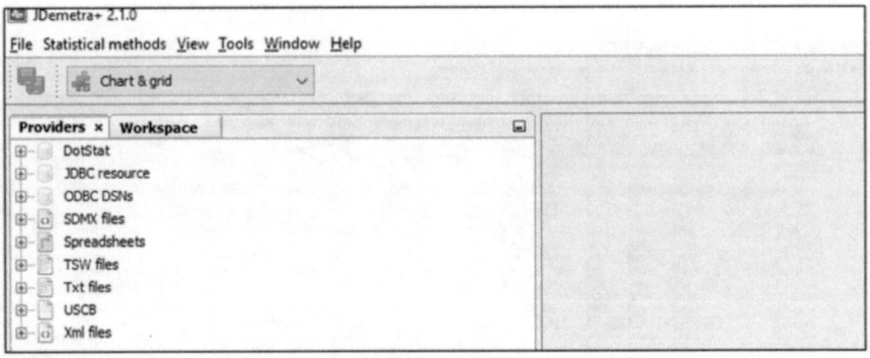

FIGURE 8.1
Source Data Format.

8.3.2 Data Format Requirements

The JD+ software has some specific requirements for the format of the input data. While the software accommodates various input formats and file types, each file must adhere to a predefined structural specification. Additionally, JD+ supports repeated seasonal adjustment of the same source data through its built-in tools. It allows the reading of revised and new charts from the (updated) original file.[5] Open JD+ and see the following view:

Figure 8.1 consists of different panels and tabs: (1) the providers option (left panel – left) provides some connection methods for uploading time series; (2) the workspace tab (left panel – hub) provides access to information used or generated in workflows; (3) the empty panel on the right may contain multiple windows created by JD+. Since it also displays the analysis, it is known as the results panel. The user can move, adjust, and close the panel as needed. Time series can be easily dragged and dropped between panels. Users can reopen closed panels via the main menu command Window. JD+ remembers the selected settings for later use.

8.3.3 Source Data Preparation

JD+ offers a user-friendly process for importing data from various file types, featuring drag-and-drop functionality and clipboard support. It also provides several alternatives for dynamic data uploading (to be explained later). JD+ allows the use of data in different formats, including (1) Microsoft Excel files (consisting of spreadsheets) with the '.xls/.xlsx' extension and text files with the extension '.txt/.csv/.tsv'; (2) X-13 software from the U.S. Census Bureau (USCB); (3) files of the TRAMO-SEATS for Windows software is TSW which is provided by the Bank of Spain; (4) Statistical Data and Metadata Exchange (SDMX) or Experience Markup Language (XML) files; and (5) files in the generic database driver, ie, Oracle Database Connection (ODBC) or Java Database Connection. These data files have specific data storage rules and are connected to JD+[6] through their own specific rules. A common file format for storing data is Microsoft Excel. Thus, Excel is used as an example to prepare the input data for JD+.[7] as shown in Figure 8.2.

Table 8.2 shows the number of observations and the start and end dates for the series in the example. The number of observations varies across the countries, but all series in the dataset are suitable for seasonal adjustment because they meet the minimum length criterion

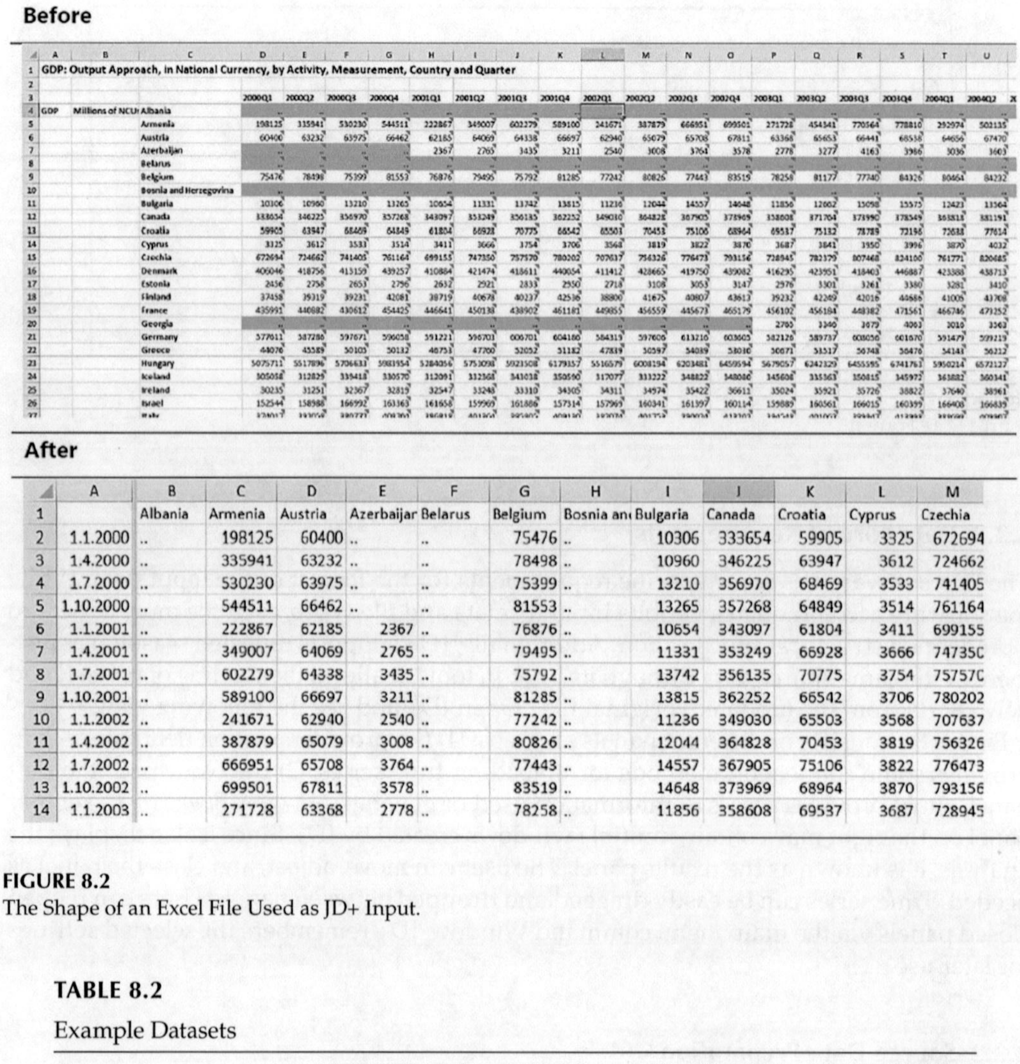

	A	B	C	D	E	F	G	H	I	J	K	L	M
1		Albania	Armenia	Austria	Azerbaijar	Belarus	Belgium	Bosnia an	Bulgaria	Canada	Croatia	Cyprus	Czechia
2	1.1.2000	..	198125	60400	..	..	75476	..	10306	333654	59905	3325	672694
3	1.4.2000	..	335941	63232	..	..	78498	..	10960	346225	63947	3612	724662
4	1.7.2000	..	530230	63975	..	..	75399	..	13210	356970	68469	3533	741405
5	1.10.2000	..	544511	66462	..	..	81553	..	13265	357268	64849	3514	761164
6	1.1.2001	..	222867	62185	2367	..	76876	..	10654	343097	61804	3411	699155
7	1.4.2001	..	349007	64069	2765	..	79495	..	11331	353249	66928	3666	74735C
8	1.7.2001	..	602279	64338	3435	..	75792	..	13742	356135	70775	3754	75757C
9	1.10.2001	..	589100	66697	3211	..	81285	..	13815	362252	66542	3706	780202
10	1.1.2002	..	241671	62940	2540	..	77242	..	11236	349030	65503	3568	707637
11	1.4.2002	..	387879	65079	3008	..	80826	..	12044	364828	70453	3819	756326
12	1.7.2002	..	666951	65708	3764	..	77443	..	14557	367905	75106	3822	776473
13	1.10.2002	..	699501	67811	3578	..	83519	..	14648	373969	68964	3870	793156
14	1.1.2003	..	271728	63368	2778	..	78258	..	11856	358608	69537	3687	728945

FIGURE 8.2
The Shape of an Excel File Used as JD+ Input.

TABLE 8.2
Example Datasets

Country	Number of observations	Starting date	Ending date
Albania	37	31.3.2009	31.3.2018
Amenia	60	31.3.2000	31.12.2014
Austria	73	31.3.2000	31.302018
Azerbaijan	69	31.3.2001	31.302018
Belarus	36	31.3.2009	31.12.2017
Belgium	73	31.3.2000	31.302018
Bosnia & Herzegovina	40	31.3.2008	31.12.2017
Bulgaria	73	31.3.2000	31.302018
Canada	73	31.3.2000	31.302018
Croatia	73	31.3.2000	31.302018
Cyprus	73	31.3.2000	31.302018
Czechia	73	31.3.2000	31.302018
Denmark	73	31.3.2000	31.302018
⋮	⋮	⋮	⋮

for time series. To be processed in JD+, a spreadsheet must meet the following requirements. The date entries in the first column must be provided in an Excel-recognisable format (eg, 'ddmmyyyy'). Monthly data entries can be on the first day of each month, for example, '1.1.2000, 1.2.2000' (dd.mm.yyyy). Quarterly data can be specified by the date of the first day of the quarter, for example, '1.1.2000, 1.4.2000, 1.7.2000' (dd.mm.yyyy). There must be a gap of three months between the dates.

- The first line contains the name of the variable;
- Time series data is in numeric format;
- The cell [A1] in the upper left corner is empty;
- The missing value to be estimated is manually set to -99999.

JD+ uses the additive outliers method to estimate missing observations. JD+ also accepts transposed tables, following the same layout, with the first column being the variable name and the first row being the date.

8.3.4 Data Import Methods

There are two ways to import data into JD+: (1) the direct method, and (2) the indirect method.

The direct way to do this is through the Providers tab (Figure 8.3). Data is used by calling the 'Provider', and dynamic data updates are also allowed. More precisely, when a user needs to rerun an existing seasonal adjustment with new data, the updated dataset can be dynamically refreshed (reread) from the file. JD+ then automatically updates the SA results for the latest data.

The indirect method is suitable for one-time analysis: (1) copying data (with date columns and header rows) in the source file; (2) clicking 'Tools → Container → Grid' in JD+; (3) right-click and paste JD+ into the Grid/List panel (Figure 8.4). This method is only

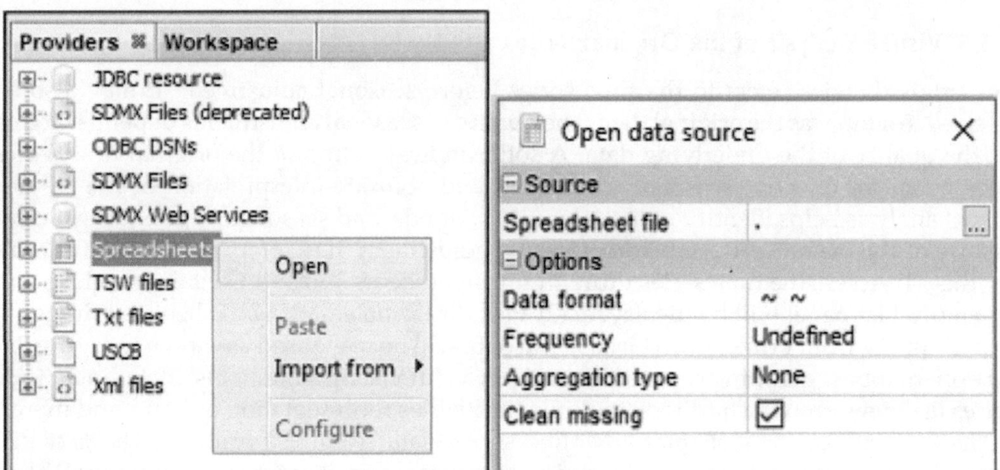

FIGURE 8.3
Importing Data into JD+ by Reading an Excel File.

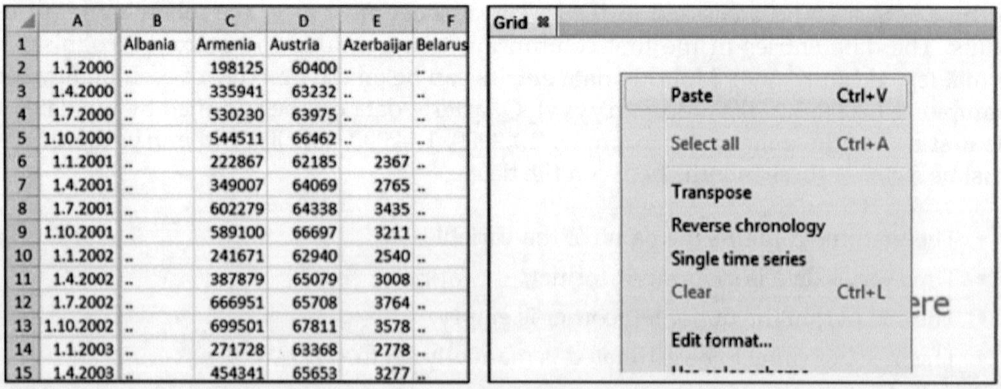

FIGURE 8.4
A Shortcut to Load Data into JD+.

suitable for one-time seasonal adjustments and rapid data analysis, and is not suitable for generating official statistics during the official production process.

Direct method: Right-click on Spreadsheet in the Providers tab and select Open (Figure 8.3, left); Select the source file by clicking '. . . .', move right to the Source file under the Spreadsheet file (Figure 8.3, right). Select the desired source file from the local folder (Excel). Once imported, right-click on the source file in the Providers tab, allowing you to edit, close, or rename the file. The file can be found using the Show in Folder option (See Figure 8.4).

Indirect method (copy and paste): Select the entire time series data, including the column containing the date in the source file and the row containing the data header, and copy it (right-click or Ctrl+C). Switch to JD+ and click Tools → Container → Grid, right-click on the open Grid panel, and paste (or Ctrl+v). With this indirect approach, dynamic updates are not possible because the data is not linked to the source file. On the other hand, it's the fastest way to view data and the one-time seasonality adjustment process.

8.3.5 Visual Analysis of the Original Series

The original series refers to the time series before seasonal adjustment. It may also be referred to simply as the original data. The quality of seasonal adjustments depends largely on the quality of the underlying data. A solid understanding of the original time series is essential for effective seasonal adjustment and accurate interpretation of the results. Visual analysis helps identify potential outliers, trends, and seasonality, providing a clear picture of the overall pattern before seasonal adjustment. JD+ offers several useful tools for this. (1) Access the time series through the menu items Tools → Container → List, and an empty List panel will be displayed. (2) Visualise problematic worksheets by dragging and dropping from Providers to List; (3) The data will be visualised as shown in Figure 8.5. JD+ offers robust graphing capabilities and powerful visualisation tools, including a time series list view, growth chart tool, multiple time series representation options, and more.

The example consists of quarterly time series data from 52 countries. The List displays the start and end dates of the series, the number of observations, and a small line chart illustrating the series' overall shape. Any action to visualise the analysis is the Tools→Container→Chart or GrowthChart option. More visualisation tools are available under Tools→Spectral Analysis. All of this becomes active when you drag and drop data

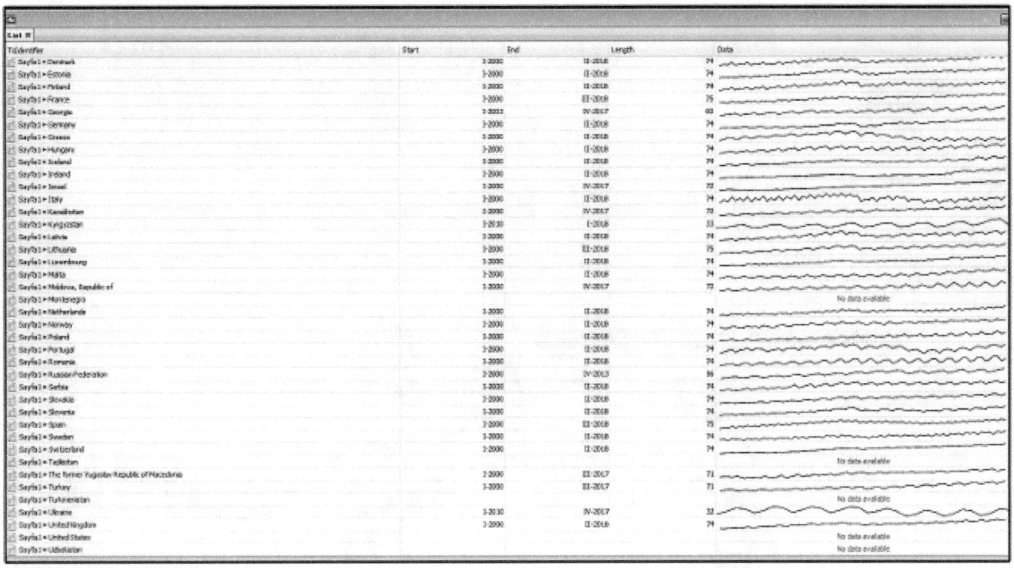

FIGURE 8.5
Time Series List View.

from Providers, List, or Grid. Figure 8.6 illustrates the Providers, List, Grid, Chart, and GrowthChart for some of the countries in the example.

The chart is shown in the upper right corner of Figure 8.6. From this example, it is already possible to determine the existence of seasonal and trend patterns. The lower-right corner shows the Growth Chart, illustrating the year-quarter-on-year-quarter (different year) growth rate. You can also change the growth rate type to Quarter:-quarter-on quarter (same year) growth rate by right-clicking on the chart and selecting Kind → Previous period. The presence and reversal of the trend (bars) can be confirmed from this chart. The chart also allows you to see data broken down by year, with each year represented by a different colour. With this type of chart, it is possible to quickly highlight level differences between months or quarters. These differences indicate significant seasonality in the data. To select this option, right-click on the chart and choose *Split into Yearly Components* (see Figure 8.7).

Periodic graphs, autoregressive spectra, and Tukey spectra are three key graphical tools used in time series spectral analysis, which can be found in the menu Tools→Spectral analysis. In all three chart types, the seasonal frequency is marked by a blue vertical line, while the purple line corresponds to the trading day frequency. For a single series, the peak of one or more of the lines indicates a seasonal or trading day effect. The Periodogram and Autoregressive Spectrum methods do not require a minimum number of observations, whereas the Tukey Spectrum requires at least 40 quarters or 80 months of data. They detect cyclical components in a time series.

A periodic chart is a mathematical tool used to point out periodic movements in a time series that cannot be directly observed (Jenkins and Watts, 1968). The cycle chart (Figure 8.8) shows a strong seasonal component but no calendar component. A peak with a frequency of $\pi/2$ (in blue, ie, it represents an annual cycle) indicates this. You don't see the spike in the calendar frequency (purple). The interpretation of the frequency depends on the nature of the data (monthly/quarterly) (See Figure 8.8).[8]

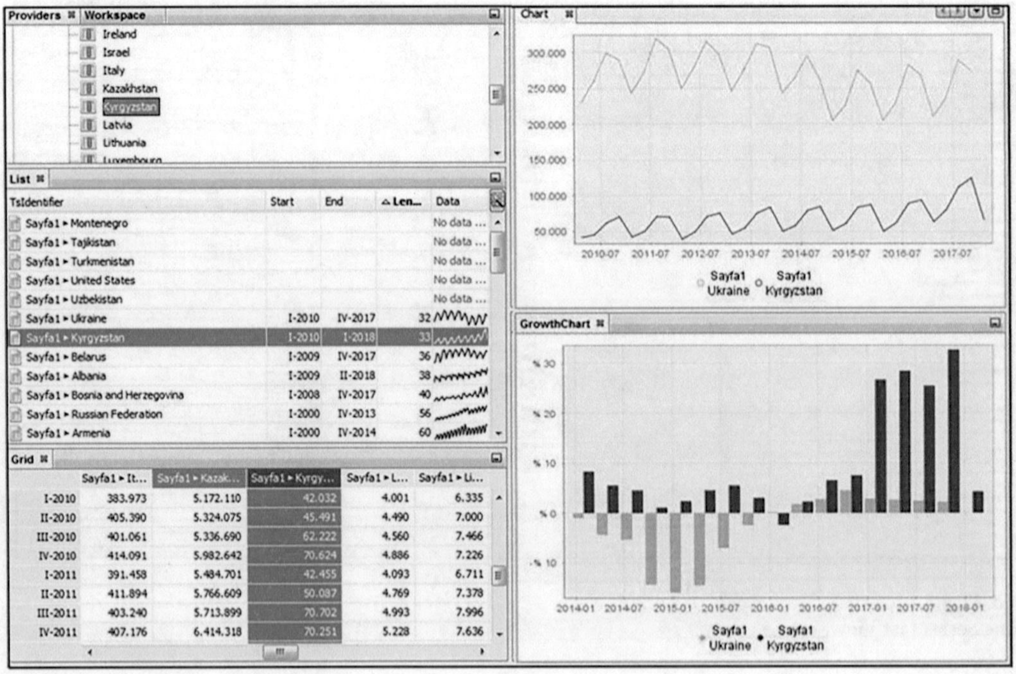

FIGURE 8.6
Charting and Growth Charting Tools.

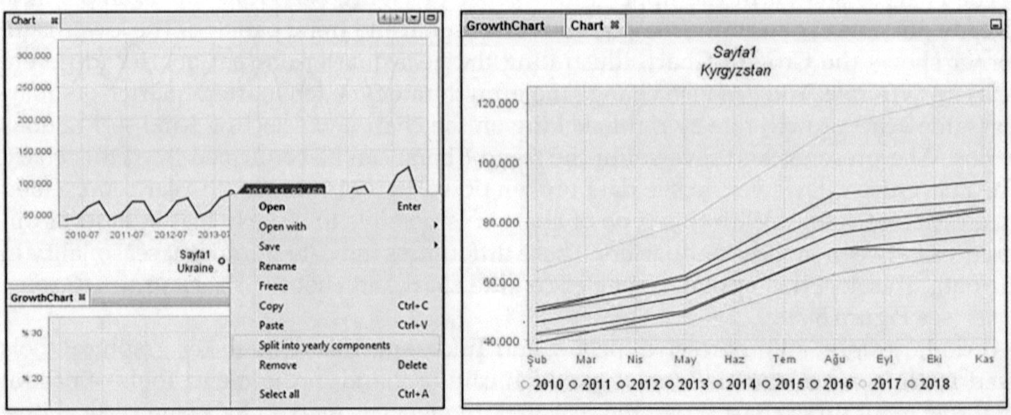

FIGURE 8.7
Time Series Presentation Types.

The Difference tool provides both further visual analysis and information about the autocorrelation structure of the first differential series and its periodic plot. This tool can be important for understanding the variance series of potential movements. It can be accessed from the menu Tools→Differencing. Differentials are common transformations that eliminate trends in a time series. Thus, a differenced series can be stationary, meaning its mean, variance, and distribution remain constant over time, which simplifies the modelling process.

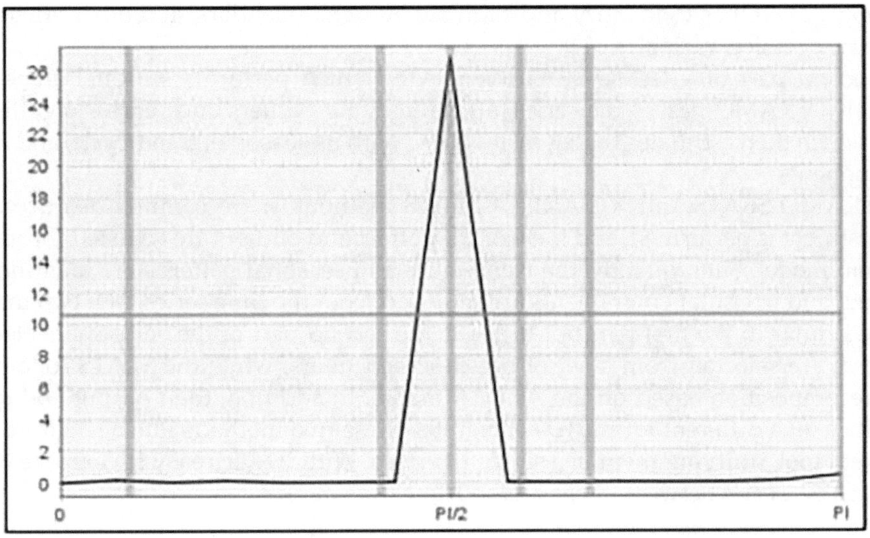

FIGURE 8.8
Example of a Periodic Chart.

8.4 JD+ Seasonal Pre-Adjustment

Currently, the most popular seasonal adjustments are the X-13ARINMA-SEATS and TRAMO-SEATS. JD+ provides statisticians with an interface to both approaches. In the econometric literature, SA series, trends, and seasonal components are often referred to as 'unobserved components' because they cannot be directly observed in time series. Statisticians have developed several methods for estimating unobserved components. X-13ARIMA-SEATS applies the X-11 moving average and performs seasonal adjustments by performing multiple iterations to derive trend periods and seasonal components.

X-11-ARIMA, introduced by Statistics Canada, adds estimates at the beginning and end of the series before making seasonal adjustments. These estimates are done by using the ARIMA model for prediction and back-transformation. The predicted end of the series significantly reduces the size of the revision and the estimated time lag. This improves the accuracy of seasonal portions. In addition to applying the basic ideas of its predecessor, the X-12-ARIMA provides a flexible tool for pre-tuning time series (regARIMA). It detects and corrects for calendar effects and outliers using regression ARIMA-type models before identifying seasonal components.

TRAMO-SEATS is a seasonal adjustment method based on 'model-based signal extraction'. The approach characterises the behaviour of the series and derives its components using an ARIMA model. As a result, the ARIMA model estimated from the observed series is made consistent with the underlying component model.[9]

The first part of TRAMO-SEATS is TRAMO, which is responsible for the pre-adjustment. TRAMO refers to a time series regression with ARIMA noise, missing observations, and outliers. It is usually a program for automatically identifying, estimating, predicting, and interpolating regression ARIMA models. This model is used to extend the series with predictions, as well as to identify and estimate outliers, calendars, and other effects with sufficient regression variables.

The second part of TRAMO-SEATS is SEATS, which performs seasonal adjustment. Specifically, TRAMO passes the series, after removing outliers and regression effects, to SEATS, which then estimates the SA series along with its seasonal, trend-cycle, and irregular components.

The TRAMO-SEATS and X-13ARIMA-SEATS methods share common features. First, pre-adjustment is performed, and the workday effect and outliers are adjusted through the regression model. Subsequently, the trend-cycle and seasonal patterns are identified and estimated, and irregular components are removed from the time series. The two methods are very similar in the first part but differ in the second part of the adjustment step. The X-13 filters are selected from a set of pre-designed filters, while the SEATS follow a so-called decomposition based on the ARIMA model. In addition, their output and quality diagnostics take different forms. JD+ integrates these two methods into a single seasonal adjustment tool, unifying them in a single interface. With JD+, it's easy to compare the differences in SA series between these methods.

8.4.1 Design and Application of Calendar Regressors

JD+ includes predefined regression variables, specifically working day and trading day variables, which reflect the relative advantage of weekdays over weekends within a given month. JD+ also offers a calendar tool for designing a more comprehensive calendar regressor. The pre-adjusted methods available in JD+, TRAMO (the modelling part of TRAMO-SEATS), and regARIMA (the modelling part of X-13) can often detect differences between working and non-working days. For certain series, a trading day effect may be observed, indicating that activity levels vary across days of the week. For example, depending on the branch, sales on Fridays may be higher than on Tuesdays. Calendar regression is a variable that contains information about the relative importance of a monthly weekday, holiday, or weekend. It may also include information about fixed or moving holidays in a country or industry. Regression quantities that include these types of national days or holidays can improve the modelling of time series. Using the appropriate amount of calendar regression, when it is necessary, will ultimately improve the overall quality of seasonal adjustments. Creating country-specific regression quantities can be time-consuming, and you can use this tool when you need to model country-specific national holidays.

To improve seasonal adjustment, TRAMO and regARIMA estimate and remove calendar effects during the pre-adjustment of the series. The pre-adjustment phase in JD+ is based on the capabilities of TRAMO and regARIMA.

8.4.1.1 Calendar Variables

JD+ provides a simple tool for creating calendar variables using the Calendars module. Calendar variables can include trading days (six variables plus leap year effect) or weekday variables (one variable plus leap year effect), as well as country or industry-specific holidays. These holidays may occur on fixed or moving dates. Several options in JD+ can be used to design specific country calendars, including predefined options. When a national calendar contains one or more movable holidays that commence or cease at defined points in time, you may specify their effective period by setting start and end dates. However, expiration dates should be used with caution. When using a national calendar, JD+ includes a long-term correction to the trading day variable. The correction does not account for expiration dates, which may introduce residual seasonal effects in the variable. For more complex holidays, you can also prepare the national calendar as an external user-defined variable, ie, as a time series to be imported into JD+.

In JD+, both TRAMO-SEATS and X-13ARIMA-SEATS automatically generate the appropriate trading day, weekday, leap year, and Easter variables according to the selected specifications. However, users may need to change the following automatic options: (1) modify the trading day, business day, and leap year variables to match national holidays that are different from JD+'s pre-specified options; (2) link two calendars with different periods; (3) combine two or more calendar variables with appropriate weights; (4) import user-defined regression variables, including holidays with varying frequency, duration, start and end dates. JD+ provides predefined regression quantities, including the following:

- Variables for the six-day effect: n{Monday}–n{Sunday}, n{Tuesday}–n{Sunday}, . . . , n{Saturday}–n{Sunday}, where n is 'the number of days in a month';
- Variables for weekday and weekend effects: N (M, T, W, Th, F) - N (Sat, Sun) × 5/2;
- Leap year variable: The leap year variable measures the impact of the difference in the number of days in February, ie, 28 or 29. This variable is set to zero for all months except February. However, in February, it adopted a value of –0.25. If the month includes 28 days of the current year and 29 days of –0.75, it is a leap year;
- Easter variable: TRAMO-SEATS assumes that the Easter effect lasts for six days. For X-13, the RSA5(c) spec tests three different lengths of the Easter effect and selects the optimal length from 1, 8, or 15 days. The user can also define the duration. Since Easter occurs in March or April, this variable takes the value zero for all months except March and April.

8.4.1.2 Create a Calendar

You can create a new calendar by right-clicking Calendars under Workspace→Utilities, and selecting Add Calendar→National[10] (see Figure 8.9) to add a national calendar:

The Add National Calendar window appears, providing editing options for the national calendar (Figure 8.9, right). You can add a new national holiday by clicking on the nearby arrow. This will open a drop-down list with options: Fixed, Easter-Related, Fixed Week, and Special Date. When all holidays are added to the national calendar, it can then be

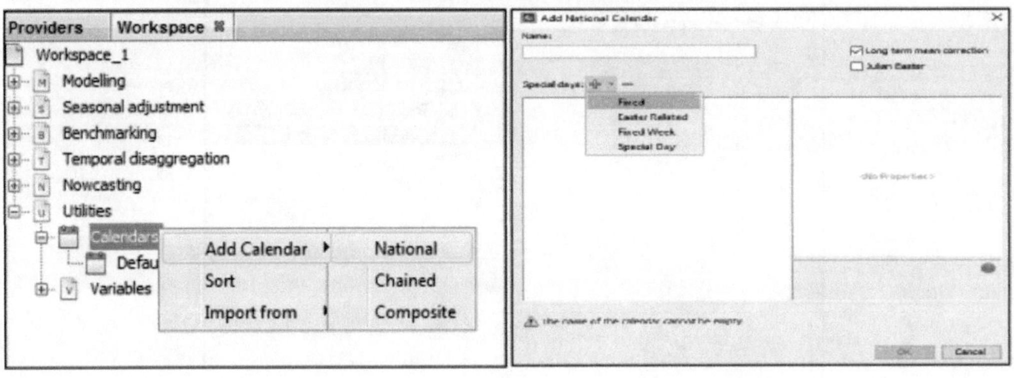

FIGURE 8.9
Adding a National Calendar.

saved by clicking OK. To demonstrate the different options and how to add a national calendar to JD+, this book will use the Belgian national calendar as an example (Table 8.3).

8.4.1.3 Special Day

JD+ offers 17 pre-specified holidays based on the most common European holidays. Since these holidays can differ across countries, it is essential to ensure the accuracy of the specified dates (See Figure 8.10).

They can be added by clicking on the special day (Figure 8.10). Once you add a holiday from the list, you can configure its offset, validity period (start and end dates), and weight. The nine holidays in the Belgian calendar are special days. Pentecost is explained in the holiday section related to Easter. National Day, Assumption Day, and Armistice Day are excluded from the special days list and must be manually added as fixed dates when required. Table 8.4 provides definitions of special dates specified in JD+.

TABLE 8.3
List of Holidays in Belgium

Name of holidays	Date
New Year's Day	Special holiday in JD+
Easter	Special holiday in JD+
Easter Monday	Special holiday in JD+
Mayday	Special holiday in JD+
Ascension	Special holiday in JD+
Pentecost	Special holiday in JD+
Pentecost Monday (WhitMonday)	Special holiday in JD+
National day	21th of July
Assumption of Mary	15th of August
All Saints	Special holiday in JD+
Amistic Day	11th of November
Christmas	Special holiday in JD+

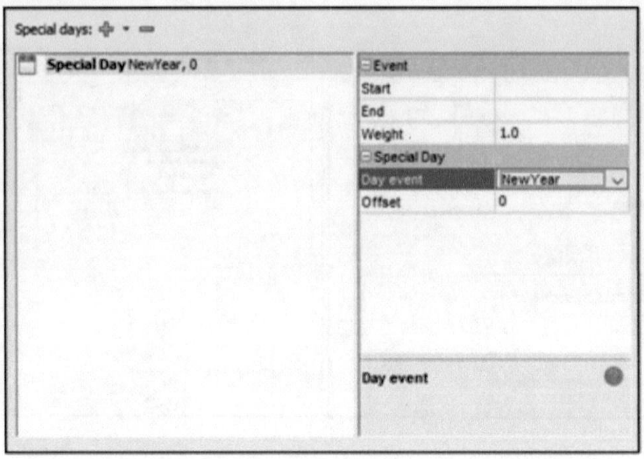

FIGURE 8.10
Adding a Special Date.

TABLE 8.4
Definitions of the Special Holidays in JD+

Holidays	General Definitions
New Year	It is a fixed holiday, which falls on the 1st of January
Shrove Monday	It is a moving holiday, which falls on the Monday before Ash Wednesday
Shrove Tuesday	It is a moving holiday, which is the preceding day of Ash Wednesday
Ash Wednesday	It is a moving holiday, which occurs 46 days before Easter
Easter	It is a moving holiday which varies between the 22nd March and the 25th of April, depending on the offset time. It always falls on a Sunday
Maundy Thursday	It is a moving holiday which falls on the Thursday before Easter. It is the fifth day of Holy Week and is preceded by Holy Wednesday and followed by Good Friday
Good Friday	It is a moving holiday, which refers to the Friday in Easter week
Easter Monday	It is a moving holiday, which is the day after Easter Sunday
Ascension Day	It is a moving holiday, which is celebrated on a Thursday, the 40th day from Easter Day
Pentecost	It is a moving holiday, which is celebrated seven weeks (50 days) after Easter Sunday
Corpus Christi	It is a moving holiday, which is two months after Maundy Thursday
Whitmonday	It is a moving holiday, which is celebrated the day after Maundy Thursday
May Day	It is a fixed holiday, which falls on May 1st. It is a Spring Festival and usually a public holiday
Halloween	It is a fixed holiday, which is observed on the 31st of October
Assumption	It is a fixed holiday, which is celebrated on the 15th of August
All Saints Day	It is a fixed holiday, which is celebrated on the 1st of November by parts of Western Christianity, and on the first Sunday after Pentecost in Eastern Christianity
Christmas	It is a fixed holiday, which is observed on the 25th of December in many countries

8.4.1.4 Fixed Holidays

Countries usually have fixed holidays in their national holiday calendars. These holidays are on the same day every year. Thus, their effects are likely captured in the seasonal component. However, fixed holidays may fall on weekdays or weekends, making seasonal factors irregular. It is recommended to include fixed holidays in the JD+ calendar regression variables. In the case of Belgium, National Day, Assumption Day, and Armistice Day are not included in the pre-designated holidays in JD+. Since they occur on a fixed date every year (see Table 8.4), they can be added as a fixed holiday. This is done by clicking and then pinning. You can then enter the appropriate spec, as shown in Figure 8.11. When fixed or special holidays are moved to a trading day (typically a Monday), a custom user-defined regressor may be needed, as the trading activity on that day reflects weekend patterns. On the other hand, this type of vacation may be captured by the seasonal component and may not require any user-defined regressors. The decision depends on the kind of holiday (See Figure 8.11).

8.4.1.5 Easter-Related Holidays

Easter and related holidays, such as Whitsunday/Whit Monday, vary in timing and duration internationally. In addition, the number of holidays around Easter may vary (offset days). As a result, JD+ provides an option called Easter-related days for users to design such a holiday (Figure 8.12). This is done with a single click, and the relevant date is determined by offset.

Since Easter-related holidays can vary from country to country, such as the Belgian calendar, it is assumed that Pentecost and Whitsunday Monday do not apply on special days.

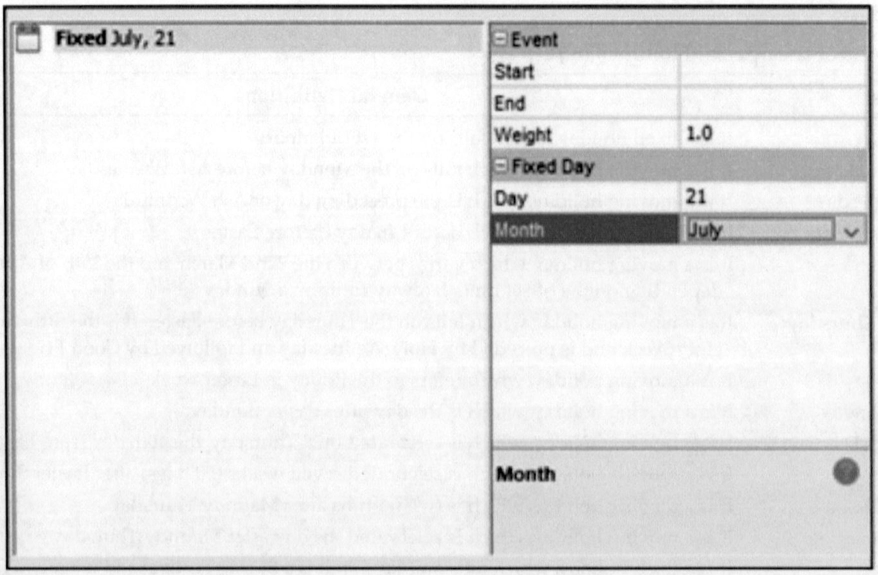

FIGURE 8.11
Fixed Holidays in the Calendar Module.

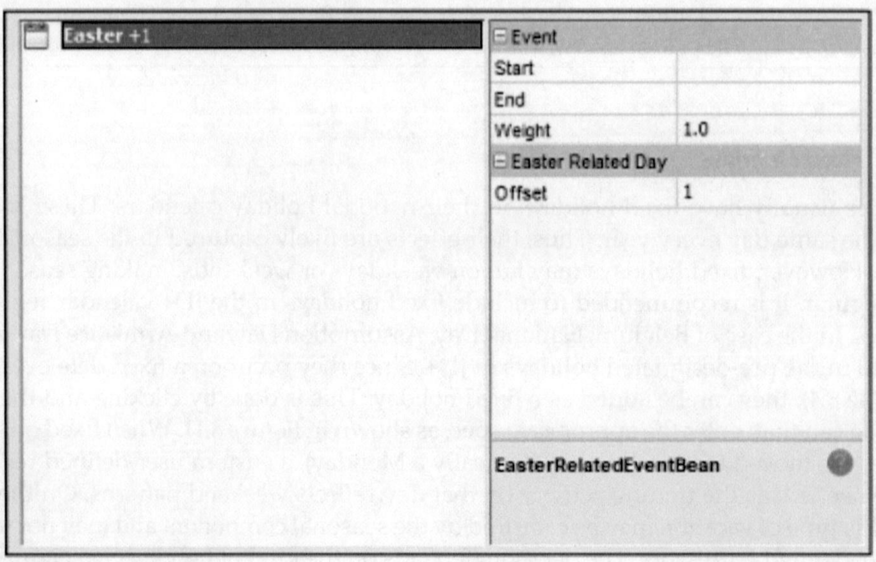

FIGURE 8.12
Choosing a Date Associated with Easter.

These holidays are defined in Table 8.4. Since the Pentecostal holiday occurs 50 days after Easter, the offset is set to 50. If it falls 50 days before Easter, the user will enter 50 with a minus sign (–). For Pentecost Monday, the offset value will be 51 because the festival is celebrated one day after Pentecost. Figure 8.13 illustrates the full holiday calendar for Belgium used in this example. It includes 7 pre-designated holidays, 2 Easter-related holidays, and 3 fixed annual dates *(eg, "Configure fixed annual dates like Christmas Day")*.

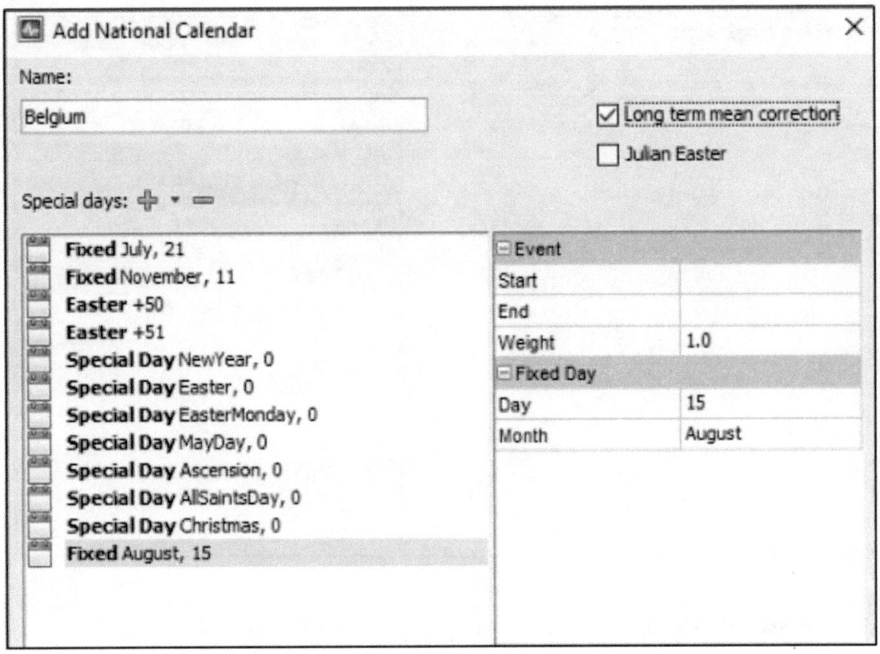

FIGURE 8.13
An Example Holiday Calendar for Belgium.

8.4.1.6 Fixed Week

In some countries, holidays occur on specific weekdays within particular weeks of the month. In the case of the Belgian calendar example, there is no fixed weekly holiday. Figure 8.14 shows an example (See Figure 8.14).

8.4.1.7 Composite Calendars and Chained Calendars

JD+ provides the option to define a composite calendar. Sometimes it can be useful, for example, this can be applied to the preparation of a calendar for the total industrial production index of EU countries. Using this application, users can combine previously created calendar variables by defining appropriate weights for each calendar. In rare cases, users of JD+ may need to link two different calendar variables. This request may appear because it was deleted after a certain date. You can create a linked calendar variable by selecting a calendar and a break date. When choosing a Chained or Composite calendar instead of a National one (see Figure 8.9), you can specify both the desired calendar and the breakpoint date. The calendar you want to use needs to be created by the user first, as shown in the Belgian example above.

8.4.1.8 Viewing Calendar Variables

JD+ displays calendar variables created for Belgium by double-clicking the calendar named "Belgium" under Workspace→Utilities→Calendars. Figure 8.15 shows the regression amount. The upper left panel of Properties displays the properties of the calendar variable. The panel at the top right graphically displays the selected regressor. The Properties

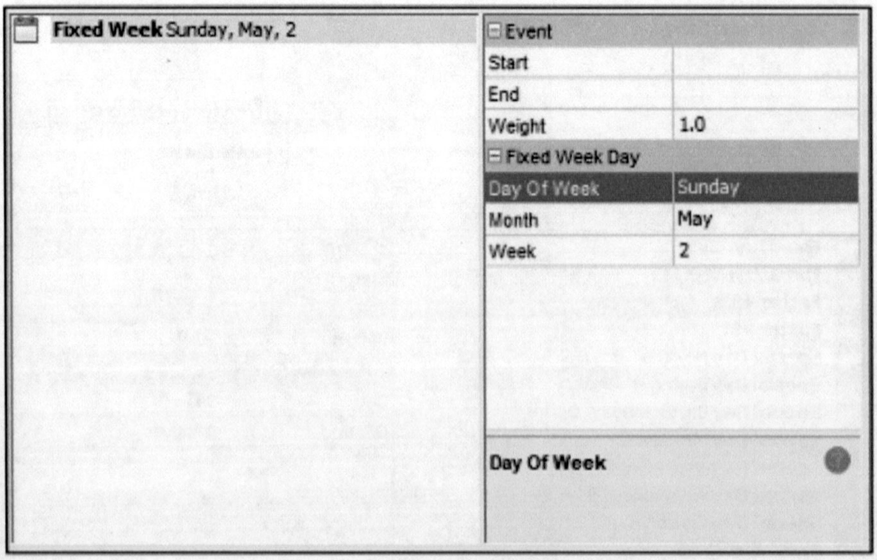

FIGURE 8.14
Adding a Fixed Week to the Calendar Module.

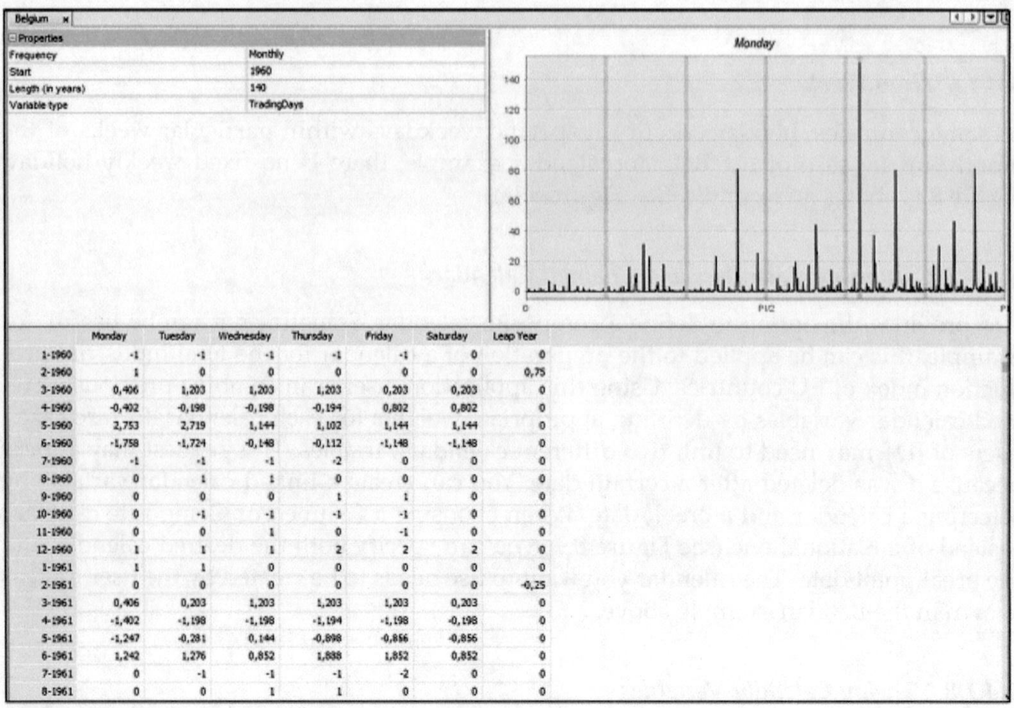

	Monday	Tuesday	Wednesday	Thursday	Friday	Saturday	Leap Year
1-1960	-1	-1	-1	-1	-1	0	0
2-1960	1	0	0	0	0	0	0,75
3-1960	0,406	1,203	1,203	1,203	0,203	0,203	0
4-1960	-0,402	-0,198	-0,198	-0,194	0,802	0,802	0
5-1960	2,753	2,719	1,144	1,102	1,144	1,144	0
6-1960	-1,758	-1,724	-0,148	-0,112	-1,148	-1,148	0
7-1960	-1	-1	-1	-2	0	0	0
8-1960	0	1	1	0	0	0	0
9-1960	0	0	0	1	1	0	0
10-1960	0	-1	-1	-1	-1	0	0
11-1960	0	0	1	0	-1	0	0
12-1960	1	1	1	2	2	2	0
1-1961	1	1	0	0	0	0	0
2-1961	0	0	0	0	0	0	-0,25
3-1961	0,406	0,203	1,203	1,203	1,203	0,203	0
4-1961	-1,402	-1,198	-1,198	-1,194	-1,198	-0,198	0
5-1961	-1,247	-0,281	0,144	-0,898	-0,856	-0,856	0
6-1961	1,242	1,276	0,852	1,888	1,852	0,852	0
7-1961	0	-1	-1	-1	-2	0	0
8-1961	0	0	1	1	0	0	0

FIGURE 8.15
Viewing Calendar Variables.

panel allows users to access and configure the regressor frequency, start date, year length, and variable type. The user can convert the regressor. The None option for variable type causes JD+ to apply only the leap year variable. Option trading days refer to the six-day variable and the leap year variable. The last option, working days, represents the calendar variable, which models the difference in activity between weekends and weekdays, as well as the leap year variable (Figure 8.15). The resulting regressor can be copied to Excel by dragging and dropping.

8.4.1.9 External User-Defined Variables

Users may want to define specific calendar variables themselves outside of JD+, for example, some moving holidays such as Ramadan or Sacrifice Day. To do this, you can prepare regression quantities in Excel and import them into JD+ as user-defined variables.

First of all, the dates of Ramadan and the Feast of Sacrifice should be determined for a specific length each day. In this example, the length covers the period from 1 January 1974 to 31 December 2023. Table 8.5 shows the design of the required Excel file. The columns from A to F represent the official holidays in Turkey. Saturdays and Sundays are also listed in the document to see when they coincide with the dates of Ramadan or the festival of sacrifices. For each moving-holiday variable, when the holiday falls on a given date, the indicator is coded in two ways: first, a value of 1 for each holiday (excluding Saturdays, Sundays, and public holidays), and 0 for all other days.

Then construct this list for the desired period for which the regression variables are required. As shown in Table 8.6, the second step is summarising the results for each month.

The third step is to extract the variables based on the monthly averages (Table 8.7), as JD+ assumes that user-defined regressors are properly centred. Specifically, the extracted variable is calculated by subtracting the mean of each user-defined regressor from its value in each calendar period. The calculation of the centred regressor is important because the change in the number of public holidays affects the monthly average over the entire period.

First, the calendar variable should be centred so as not to have any horizontal impact on the series. Second, monthly averages should be extracted from calendar variables, as calendar variables may contain seasonality. Calendar variables should only measure calendar effects, not seasonality. The average of the regression amounts is calculated for each month. Users can deduct the sum of the Ramadan and Sacrifice holiday variables directly from the monthly average to get the total holiday variables.

To import user-defined calendar variables, you can drag and drop them into JD+ to create static variables, or you can read them directly from an Excel file to create dynamic variables. For the first option, create a new variable by selecting New after right-clicking Workspace/Utilities/Variables. After double-clicking the newly created 'Vars-1', you can import the time series by dragging and dropping it from Excel, or by copying and pasting using Ctrl+C and Ctrl+V. To apply this imported regressor, you need to change the spec type of the calendar effect to user-defined (Figure 8.16).[11]

8.4.2 Seasonal Adjustment Method and Spec Selection

The seasonal adjustment process separates seasonal effects from the time series to reveal underlying movements. JD+ first determines the structure of the components, then estimates the components, and decomposes the time series into trend, seasonality, and irregular components. Decomposition can be in additive or multiplicative form. JD+ provides seasonal adjustments for single analysis or multiprocessing.

TABLE 8.5

An Example of Structure for Constructing External Calendar Variables in Excel File

			A	B	C	D	E	F			G	H	I	J
Date	Sacrifice	Ramadan	New Year	23-Apr	01-May	19-May	30-Aug	29-Oct	Saturday	Sunday	Sacrifice=1; otherwise=0 (Excl. Sat, Sun, and official holidays	Sacrifice=1; otherwise=0 (Excl. Sun and official holidays	Ramadan=1; otherwise=0 (Excl. Sat, Sun, and official holidays	Ramadan=1; otherwise=0 (Excl. Sun and official holidays
1.1.1974	0	0	1	0	0	0	0	0	0	0	0	0	0	0
2.1.1974	0	0	0	0	0	0	0	0	0	0	0	0	0	0
3.1.1974	0	0	0	0	0	0	0	0	0	0	0	0	0	0
4.1.1974	1	0	0	0	0	0	0	0	0	0	1	1	0	0
5.1.1974	1	0	0	0	0	0	0	0	1	0	0	1	0	0
6.1.1974	1	0	0	0	0	0	0	0	0	1	0	0	0	0
26.12.2023	0	0	0	0	0	0	0	0	1	0	0	0	0	0
27.12.2023	0	0	0	0	0	0	0	0	0	1	0	0	0	0
28.12.2023	0	0	0	0	0	0	0	0	0	0	0	0	0	0
29.12.2023	0	0	0	0	0	0	0	0	0	0	0	0	0	0
30.12.2023	0	0	0	0	0	0	0	0	0	0	0	0	0	0
31.12.2023	0	0	0	0	0	0	0	0	0	0	0	0	0	0

TABLE 8.6

An Example of Aggregated Monthly Values

Date	Sacrifice Holidays (Excl. Sat, Sun, and official holidays)	Ramadan Holidays (Excl. Sat, Sun, and official holidays)	Sacrifice Holidays (Excl. Sun and official holidays)	Ramadan Holidays (Excl. Sun and official holidays)
1.1.1974	2	0	3	0
1.2.1974	0	0	0	0
1.3.1974	0	0	0	0
1.4.1974	0	0	0	0
1.5.1974	0	0	0	0
....	....	...	...	...
1.5.2023	0	0	0	0
1.9.2023	3	0	4	0
1.10.2023	0	0	0	0
1.11.2023	0	0	0	0
1.12.2023	0	0	0	0

TABLE 8.7

An Example of Feast of Sacrifice Holiday Variable

Date	Sacrifice Holidays (Excl. Sat, Sun, and other holidays)	Average Values for 1974–2015	Sacrifice Regressors (Excl. Sat, Sun, and other holidays)
1.1.1974	2	0.2381	1.7619
1.2.1974	0	0.2143	–0.2143
1.3.1974	0	0.2143	–0.2143
1.4.1974	0	0.1905	–0.1905
1.5.1974	0	0.1429	–0.1429
1.6.1974	0	0.2143	–0.2143
....	...	...	...
1.7.2023	0	0.2143	–0.2143
1.8.2023	0	0.2143	–0.2143
1.9.2023	3	0.2619	2.7381
1.10.2023	0	0.3333	–0.3333
1.11.2023	0	0.4048	–0.4048
1.12.2023	0	0.2857	–0.2857

8.4.2.1 Choose a Method

JD+ offers several different methods for adjusting time series. Firstly, you need to choose one of two methods: TRAMO-SEATS or X-13ARIMA-SEATS. You can use both methods to perform seasonal adjustments to see the difference. Both methods are endorsed by Eurostat, the European Central Bank, the OECD, and the International Monetary Fund as official recommendations. Both produce similar but not identical results. The choice between X-13 and TRAMO-SEATS should ensure continuity and standardisation in the production of official statistics. Secondly, the method selection can be done from the menu Statistical Methods→ Seasoning panel (Figure 8.17) for a single analysis; the preferred method (top

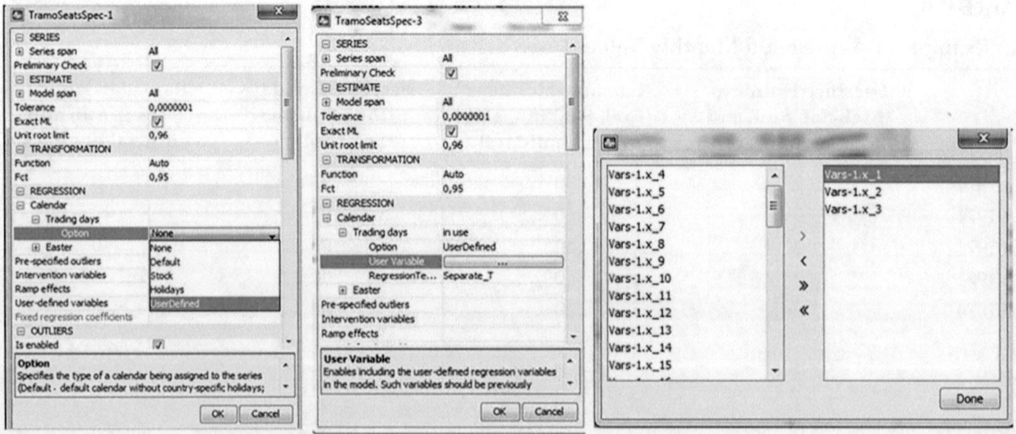

FIGURE 8.16
User-defined Regressor in the Spec.

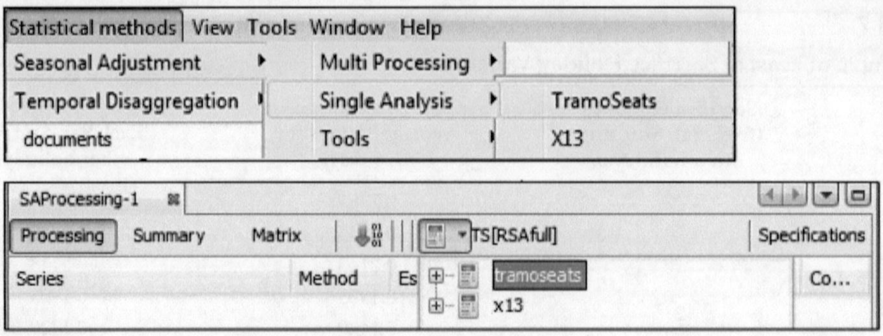

FIGURE 8.17
Method Selection.

half of Figure 8.17) or the new → for multiprocessing, and then users can find the method in the multiprocessing panel (bottom half of Figure 8.17).

For simplicity's sake, the example method chosen would be TRAMO-SEATS. It is also possible to use the X-13 to replicate all applications and exercises.[12]

8.4.2.2 Spec and User Regressors

Seasonal adjustment implies a series of mathematical and statistical tasks (outlier detection, model identification, decomposition, etc.), each of which can be specified in the software. This can be done under Workspace→Seasonal Adjustment→Specification. JD+ provides 7 predefined specs for TRAMO-SEATS and X-13 and uses some default specs for each method (Figure 8.18). Options range from simple airline models (RSA0) to more complex specifications (RSAfull), corresponding to those used in TRAMO-SEATS and X-13ARIMA-SEATS (see Table 8.8). Users can select from these predefined specifications and modify them if needed, as explained later in this section.

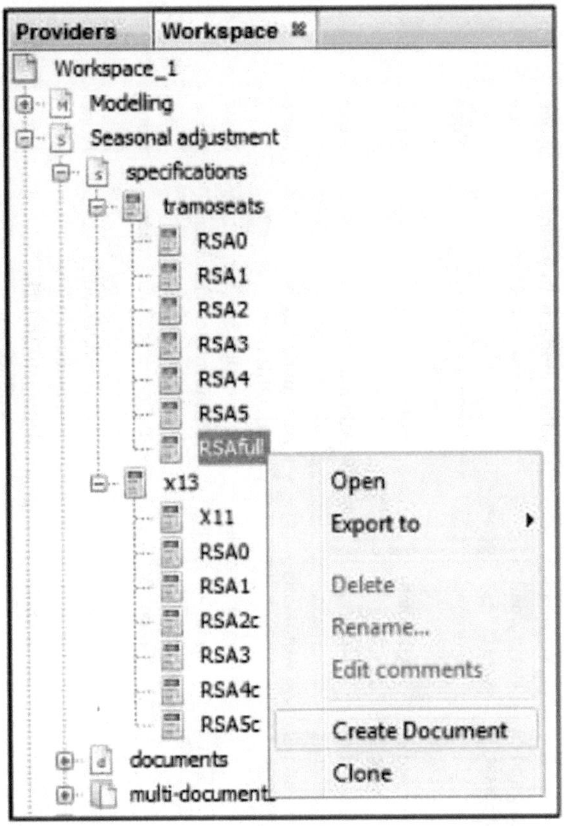

FIGURE 8.18
Spec.

The selected spec defines whether TRAMO and regARIMA should be tested for trading day and weekday impacts, leap year effects, etc. If the effect is significant, they run the test and apply the regression variables accordingly. If included in the spec, TRAMO and regARIMA will also test the Easter effect, but not all moving holidays are tested by default, as these changes vary from country to country (See Figure 8.18).

First, the RSAfull spec in TRAMO-SEATS or the RSA4c or RSA5c spec for X-13ARIMA-SEATS is recommended for time series with significant calendar effects. The RSA3 spec for both methods is a good benchmark for time series without calendar effects. While the RSA3 spec does not test for calendar effects, TRAMO-SEATS' RSA4 and RSA5 specs pre-test and determine the weekday/leap year effect and the trading day/leap year effect, respectively. Although the working day effect may be valid for financial time series (ie, monetary aggregates such as M1, M2, or deposit amounts), and the trading day effect for real sector time series (ie, a measure of industrial production, construction, or services), this is not the case in all cases. The RSAfull spec of TRAMO-SEATS tested all effects and selected the appropriate calendar effect based on its significance. The RSAfull and RSA4c/RSA5c specs for TRAMO-SEATS and X-13ARIMA-SEATS also tested the Easter holiday and leap year effects. This means that you don't need to create any additional regressors

TABLE 8.8
Definitions of the Special Holidays in JD+

Method	Name	Transformation	Calendar effect	Outlier detection	Modelling	Benchmarking
TRAMO/SEATS	RSA0	Level	None	None	Alline model with mean	None
TRAMO/SEATS	RSA1	Test for log/level	None	Automatic	Alline model with mean	None
TRAMO/SEATS	RSA2	Test for log/level	Test for working day and or leap year and/or Easter	Automatic	Alline model with mean	None
TRAMO/SEATS	RSA3	Test for log/level	None	Automatic	Automatic Model Identification	None
TRAMO/SEATS	RSA4	Test for log/level	Test for working day and or leap year and/or Easter	Automatic	Automatic Model Identification	None
TRAMO/SEATS	RSA5	Test for log/level	Test for working day and or leap year and/or Easter	Automatic	Automatic Model Identification	None
TRAMO/SEATS	RSAfull	Test for log/level	Test for working day and or leap year and/or Easter	Automatic	Automatic Model Identification	None
X-13	X-11	Non-applicable	Non-applicable	Non-applicable	Non-applicable	None
X-13	RSA0	Level	None	None	Airline model with mean without mean	None
X-13	RSA1	Test for log/level	None	Automatic	Airline model with mean without mean	None
X-13	RSA2c	Test for log/level	Test for working day and or leap year and/or Easter	Automatic	Airline model with mean without mean	None
X-13	RSA3	Test for log/level	None	Automatic	Automatic model identification	None
X-13	RSA4c	Test for log/level	Test for working day and or leap year and/or Easter	Automatic	Automatic model identification	None
X-13	RSA5c	Test for log/level	Test for working day and or leap year and/or Easter	Automatic	Automatic model identification	None

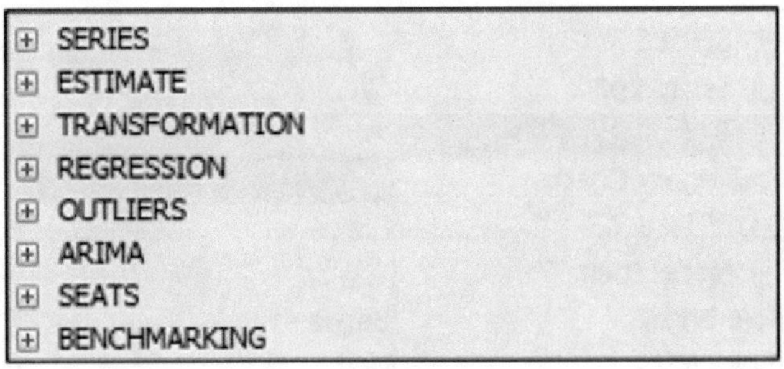

FIGURE 8.19
Dialogue Box.

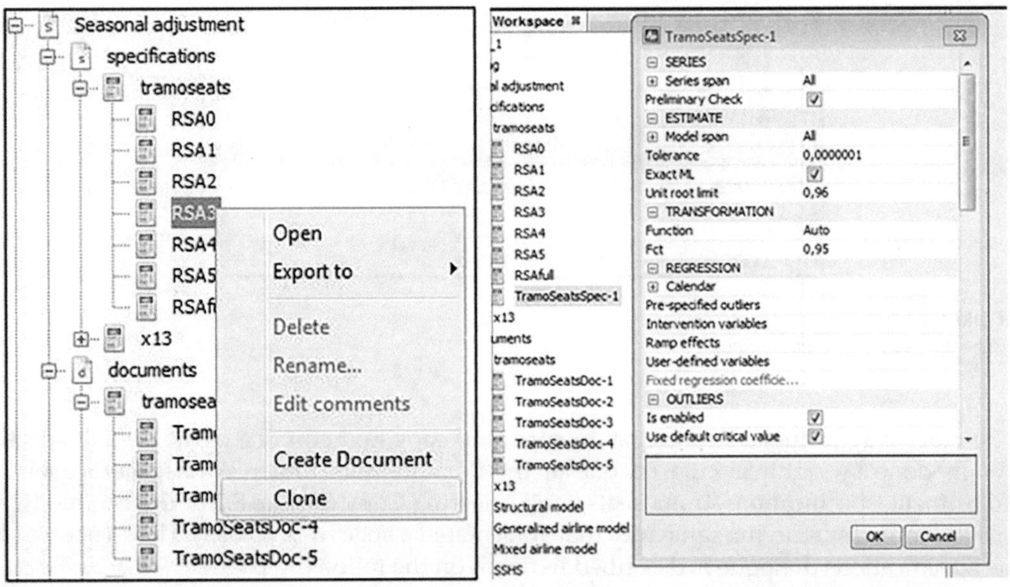

FIGURE 8.20
(a) User-Defined Spec Steps.

for these calendar effects. If additional calendar effects, such as country-specific moving holidays, need to be included in the pre-adjustment process, you can import an external regressor and select it in the specification. If user-defined specs are frequently used, such as including national calendar effects or excluding certain types of outliers, the best solution is to define a spec and add it to the workspace.

8.4.2.3 The General Structure of the Spec Dialogue

This can be done by right-clicking TRAMO-SEATS or X-13 in Workspace/Seasonal Adjustment/specifications, then selecting New. The new spec will automatically appear in the Workspace panel. Double-click it, or right-click to open a dialogue box that lets you specify its properties directly (Figure 8.19).

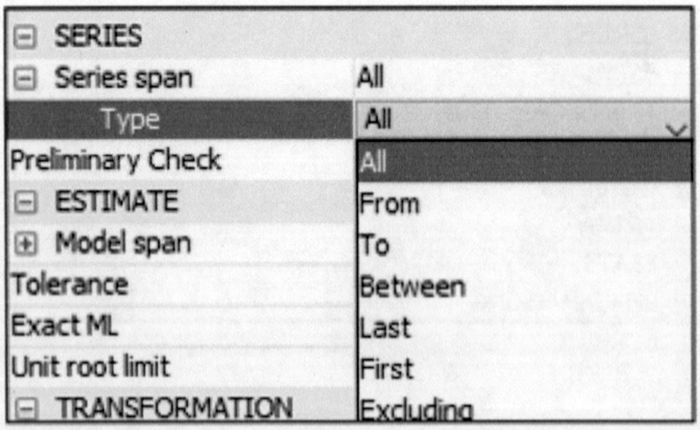

FIGURE 8.21
(b) The Series Section of the Spec.

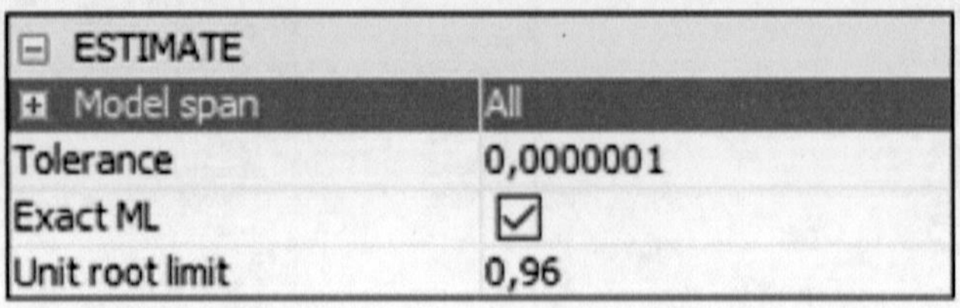

FIGURE 8.22
(c) The 'Estimates' Section of Spec.

Another way to create a new custom spec is to copy and edit one of the default specs. This is done by right-clicking on one of the default specs under Workspace/Seasonal Adjustment/specifications/tramoseats/ and selecting Copy (Figure 8.20). You can modify a newly copied spec in the same way that you created a spec from scratch. The structure of the Specifications dialogue is described in detail on the following pages.

8.4.2.4 Steps of User-Defined Spec Steps

Saving the workspace also saves any newly created specifications. They can be invoked by reloading the Workspace and then made available for use. The first part of the Series spec box is the SERIES, which contains the Series span and the Preliminary check (See Figures 8.20 & 8. 21).

The user can set the span of the original time series to be analysed/SA by selecting one of several options, such as From, To, Between, and so on. When the Preliminary Check option is selected, the quality of the input series is evaluated, and highly problematic series, such as those with multiple identical observations or missing values exceeding a predefined threshold, are excluded. When unchecked, thresholds are ignored, and the process is performed where possible. When 'type All' is selected, a series analysis of its full length is performed (Figure 8.21). The next section, ESTIMATE, contains the Model Span, Tolerance, Exact ML, and Unit root limit options. Typically, these options don't need to be changed or set to other values (Figure 8.22). However, the Model Span option can be used to limit the data span used to determine the regARIMA model. However, the selected regARIMA model will be applied to the entire family.

FIGURE 8.23
(d) The 'Transformation' Section of the Spec.

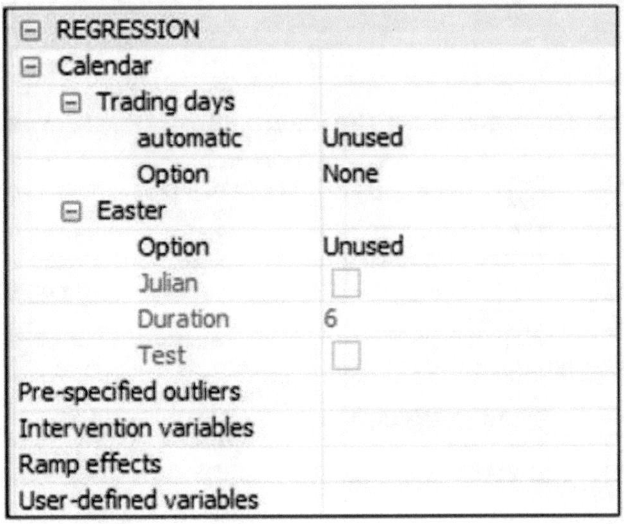

FIGURE 8.24
The Regression Section of the Spec.

The TRANSFORMATION section includes Function and FCT (Figure 8.23). The options for the function include (1) None, which does not convert the original time series; (2) Log, which takes the logarithm of the original time series; (3) Automatic, which performs a test that selects a logarithmic or level specification to automatically select the conversion spec. FCT controls deviations in testing. If FCT > 1, if skewed, it is horizontally skewed; otherwise, it is logarithmically skewed.

8.4.2.5 Regression

Regression consists of five main components: calendar, pre-specified outliers, intervention variables, slope effects, and user-defined variables (Figure 8.24).

(1) The calendar consists of two separate sections: Trading Day and Easter. The option is used to define the type of calendar regressor, such as None, Default, Stock, Holidays, or 'UserDefined'. Automatic is used to define the type of test, eg, 'F-Test' for multiple tests and 'WaldTest' for a single test (Figure 8.25). The default options include selecting the calendar type ('None', 'TradingDays' (6 regressions), or 'WorkingDays' (1 regression)) and the leap year effect regression. You can specify the type of test for the regression of the trading day effect.

When the F-test or t-test is selected, JD+ will perform the test of the trading day variable. Otherwise, testing is not necessary before using the trading day variables. The

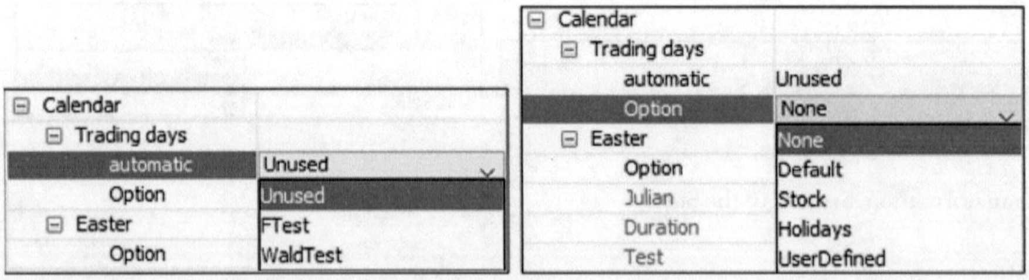

FIGURE 8.25
The Calendar Section of the Spec.

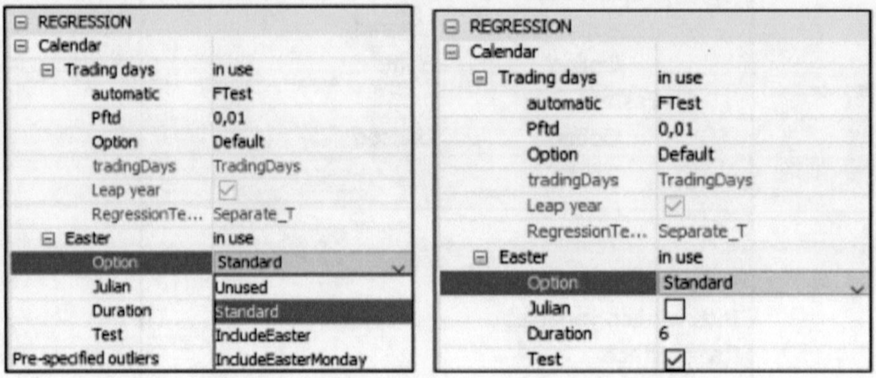

FIGURE 8.26
The 'Easter' Section of the Spec.

week-on-week effect of stock variables can be described and measured using the Stock option. Holidays can be used when using a user-defined calendar with country-specific holidays. The 'UserDefined' option is used for the uploaded calendar variable. Finally, 'None' means that calendar variables are not used for regression models.

Calendar effects can be added to the model manually (by setting the automatic parameter to Unused) or automatically, where the selection of the number of calendar variables is based on 'F-test' or 'Waldtest'. In this example, Option is set to Default, automatically set to 'F-test', and has a 'Pftd' value of 1% (Figure 8.26). Figure 8.26 also shows the main attributes of Easter. The standard option is recommended because the impact of Easter and Easter Monday is often accounted for in user-defined regression models. If necessary, the duration of this effect can be modified from the default six-day length (for TRAMO-SEATS) if the impact of the country or economic activity in question may be different. If the Test box is checked, a statistical t-test is performed on the Easter effect. Otherwise, it is added to the model without prior testing.

(2) Pre-specified outliers and intervention effects are not used at the beginning of the seasonal adjustment process. A pre-specified outlier is an outlier known before adjustment. Intervention variables refer to known special events that affect the data, such as strikes, political decisions, etc. Not only can it create outliers, but it can also make more complex interventions to match specific events. They may be needed in cases where seasonal adjustments are problematic. The remaining two options, Ramp effects and User-defined variables, are beyond the scope of this basic seasoning process. The slope effect

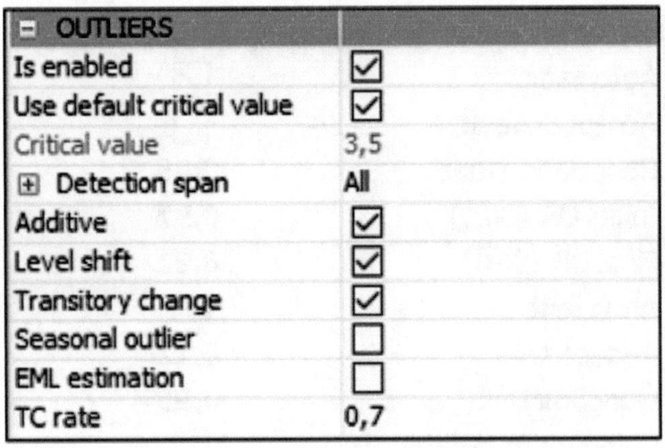

FIGURE 8.27
The Outliers Section of the Spec.

refers to a linear increase or decrease in the level of a series that persists over a certain time. They are more relevant for advanced time series analysis or seasonal adjustments.

Outliers (see Figure 8.27) indicate that automatic outlier detection and correction will be performed during the seasonal adjustment process, and this option is active for all default specs.

Defining thresholds for detecting outliers using default values is the most important fine-tuning parameter that affects the quality of the seasoning process. If this value increases, the number of outliers decreases, and vice versa. The default value is recommended for the initial adjustment, but adjustments may improve results for the given series. However, caution should be exercised when adjusting the default values. Sometimes it can be useful to limit the detection span of automatic outlier detection to minimise revisions. It allows for the selection of four main types of outliers: additive outliers (AO), horizontal shifts (LS), transient changes (TC), and seasonal outliers (SO). In the case of SA standards, the remaining two options, namely EML estimates and TC rates, are generally not considered.

8.4.2.6 ARIMA and SEATSA

ARIMA and SEATS are the core parts of the spec when using the TRAMO-SEATS method. In general, it is not necessary to modify their parameters. The only exception is the ARIMA option, which should be unchecked when the 'Automatic' parameter is selected to apply the default ARIMA model. However, this is only needed if there is a problem with seasonal adjustments. By default, when set to Automatic, TRAMO-SEATS performs automatic model recognition (Figure 8.28).

8.4.2.7 Benchmarking

For many economic indicators, such as compiling annual summaries for comparison, the annual aggregation of SA data rarely equals that of the original time series due to the non-additive nature of seasonal adjustment. However, this is often not the case because seasonal components are random, so seasonality is not necessarily cancelled out over the year. JD+ offers a benchmarking feature that enforces consistency between annual sums:

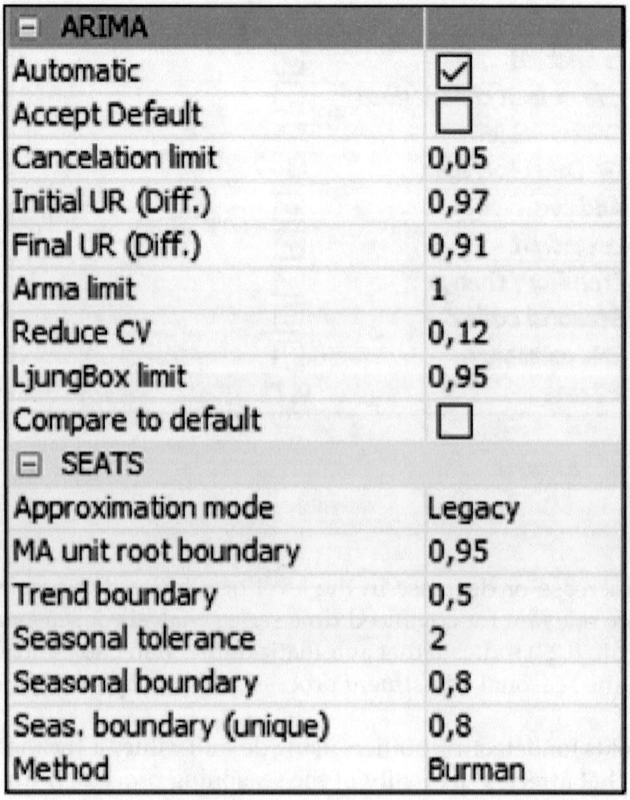

FIGURE 8.28
The 'ARIMA' Section of the Spec.

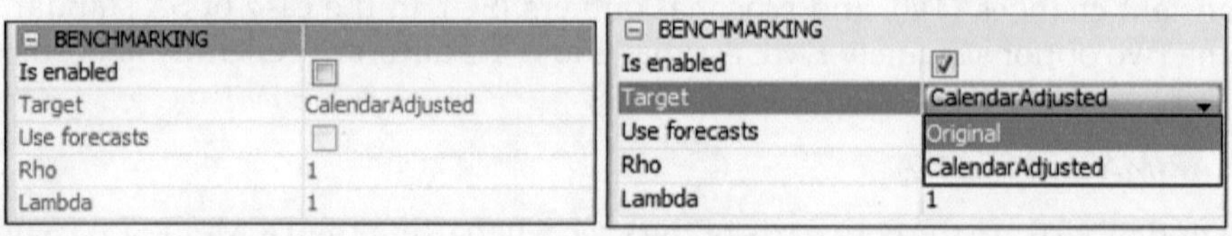

FIGURE 8.29
Benchmark Spec.

the year-over-year totals of SA data will match those of the original (or calendar-adjusted) source data. Activate this option by checking 'Is enabled' in benchmarking (Figure 8.29). Target specifies the target variable for the benchmarking process, which can be the original series or the calendar-adjusted series. When the Use forecasts are checked, the benchmark is also applied to the SA series and forecasts for the target variable. Rho is the value of the first autoregressive (AR 1) parameter (set between 0 and 1) in the function used for benchmarking. Lambda is a parameter in the same function that is related to the weights in the regression equation; It is usually equal to 0, 1/2, or 1.

8.4.3 Single-Series Processing Process

Once the spec is determined, there are two ways to initiate a single process of seasonal adjustment. In the case of using a default spec such as 'RSAfull' for TRAMO-SEATS, the user can start the process from the main menu by selecting Statistical Methods → Seasonal Adjustment→Single Analysis→TramoSeats. Create a new blank document under Workspace/Seasonal adjustment/documents, and you can drag and drop the time series to which you want to adjust the season.

To use a pre-modified spec, a second method of initiating seasonal adjustment must be applied. To achieve this: Right-click Workspace → Seasonal Adjustment → Specifications → tramoseats, then select your desired specification under 'Create Document' (Figure 8.30). Similarly, a new blank document is created, to which the relevant time series can be dragged and dropped.

Figure 8.31 shows and explains step-by-step the entire process of a single process, starting with creating a national calendar using the Belgian example in Figure 8.31, as explained in the following figure.

8.4.4 Multi-Series Batch Processing

JD+ can seasonally adjust a set of time series for each time series using a different spec approach.

(a) To create a new multi-series processing document, select Statistical Method → Seasonal Adjustment → Multi-Series Processing → New from the main menu (Figure 8.32, left). Open the new document SAProcessing-1 and drag and drop the time series to be SA as you would in a single-series process (Figure 8.32, right).

As a result, the selected time series is listed in SA Processing-1 (Figure 8.33). Here, each time series displays the selected method, estimate type, time series status, and tuning quality. In the example, the method is 'RSAfull', and the entire time series is unprocessed.

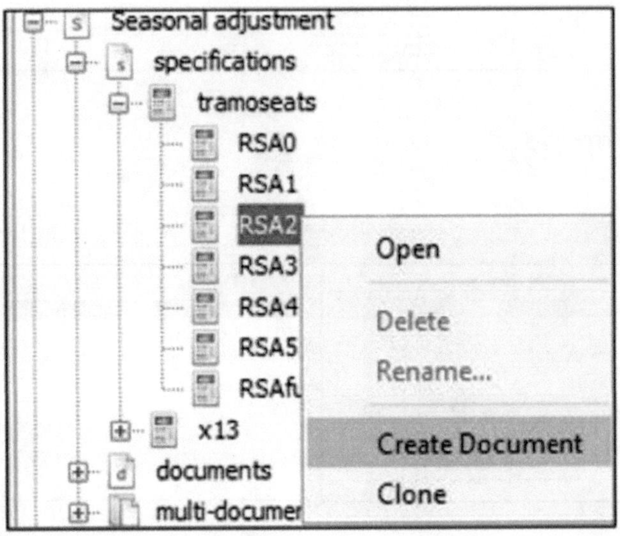

FIGURE 8.30
Create a Single-Processing Document.

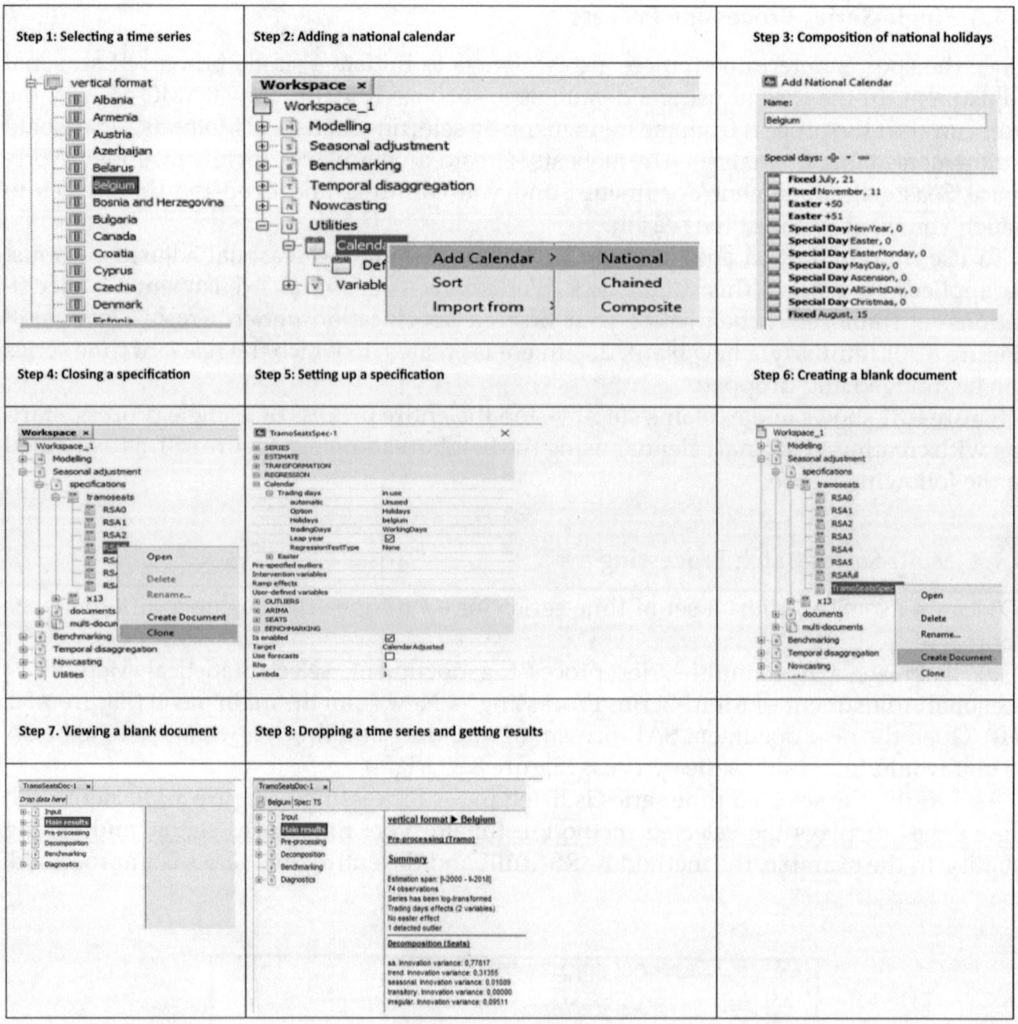

FIGURE 8.31
Step-by-Step Order Processing.

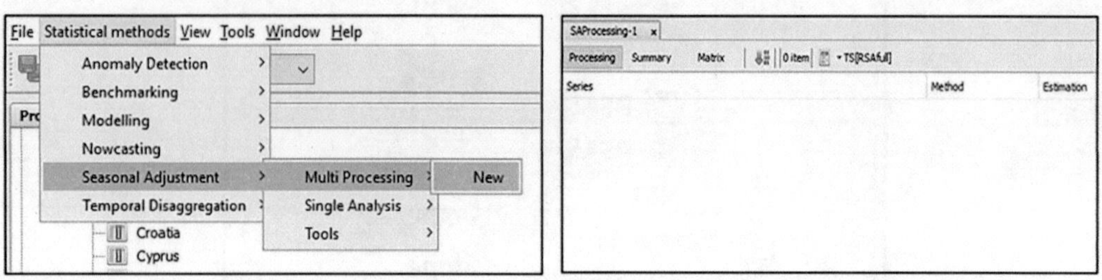

FIGURE 8.32
Creating a Multi-Series Process.

TABLE 8.9

Description of Individual Processing Steps

Steps	Explanation	Details
Step 1	The data set is called by providers and the time series of Belgium is identified.	
Step 2	A user-defined calendar is created.	
Step 3	All 11 Belgian calendar events are entered into the calendar module.	Using the button '+' sign, one defines the type of the calendar effect depending on whether it is fixed, a special day, or another
Step 4	A default specification is cloned as the basis for user-defined specifications.	
Step 5	The regressor and benchmarking parts of the specification are modified.	REGRESSOR: Automatic='Unused', Option= 'Holidays',Holidays='belgium', TradingDay='workingDay',Leap year='True', RegressionTestType='None' BENCHMARKING: Is enabled='True',Target= 'CalendarAdjusted'
Step 6	The new seasonal adjustment document has been created.	
Step 7	The blank document is ready to have the time series dragged and dropped on.	
Step 8	View of the results after performing the single process.	

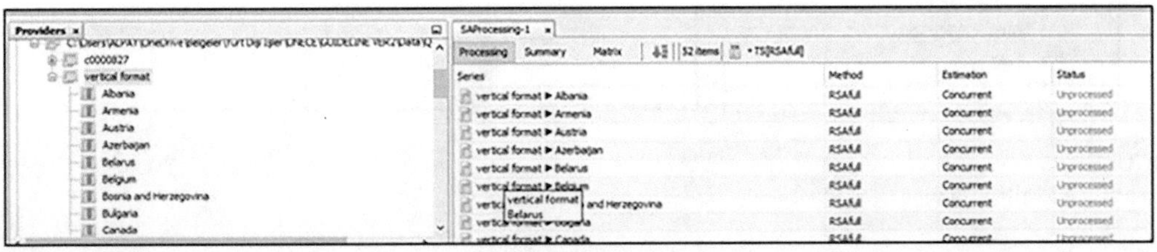

FIGURE 8.33
Multi-Series Processing View.

Once all required specifications have been set, perform the seasonal adjustment simultaneously by clicking the green arrow (Figure 8.34, left). Multi-series processing documents can be found under Workspace/Seasonal adjustment/multi-documents for future use (periodic generation of seasonal adjustment data), as shown in Figure 8.34 on the right.

Another way to create a multi-series process is to right-click Workspace/Seasonal adjustment/ multi-documents and select New (Figure 8.35).

Multi-Series Processing: When performing multi-series processing in a step-by-step manner, it is recommended to start with the default, most comprehensive spec. They are RSAfull for TRAMO-SEATS and RSA5c for X-13. Then, if necessary, the user can change the spec for each time series. You can also change the default settings under Tools → Options in the main menu.

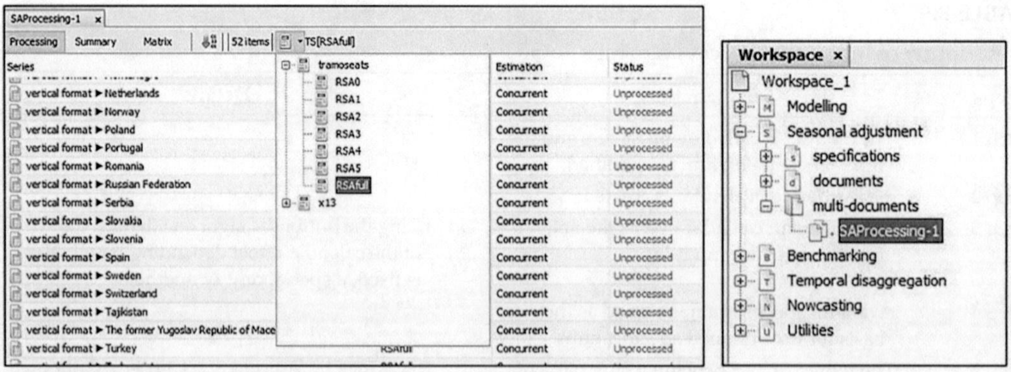

FIGURE 8.34
Multiprocessing in the Workspace.

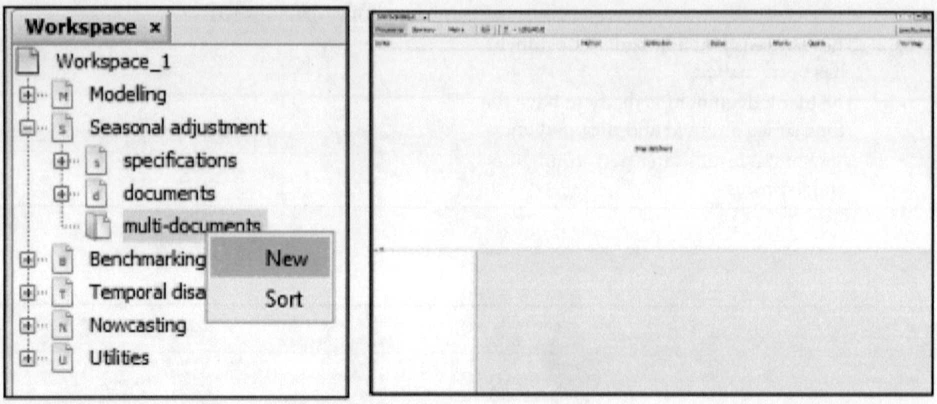

FIGURE 8.35
Creating a Multi-Process.

(b) *The main step of multi-series processing*

Step 1: Creating a blank multi-processing document	Step 2: Dragging and dropping the set of time series into the blank area called 'Drop data here'	Step 3: Pressing the green button and performing seasonal adjustment with the default specification
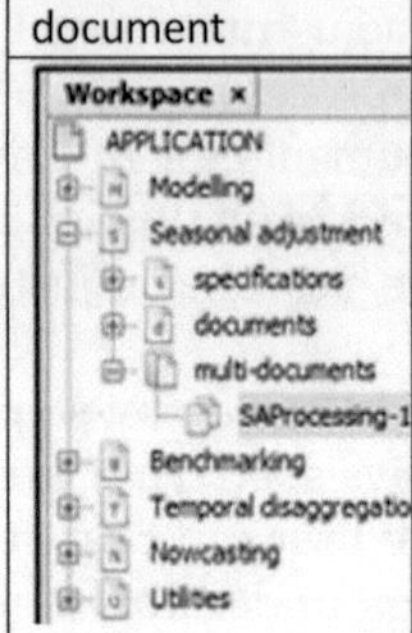		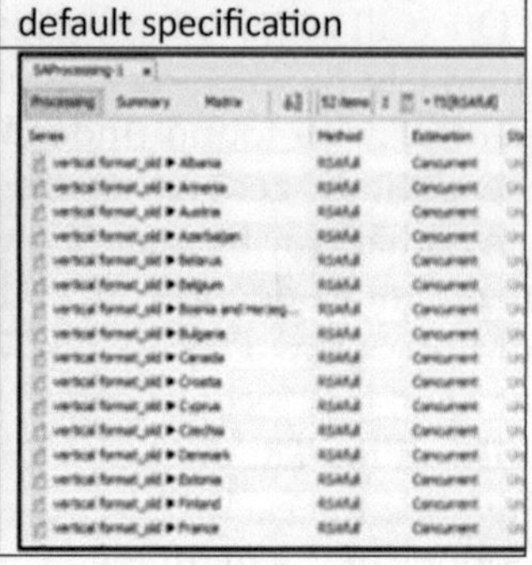

FIGURE 8.36
The Main Steps of Performance in a Multi-Series Processing Seasonality.

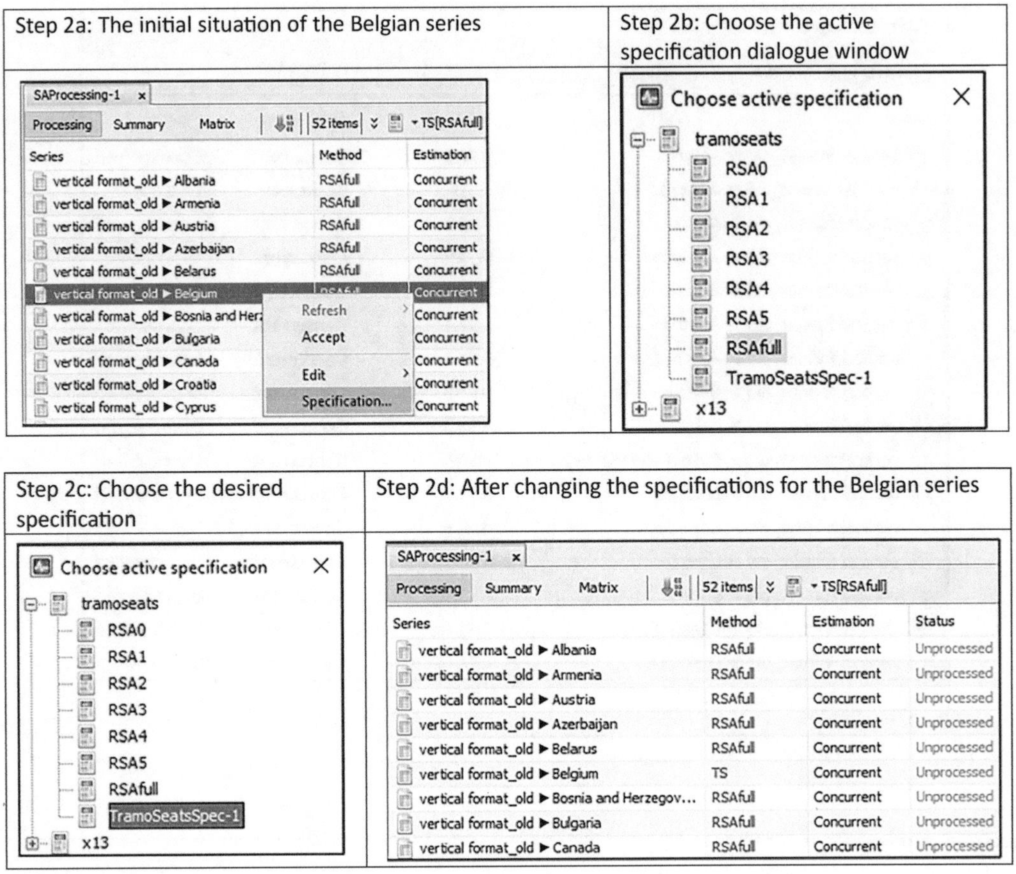

FIGURE 8.37
Set Up Multi-Process Spec.

To modify specifications (eg, adding the Belgian calendar), complete intermediate steps between steps 2 and 3 as shown in Figure 8.37. In this example, the spec for the Belgian series is changed from 'RSAfull' to a user-created TS.

(c) *Set up multi-process spec*

As can be seen from step 2c (Figure 8.37), the SA method can also be changed to X-13. Thus, even within the same series, multiple machining operations may use different methods or specs for seasonal adjustment. To do this, you need to create multiple copies of the same original series (copy + paste). This provides a great way to compare different methods and specs at the same time.

(d) *Multiple specs for multiple series*

Figure 8.38 shows an example of multiple specs in a more complex multiprocessing in JD+.

The multi-series processing is then completed through step 3 (Figure 8.38). When the green button is pressed, a seasonal adjustment is performed, including the creation of a report with relevant statistical diagnostics. If you change any of the specs of multiprocessing, you must repeat this step.

FIGURE 8.38
Multiple Specs for Multiple Series.

8.5 Result Analysis and Quality Diagnosis

JD+ delivers core quality diagnostics through its integrated TRAMO-SEATS methodology. Its purpose is to help open the 'black box' of the SA process performed by the software, thereby supporting the acquisition of robust SA data. Quality issues include (1) visual quality; (2) statistical diagnosis; (3) model (preprocessing); (4) the residuals of the model; and (5) component (decomposition). Baseline seasonality is not a reliable and precise fact, but is determined based on assumptions about underlying conditions and models. The purpose of the quality diagnosis is to reveal any fundamental weaknesses in the SA results to prevent the use of misleading results that are prone to large corrections and may provide false signals about the economy. The automatic programs of TRAMO-SEATS and X-13ARIMA-SEATS are very reliable and can therefore be used to adjust many series, especially important series with limited data. A critical interpretation of diagnostics is imperative when evaluating model performance.

The output of JD+ shows the statistical and mathematical properties of the identified models and components. It includes a wide range of quality diagnostics that reflect the different approaches of the TRAMO-SEATS and X-13ARIMA-SEATS. However, they also have several common quality diagnoses. In addition to digital diagnostic tests, JD+ provides users with a variety of ready-to-use illustrative charts. The smoothness of a SA series is not a measure of quality. Conversely, the irregular components are part of the seasonal adjustment series. Ideally, the number of outliers is relatively small, and they should not be unevenly distributed across the series. Modelling problems are more likely to occur in short time series with many outliers. Longer time series are particularly useful for identifying seasonal patterns from highly volatile time series. However, high volatility may be a natural attribute of many time series in emerging economies or fast-growing industries.

JD+'s careful evaluation of SA data includes an analysis of the stability of the seasonal components. Diagnostics include both statistical tests and graphical outputs, contingent on the selected seasonal adjustment method. The M-statistic for X-13 is explained in Chapter 6, with a note on interpreting the SEATS results.[13] The residuals of the TRAMO process provide a useful tool for verifying that seasonal adjustments are satisfactory. Theoretically, the residuals should be random and free of seasonality.

JD+, on the other hand, also provides a test to confirm that there is no residual seasonality in the SA data either. Quality diagnostics (eg, X-11's M-statistics, TRAMO-SEATS' spectral tests) quantify residual seasonality; values <1 indicate acceptable reduction. The benchmark assesses the equality of different frequency series between the adjusted series and the calendar-adjusted/original series. Benchmarking is sometimes applied if the difference between the aggregation of the original series and the adjusted series (which should be low in theory) is too large. However, benchmarking is generally not recommended because it may not yield the best SA series and may compromise its random structure.

The overall quality metrics displayed by JD+ help draw attention to the most problematic series. When diagnostics indicate potential issues with result validity, modify your specifications and rerun the seasonal adjustment. Given the resource-intensive nature of specification tuning, focus manual interventions on high-stakes series where diagnostic risks directly impact core objectives. A quality diagnostic indicates the characteristics of a problem with the standard seasonal adjustment process. Highly nonlinear series saturated with outliers often resist standard modelling, even with series truncation, resulting in irreconcilable diagnostics. In such cases, alternative approaches are required. A dominant irregular component or a large number of outliers may mask a seasonal component. Inconsistent adjustments across overlapping periods, for example, revisions in successive runs often indicate fundamental seasonal instability that requires methodological intervention rather than simple specification tweaks.

8.5.1 Interpretation of Single-Series Processing Results

JD+ provides a comprehensive and detailed test for analysing the quality of seasonal adjustments. The tests available vary depending on the tuning method chosen. The results of the seasonal adjustment and its testing are displayed in the document created when the seasonal adjustment is performed. These documents are located under Workspace/Seasonal adjustment/documents and can be found under the selected method. The document consists of six main sections: Inputs, Primary Results, Preprocessing, Decomposition, Benchmarking, and Diagnostics. For the X-13 adjustment method, the decomposition part is the so-called Decomposition (X11).

8.5.1.1 Main Results

The main results include a brief description of the model used and an assessment of the quality of the seasonal adjustment. Figures 8.39 and 8.40 show examples of the SA results of Belgian GDP at constant prices.

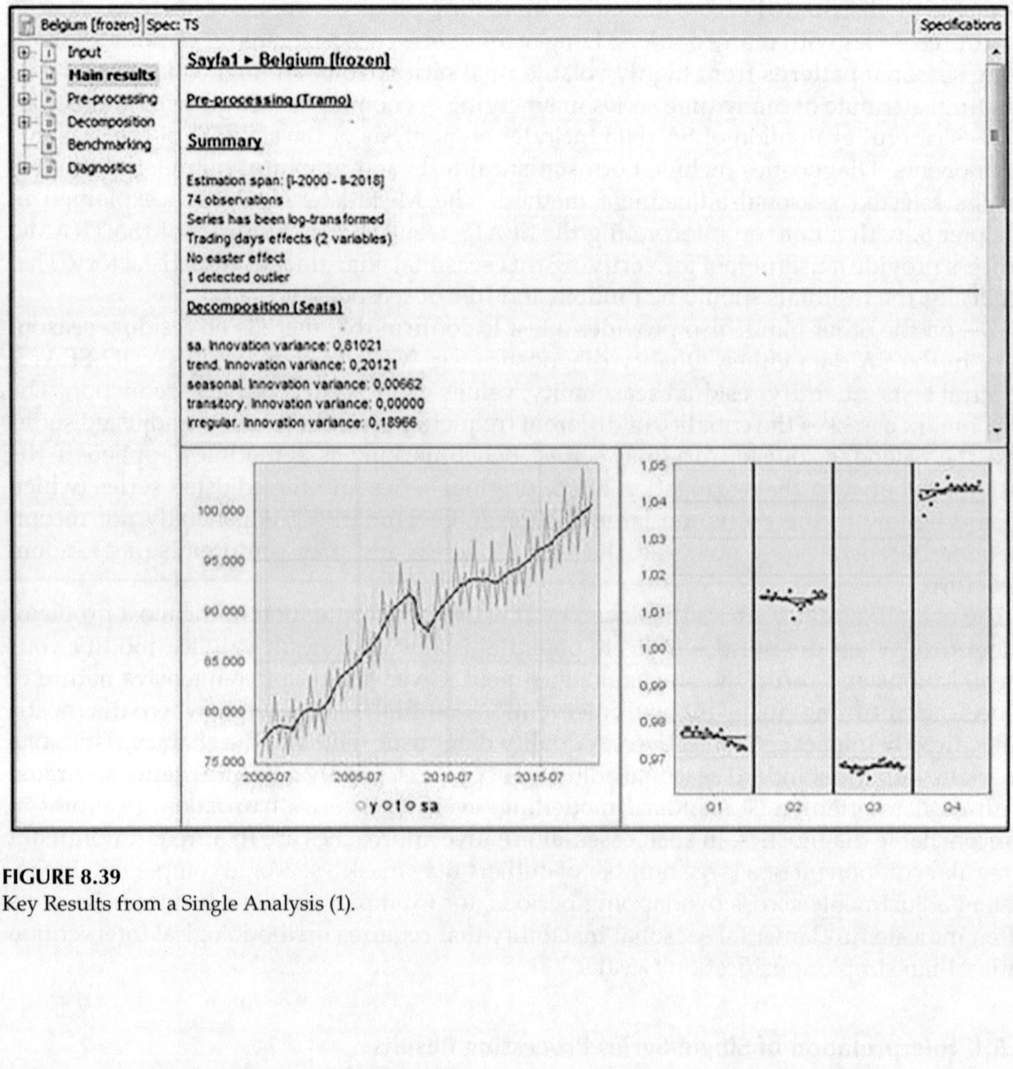

FIGURE 8.39
Key Results from a Single Analysis (1).

8.5.1.1.1 Primary Results from a Single Analysis (1)

The main results (Figure 8.39) include the following information on the adjustment process: (1) the estimated span used to identify seasonal patterns is from the first quarter of 2000 to the second quarter of 2018 and contains 74 observations; (2) TRAMO-SEATS applied a logarithmic transformation to the original time series so that the data was closer to the statistical assumptions and to help fit the model. During additive decomposition, the model is directly fitted, but in the case of multiplicative decomposition, the original time series is logarithmically transformed to convert it into an additive form before fitting the model; (3) estimated trading day effect with two variables, which have been defined in the spec without any prior testing. No significant Easter effect was found: (4) the outlier detection process detected an outlier in the original time series; (5) the variance of seasonal and trend-cyclical component innovation is higher than that of irregular components. In summary, all quality diagnostics were marked as 'good', except that spectral seasonal peaks were marked as 'inconclusive'.

8.5.1.1.2 Primary Results from a Single Analysis (2)

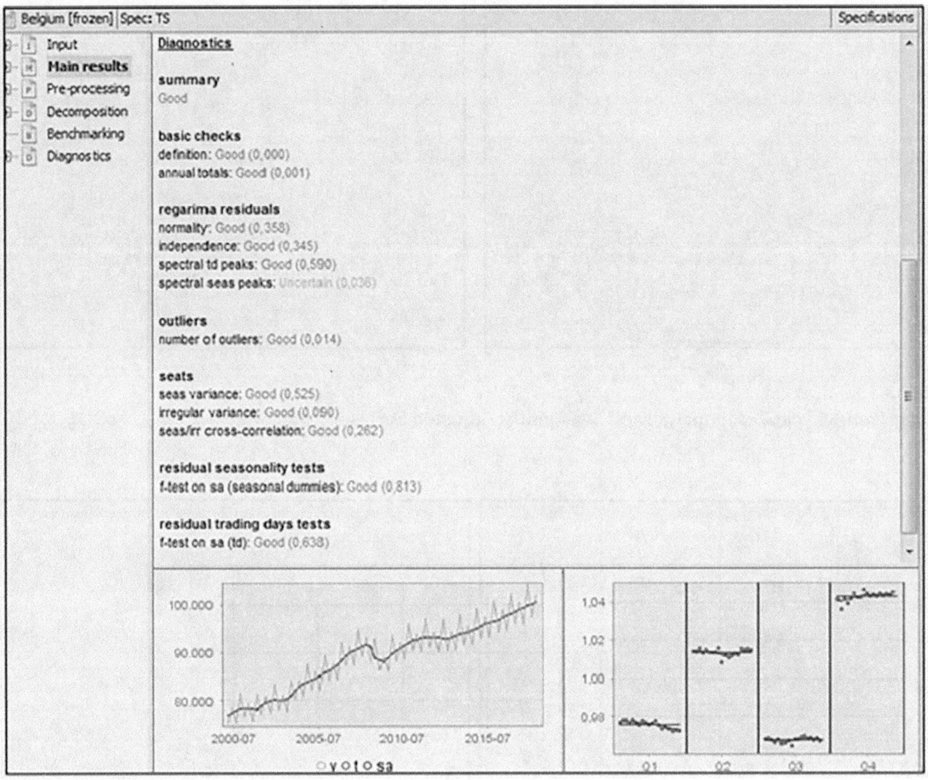

FIGURE 8.40
Key Results from a Single Analysis (2).

8.5.1.2 Visual Inspection

Viewing JD+'s visual tools, you can get a first impression of the SA results. The software displays multiple charts and tables under the 'Main Results'. Raw data, trend-cycle components, and SA series are shown in the left figure, and the S-I (seasonal-irregular) ratio is shown in the right figure (Figure 8.40). Double-clicking on a series in the chart on the left will only open a new window for that series, while for S-I charts, double-clicking on a specific quarter will open a more detailed view of that individual quarter (or 'month', if it's monthly data). When you open the Main Results node, you can find a detailed view of two charts, Charts/Sa, trend, and SI ratio on the left and SI ratio on the right. Under the Charts node, there is a second useful visualisation that can be accessed by clicking 'Cal.', 'sea.', and 'irr.'. It provides a breakdown of factors into calendar ("calendar effect"), seasonal ("season component"), and irregular ("irregular") components (see Figure 8.40). By visual analysis of charts, users can find preliminary answers to questions about whether the original series shows a seasonal, trending, or calendar effect. Visual inspection of the original series plot enables preliminary outlier identification, though statistical confirmation remains essential for robust adjustments. The Chart Details node is shown in Figure 8.41.

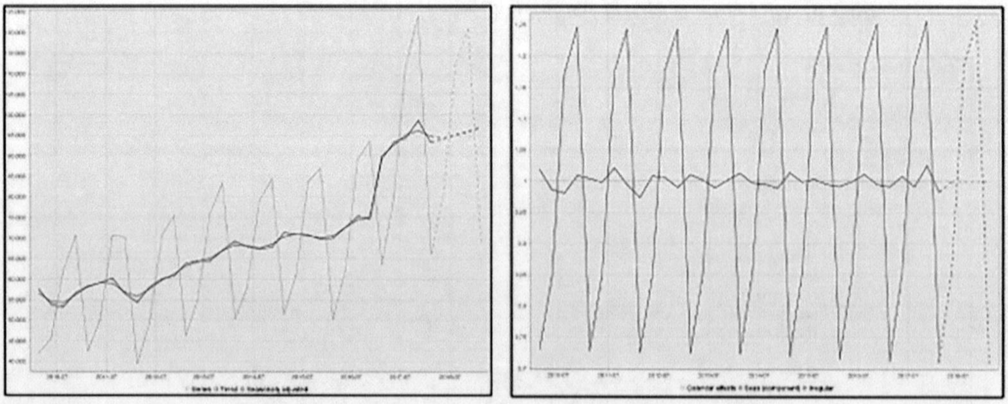

FIGURE 8.41
Raw data, Trend-Cycle Components, and Seasonally Adjusted Series.

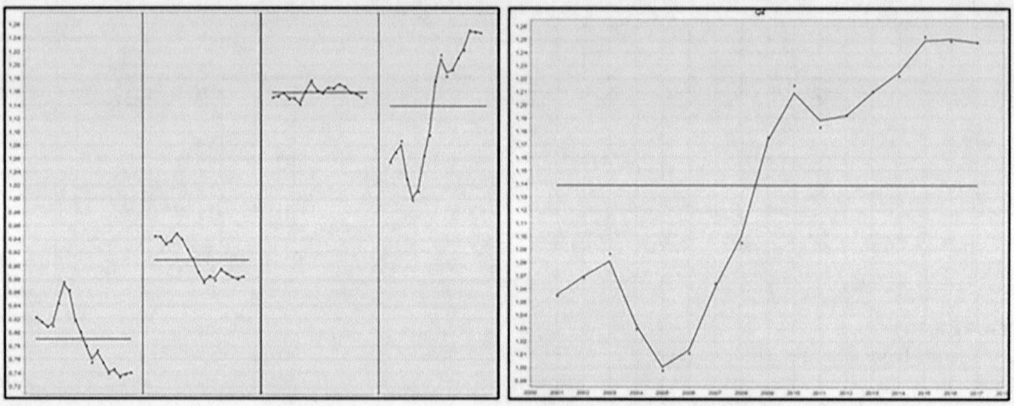

FIGURE 8.42
SI Ratio Chart.

8.5.1.3 SI Ratio Diagram

SI ratio charts can be used to analyse the development of seasonal patterns, ie, to detect seasonal breaks or shifts in seasonality. The blue line represents the detrending value for each quarter series. A gradual increase or decrease in seasonality factors can mean a change in seasonality. However, if this occurs in a specific quarter or month, it may indicate a seasonal break; for example, a time series that consistently peaks in a particular month or quarter over many years, but then begins to show a sustained decline or increase in subsequent years. One such case is Q4 in Azerbaijan's GDP compared to the RSAfull model (see Figure 8.42). The presence of seasonal outliers can also be formally tested.

Such problems can be addressed using a model-based approach to automatic seasonality, or manually by incorporating parameters into the model that account for moving seasonality.

8.5.1.4 Applied Models

JD+ provides details on seasonal adjustments in the Preprocessing and Decomposition sections of the documentation. The Preprocessing portion is the same for both methods, but the Decomposition portion varies depending on the method chosen.[14] The preprocessing section includes a summary portion of the ARIMA model estimate (eg, the estimated span used, transformation information, the estimated ARIMA model, the estimated parameters of the trading day, and/or user-defined regression variables, Easter effects, and outliers) and sub-nodes – predictions, regression variables, ARIMA, pre-adjusted series, residuals, and likelihood. Each of these nodes provides important information to help assess the quality of the ARIMA model estimated for preprocessing. In particular, the forecasts and residual diagnostics assess the validity of the ARIMA model by testing the significance of the predictions and evaluating the distribution of the residuals. Decomposition provides a summary of the applied decomposition model and more details on the variance, autocorrelation, and intercorrelation of the components obtained by SEATS. On the main Decomposition page, the new variance for all theoretical model components is shown. The new interest variance is the maximised variance of the model, with irregular components to derive trend periods and seasonal components that are as stable as possible, meaning that no additional white noise can be removed from them. This assumption is also known as 'spec decomposition'. Irregular components include random fluctuations that cannot be attributed to other components.

8.5.2 Pre-Adjustment and Model Evaluation

The pre-adjustment displays the properties of the pre-adjustment and contains the following information: (1) the statistical properties of the ARIMA model used in the seasonal adjustment; (2) regression and its coefficients; (3) pre-adjusted series listings; (4) residuals.

8.5.2.1 ARIMA Evaluation

Under Preprocessing, information is provided about the statistical properties of the ARIMA model used in seasonal adjustments. TRAMO and regARIMA determine the most appropriate ARIMA model and estimate the parameters for each time series. If they can't find a specific ARIMA model in automatic model recognition, they use the airline model. ARIMA models (p, d, q) are used to model and forecast time series data. The ARIMA model includes three types of parameters: autoregressive parameters (p), differencing order (d), and moving average parameters (q). Seasonal series typically have two sets of parameters: a regular component defined by (p, d, q) and a seasonal component (P, D, Q). Different ARIMA models can be installed in the same series. However, it is generally recommended to select the simplest model with the fewest parameters that still provides a satisfactory fit. In the case of Belgium, TRAMO chose the airline model (0,1,1) (0,1,1), as shown in Figure 8.43, where the coefficient Theta represents a stochastic (volatility) trend and 'BTheta' represents a deterministic seasonality, rather than a trend.

In general, automatic model recognition should produce satisfactory results, and the model does not need to be defined manually. Test results may indicate that the specifications need to be improved and changed accordingly.

Arima model
[(0,1,1)(0,1,1)].

	Coefficients	T-Stat	P[|T| > t]
Theta(1)	0,2759	2,27	0,0265
BTheta(1)	-0,6960	-6,95	0,0000

FIGURE 8.43
Estimated ARIMA model.

Regression model
Working days

	Coefficients	T-Stat	P[|T| > t]
Week days	0,0004	2,90	0,0051

Leap year

	Coefficients	T-Stat	P[|T| > t]
Leap year	0,0021	1,30	0,1978

FIGURE 8.44
Estimation of the Calendar Effect.

8.5.2.2 Calendar Effect Evaluation

The parameters estimated for the calendar effect are displayed at the bottom of the pre-processing master page. The ARIMA calendar regression and its estimated effects are provided in detail under the regressors and pre-adjustment series nodes. JD+ offers three options for dealing with calendar effects: (1) Trading days: six regression quantities that measure the impact of each weekday starting on Sunday; (2) Weekdays: a regression quantity used to measure the impact of weekdays and weekends; (3) Leap year: a regression measure used to measure the impact of an extra day in a leap year; (4) Easter: a regression measure of the Easter effect; (5) None: Only one leap year effect variable is included.

Figure 8.44 shows information about the estimated coefficients and their Belgian example. In this example, the weekday regression is statistically significant at the 99% confidence level, whereas the leap year effect is not even statistically significant at the 90% confidence level.

8.5.2.3 Predictions Evaluation

JD+ generates one-year forecasts for the original series, with confidence intervals derived from the ARIMA model estimated during preprocessing (see Figure 8.45). It also includes

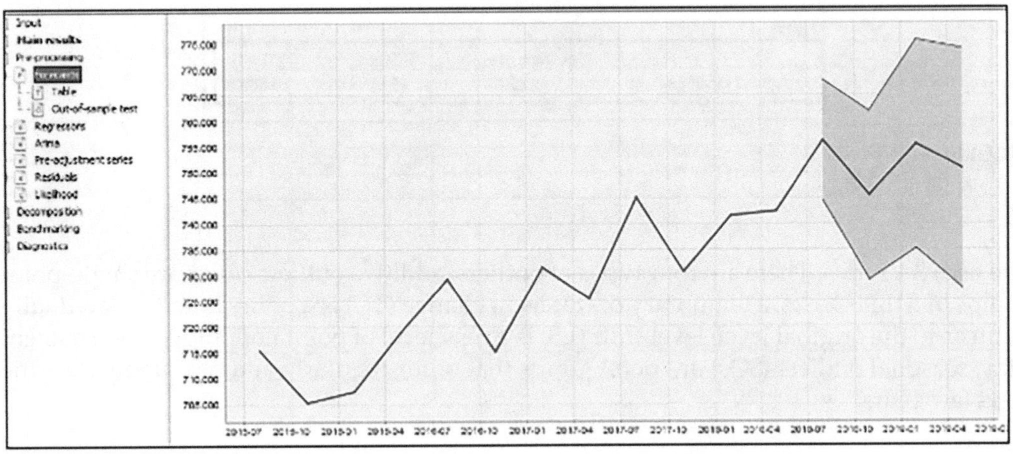

FIGURE 8.45
Prediction Section.

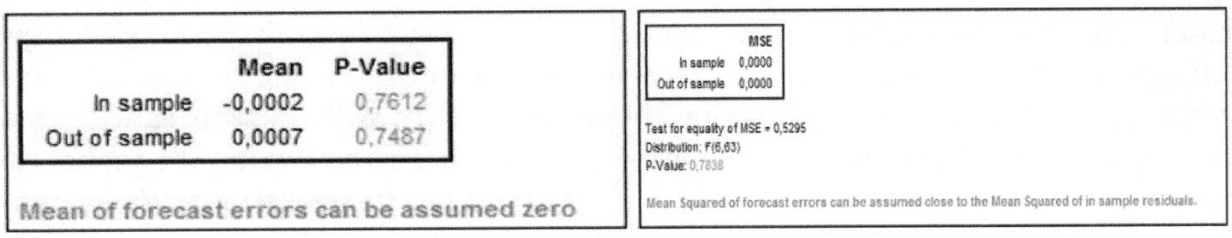

FIGURE 8.46
Tests for the Prediction Section.

an out-of-sample test for prediction, which indicates the statistical significance of the prediction.

Evaluate forecasts based on distribution attributes. The idea behind the assessment is that the distribution of data within the sample should be the same or similar to the distribution of data outside the sample. Statistical distributions are characterised by their mean and variance. Thus, an out-of-sample test is conducted by estimating the model using a restricted portion of the sample (90%) and forecasting the remaining portion with fixed model parameters. The evaluation is then based on whether the mean of the forecast errors is equal to zero and on the average squared error. Figure 8.46 shows the results of the Forecasts/Out-of-sample test if both measures meet the criteria.

8.5.2.4 Outlier Detection Evaluation

Outliers are observations that fall outside a hypothetical distribution. It is important to highlight 'what ifs' because, based on an observation, one distribution is an outlier but may not be an outlier based on another. As a result, an observation identified as an outlier based on one ARIMA model may not be an outlier determined according to another ARIMA model. Detected outliers, along with their types, locations, and coefficients, are available on the preprocessed home page. TRAMO and regARIMA automatically detect outliers and temporarily replace them with expected values before estimating seasonal and calendar components. After that, the estimated effect of the outliers is added back to

Outliers

	Coefficients	T-Stat	P[|T| > t]
LS (IV-2008)	-0,0276	-5,51	0,0000

FIGURE 8.47
Estimation of Outliers.

the final SA series. There are four types of outliers: additive outliers (AOs) are single-point jumps in a time series; a temporary or transient change (TC) is a point jump that gradually returns to the original level; level shift (LS) is a new level of point hopping and permanent stay; seasonal outliers (SOs) are point jumps that occur regularly and repeatedly over the next few years.

8.5.2.5 TRAMO and regARIMA Test Evaluation

TRAMO and regARIMA use the threshold of the test, known as the VA, to identify statistically significant outliers. The VA defines the observed value as an outlier when the unexplained portion of the observed value exceeds the VA. Otherwise, it's not an outlier. You can modify the default value of VA based on the statistical requirements described in the outliers. In the case of Belgium, a change in levels in the fourth quarter of 2008 was determined (Figure 8.47). This means that the Belgian GPD's airline model cannot account for this observation. This outlier is not surprising, as it occurred during the global financial crisis.

At the end of the time series, information that can explain the outlier is especially important when the nature of the outlier is uncertain, as the next observation is not yet available. A later change in the outlier type can result in a significant revision of the seasonal adjustment series.

8.5.2.6 Residuals Evaluation

The residuals of the model can already indicate the quality of the seasonal adjustment. The residuals should be independent and random, and follow a normal distribution with a mean of zero. JD+ displays information about residuals under Preprocessing/Residuals/Statistics. Under Distribution, major attributes such as autocorrelation and residual distribution are visualised (Figure 8.48, right). The main properties of the distribution, such as mean, skewness, kurtosis, and normality, are displayed here. JD+ also evaluates the performance of residuals. The test results are available under the Statistics section and include the Normality test, which assesses whether the residuals follow a normal distribution, as well as the Independence test, which checks for autocorrelation in the residuals.[15] The randomness test measures whether the positive and negative signs of the residuals are randomly distributed or periodic. Finally, the linearity test is a heteroskedasticity test that assesses whether the variance is constant over time.

Tests for independence and randomness are essential to assess the unbiasedness of the estimator, while tests for normality and linearity assess the estimator's efficiency. One can prioritise impartiality, thereby ensuring independence and randomness in the test results. In the Belgian example, the residual test did not indicate any statistical problems (Figure 8.47). All null hypotheses in the statistical tests were accepted at the 5% significance level. Thus, It is assumed that the residuals are randomly distributed, normally

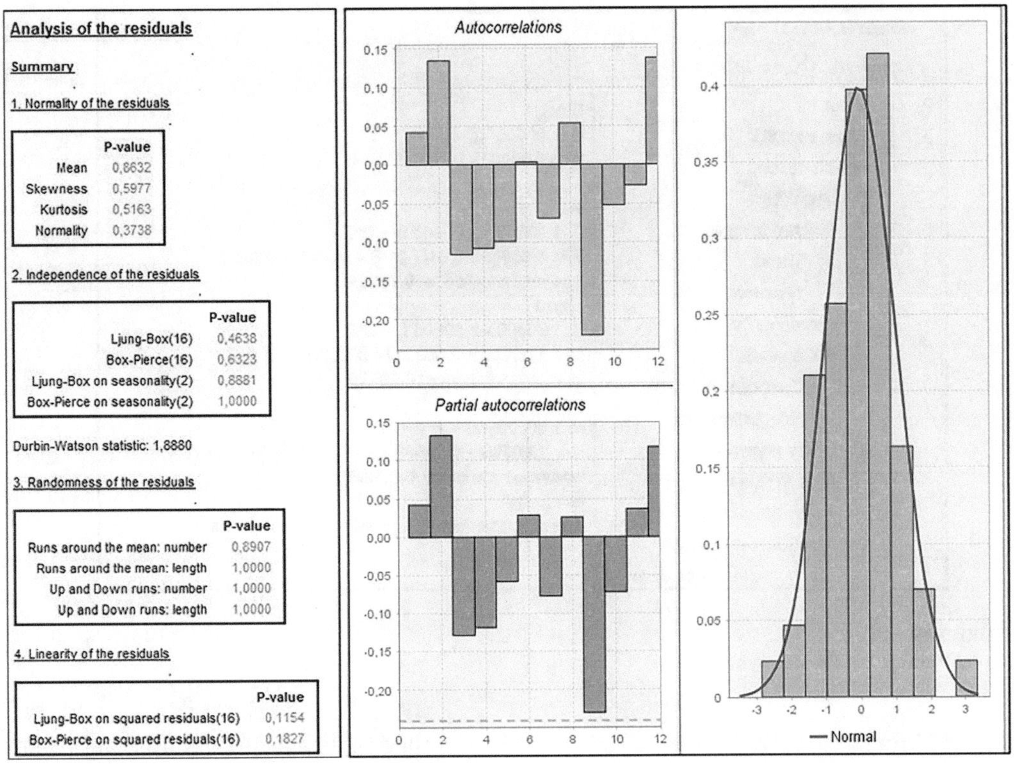

FIGURE 8.48
Test Statistic and Distribution Plot of Residuals.

distributed, and independent, implying that they contain no systematic information and satisfy the theoretical requirements. The significance level is the criterion used to test a hypothesis (ie, a null hypothesis about the data). Experienced users can modify the significance level used by JD+ via Tools > Options > Statistics.

8.5.3 Decomposition Model and Component Analysis Evaluation

The decomposition performed by SEATS assumes that all the components in the time series (ie, trend periods, seasonality, and irregularities) are independent of each other. This independence is also assumed to be a temporary component, which is rarely identified and estimated. Decomposition is performed by maximising the variance of the irregular components (Innovations Variance). The aim is to derive trend periods and seasonal components as smoothly as possible. JD+ presents a mathematical model for each component on the main Decomposition page (Figure 8.49). The decomposition consists of eight parts: the random series, components, Wiener–Kolmogorov (WK) analysis, and result error analysis. The steady-state variance decomposition includes several intermediate results from SEATS that can provide additional insights. Component decomposition contains information about the ARIMA model applied by SEATS. This section describes the characteristics of SEATS' Wiener–Kolmogorov filters, which are used to extract components from the original series. Decomposition requires that the components are not correlated. When

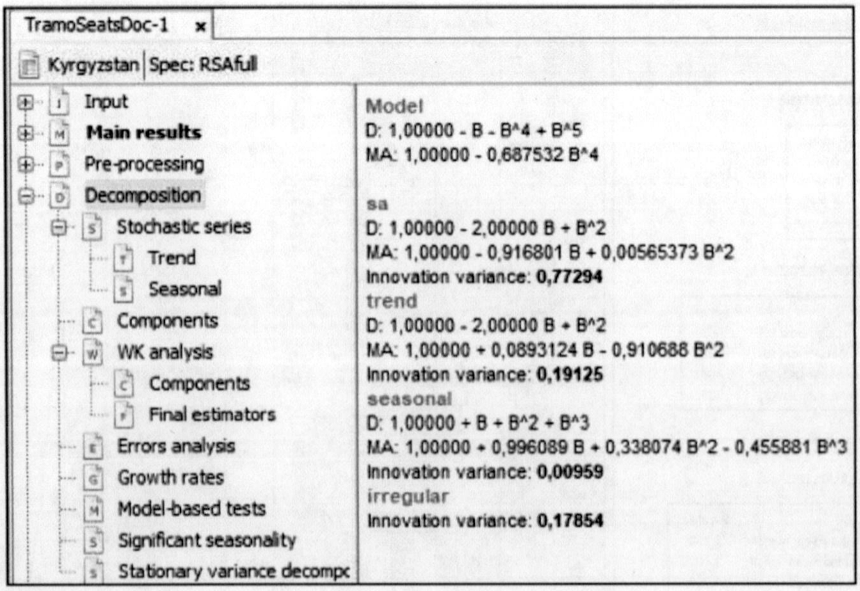

FIGURE 8.49
General View of the Breakdown.

using X-13 for seasonal adjustment, JD+ provides detailed tables (A through E) that present the results of each successive phase of the X-11 decomposition process performed by X-13.

SEATS identified an ARIMA model for each component of the Kyrgyzstan series, which includes trend cycle, seasonal, and irregular components. JD+ displays the mathematical formulas for each ARIMA model of the component on the home page of decomposition. There are several parts of Decomposition that users can use effectively. These are Growth rates, Model-based tests, and Significant seasonality. In the growth rate (Figure 8.49), the growth rate of the seasonally adjusted series (SA_t) or trend component ($Trend_t$) over the period (t, t-1) is expressed as percentiles:

$$\left(\frac{SA_1}{SA_{t-1}}-1\right)\times 100$$

$$\left(\frac{Trend_1}{Trend_{t-1}}-1\right)\times 100$$

The standard error is calculated using a linear approximation of the ratio.[16] When periods vary greatly, these standard errors should be interpreted as generalised approximations, which tend to underestimate true values (Figure 8.50).

8.5.3.1 Model-Based Testing Evaluation

Model-based tests present distributions of components, theoretical estimators, and empirical estimates (for their stationary transformations) in terms of variance, autocorrelation, and cross-correlation. The tests in this section can be used to assess whether the components are estimated (under- or over-adjusted) according to a theoretical model. The

Period to period growth rates

Seasonally adjusted series

Period	Growth	Std error (revisions)	Std error (total error)
II-2018	0,57%	0,14%	0,2%
I-2018	0,12%	0,14%	0,2%
IV-2017	0,61%	0,13%	0,2%
III-2017	0,19%	0,13%	0,2%
II-2017	0,7%	0,11%	0,18%
I-2017	0,35%	0,1%	0,18%
IV-2016	0,28%	0,1%	0,18%
III-2016	0,42%	0,1%	0,18%
II-2016	0,58%	0,08%	0,17%

Trend

Period	Growth	Std error (revisions)	Std error (total error)
II-2018	0,41%	0,15%	0,23%
I-2018	0,36%	0,06%	0,19%
IV-2017	0,38%	0,05%	0,18%
III-2017	0,42%	0,05%	0,18%
II-2017	0,49%	0,04%	0,18%
I-2017	0,42%	0,04%	0,18%
IV-2016	0,33%	0,04%	0,18%
III-2016	0,42%	0,04%	0,18%
II-2016	0,48%	0,03%	0,18%

FIGURE 8.50
Adjusting the Growth Rate of the Series.

Variance

	Component	Estimator	Estimate	P-Value
Trend	0,3785	0,1650	0,1624	0,9435
Seasonally adjusted	1,5253	1,3318	0,9066	0,1312
Seasonal	0,0145	0,0012	0,0009	0,5506
Irregular	0,1911	0,0836	0,0555	0,1069

FIGURE 8.51
Test of Variance for Components.

Cross-correlation

	Estimator	Estimate	P-Value
Trend/Seasonal	-0,2001	-0,0741	0,4198
Trend/Irregular	-0,0019	-0,1270	0,3852
Seasonal/Irregular	0,0837	0,2211	0,2589

FIGURE 8.52
Cross-Correlation of Components.

estimator and estimated variances are expected to be close to each other and smaller than the component variance. This is measured by the F-test, whose value is shown in the last column of the table in Figure 8.51.

Another important part of model-based testing is the cross-correlation between the measured components. An inconsistency between the estimator and the estimated component suggests a misidentification of the component. This agreement is assessed using the F-test, with the corresponding p-value displayed in the last column (see Figure 8.52). The cross-correlation table provides the user with useful information for testing whether the estimated or actual estimates of the components are correlated with each other. Cross-correlation results provide a means of testing the hypothesis that components are orthogonal (ie, uncorrelated)

Figure 8.52 shows an example of a cross-correlation table made by SEATS for Belgium's GDP. In JD+, the colour of the p-value represents the outcome; for example, a green p-value indicates a non-significant cross-correlation.

Finally, significant seasonality indicates the number of quarters (or months) that exhibit notable seasonal changes in different segments of the time series, evaluated at the 95% and 99% confidence levels, respectively. This is divided into three sections, namely Historical, Current, and Forecast. It is expected that the historical and current periods exhibit similar seasonality, and that the forecast period also demonstrates positive seasonality consistent with the current period.

8.5.4 Quality Diagnostic Tools

Diagnostics provide detailed information about seasonal adjustment procedures. This information is typically descriptive and is calculated identically for both TRAMO-SEATS and X-13, ensuring comparability. JD+ divides diagnostics into five main sections: seasonality testing, spectrum analysis, sliding span, revision history, and model stability analysis. To ensure the high quality of the seasonal adjustments, it is recommended to use the various quality measures offered by JD+. This book focuses on quality diagnostics using the TRAMO-SEATS method.[17] JD+ uses the classifications in Table 8.10 to report all test statistics as 'messages'. This allows for quick interpretation of test statistics.

Overall, information is uncertain, and goodness can be assessed as acceptable. The home page for diagnostics provides a summary of the test statistics and an overall summary value. This value provides a quick and overall indication of the quality of seasonal adjustment (Figure 8.53). Here, the diagnosis of regARIMA residuals, out-of-sample, and residual seasonality is the most important outcome and should be checked before deciding on a SA model.

8.5.4.1 Seasonal Test Diagnostic

JD+ information for quality measures are shown in Table 8.10. In addition, under the basic check, non-fundamental properties of seasonal adjustments will be tested, defining whether the test decomposition follows mathematical relationships between different components and effects (eg, the seasonal adjustment series should be equal to the difference between the original time series and the seasonal component, and the sum of the trend period component and the irregular component). Annual totals compare the yearly sums of the original series with those of the SA series. The difference should be as low as possible (See Figure 8.53).

TABLE 8.10

JD+ Information for Quality Measures

Test Statistics	Reasons
Undefined	The test is probably not designed or suitable for that process, or an unsatisfied condition to calculate the statistics.
Error	The test statistic is unacceptable and needs to be recalculated.
Severe	The test statistic is unacceptable and needs to be recalculated.
Bad	The test statistic is critical and needs to be recalculated.
Uncertain	The test statistic is acceptable, but fine-tuning could remove the uncertainty.
Good	The test statistic is acceptable; no more attempts needed.

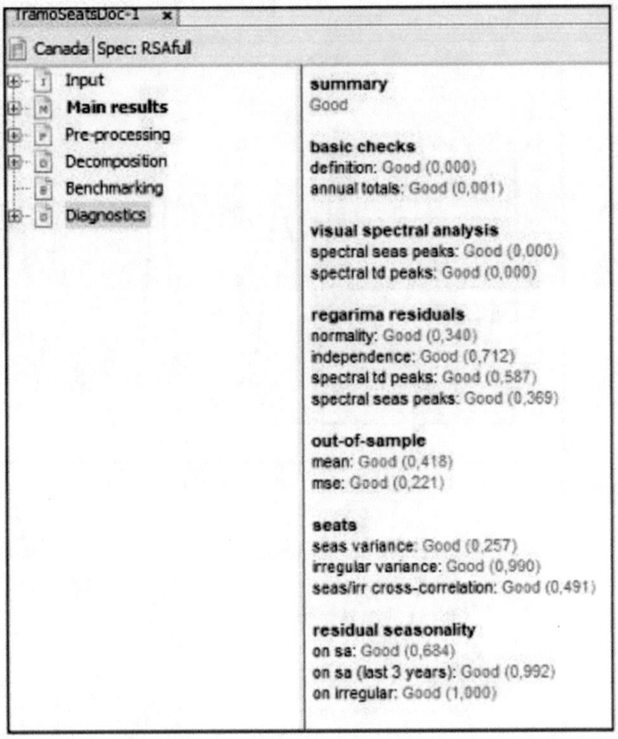

FIGURE 8.53
General View of Diagnostics.

8.5.4.2 Spectrum Analysis Diagnostic

A special diagnostic analysis section provides cycle plots and autoregressive spectra. The spectral pattern enables the examination of residuals, irregular components, and SA series for the presence of seasonal or trading day effects. The peak at the seasonal frequency of the adjustment series (Figure 8.54) indicates that the seasonality filter is insufficient in the time interval used for spectrum estimation, so different model specifications or data span lengths should be considered. The peak at the trading day frequency (Figure 8.54) indicates the extent to which the regression model may need to be re-evaluated. If some remaining seasonality is identified, you need to reconsider the model spec, regression variables, or periods used for modelling. The Belgian series showed no signs of residual seasonality or residual calendar effects. In other words, in the period portion of Figure 8.54, there are no spectral peaks at seasonality or trading day frequency. Peaks are considered only when the value is above the threshold of 10.

The Sliding spans a section of Diagnostics that provides the results of several tests for the presence of seasonality in different time series, subtracted from the original series. These series include primitive series, linearised series, residual series, and others. Table 8.11 gives an interpretation of these tests and their expected results. The different tests used to test for seasonality in correlation series are Seasonal Lag Autocorrelation, Friedman, Kruskal–Wallis, Spectral Peaks, Periodic Graphs, Seasonal Dummies, and Seasonal Dummies

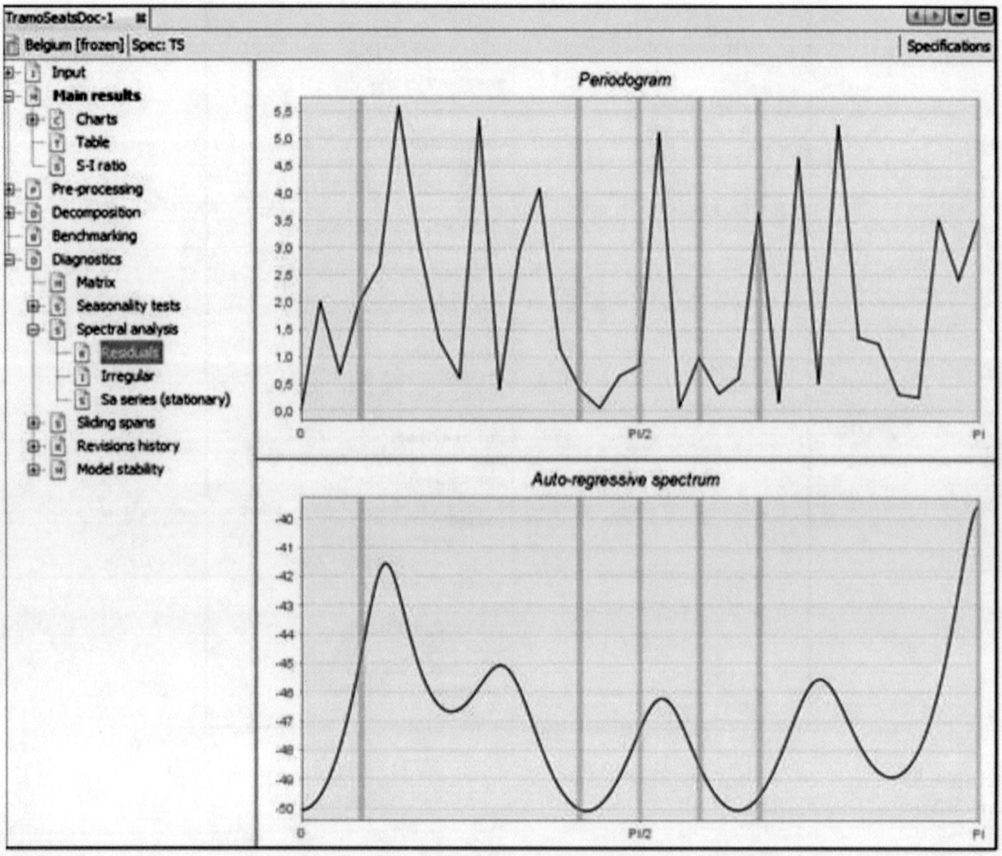

FIGURE 8.54
Spectral Plot of Residuals.

TABLE 8.11
Seasonally Adjusted Series Test

No	Type of test	Expected result	Explanation
1	Original (transformed) series	yes	There are several types of tests. It is desired that most of the test approves the existence of seasonality in the original time series to be SA.
2	Linearised series	yes	This is the original time series corrected for outliers and calendar effects, if any. The explanation is the same as the original time series.
3	Full residuals	No	This is the part of the original time series unexplained by the ARIMA model estimated. The presence of seasonality in the residuals is to be avoided.

TABLE 8.11 (*Continued*)

Seasonally Adjusted Series Test

No	Type of test	Expected result	Explanation
4	Combined test	Identifiable seasonality presented	This test consists of three separate tests which measure the stability, the presence, and the evolution of seasonality. Seasonal adjustment should only be considered if identifiable seasonality is present.
5	SA series	No	This is the SA series. According to the potency criteria, most tests shouldn't detect any remaining seasonality in the series.
6	Irregular	No	This is the unexplained part of the original time series by the SEATS decomposition part. The presence of seasonality in the irregular part is not desirable.
7	Residuals (last period)	No	Same as the 3rd row in the table, but the tests are applied to the last three years of the series in question.
8	SA series (last period)	No	Same as the 5th row in the table, but the tests are applied to the last three years of the series in question.
9	Irregular (last period)	No	Same as the 6th row in the table, but the tests are applied to the last three years of the series in question.
10	Residual seasonality	No	It is a test for residual seasonality with the same purpose as the 3rd test, but with a different statistical approach. It measures the residual seasonality in both and the restricted part of the residuals.

(AMIs). Figure 8.55 gives an overview of the seasonal test results. In this example, the results are exactly as expected, as explained in Table 8.11. There is no residual seasonality across the series, but there may be in the last three years. However, this is negligible, as the overall results indicate favourable conditions for identifying seasonal patterns. A more detailed explanation of these tests can be found in Grudkowska (2015).[18] If the test results for any series are contradictory, it is recommended to give greater weight to the parametric test than to the nonparametric test. The parametric tests are the 1st, 6th, and 7th tests.

8.5.4.3 Sliding Span Analysis

The Sliding Spans analysis portion of Diagnostics was originally used for X-13. It is particularly useful for series with strong seasonality or a large number of outliers. From the analysed time series, the program extracts a period spanning eight years. In this example, it extracts 4 time spans that are one year apart. The sliding span is calculated based on the seasonal component, trading day effects (if any), and SA series. When using additive decomposition, the sliding span analysis is based on absolute differences; otherwise, it is based on factor ratios. The threshold detection outlier is set to 3%. Any larger value is unstable. Outliers in relevant parts of the component can be analysed in detail. In Figure 8.56, the seasonal factor appears identifiable from the sliding span analysis of Belgian GDP, although the seasonal component exhibits a shifting structure.

Original series

Summary
Data have been differenced and corrected for mean

Test	Seasonality
1. Auto-correlations at seasonal lags	YES
2. Friedman (non parametric)	YES
3. Kruskall-Wallis (non parametric)	YES
4. Spectral peaks	YES
5. Periodogram	YES
6. Seasonal dummies	YES
6bis. Seasonal dummies (AMI)	YES

Linearized series

Summary
Data have been differenced and corrected for mean

Test	Seasonality
1. Auto-correlations at seasonal lags	YES
2. Friedman (non parametric)	YES
3. Kruskall-Wallis (non parametric)	YES
4. Spectral peaks	YES
5. Periodogram	YES
6. Seasonal dummies	YES
6bis. Seasonal dummies (AMI)	YES

Full residuals

Summary

Test	Seasonality
1. Auto-correlations at seasonal lags	NO
2. Friedman (non parametric)	NO
3. Kruskall-Wallis (non parametric)	NO
4. Spectral peaks	NO
5. Periodogram	NO
6. Seasonal dummies	NO
6bis. Seasonal dummies (AMI)	NO

Summary
Data have been differenced and corrected for mean

Test	Seasonality
1. Auto-correlations at seasonal lags	NO
2. Friedman (non parametric)	NO
3. Kruskall-Wallis (non parametric)	NO
4. Spectral peaks	NO
5. Periodogram	NO
6. Seasonal dummies	NO
6bis. Seasonal dummies (AMI)	NO

Irregular series (stochastic)

Summary

Test	Seasonality
1. Auto-correlations at seasonal lags	NO
2. Friedman (non parametric)	NO
3. Kruskall-Wallis (non parametric)	NO
4. Spectral peaks	NO
5. Periodogram	NO
6. Seasonal dummies	NO
6bis. Seasonal dummies (AMI)	NO

Residuals (last periods)

Summary

Test	Seasonality
1. Auto-correlations at seasonal lags	NO
2. Friedman (non parametric)	NO
3. Kruskall-Wallis (non parametric)	NO
5. Periodogram	NO
6. Seasonal dummies	NO
6bis. Seasonal dummies (AMI)	NO

Seasonally adjusted series (last periods)

Summary
Data have been differenced and corrected for mean

Test	Seasonality
1. Auto-correlations at seasonal lags	NO
2. Friedman (non parametric)	NO
3. Kruskall-Wallis (non parametric)	NO
5. Periodogram	NO
6. Seasonal dummies	NO
6bis. Seasonal dummies (AMI)	NO

Irregular series (last periods)

Summary

Test	Seasonality
1. Auto-correlations at seasonal lags	NO
2. Friedman (non parametric)	NO
3. Kruskall-Wallis (non parametric)	NO
5. Periodogram	NO
6. Seasonal dummies	NO
6bis. Seasonal dummies (AMI)	NO

Combined seasonality test

Identifiable seasonality present

Residual seasonality test

No evidence of residual seasonality in the entire series at the 10.0 per cent level: F=0,2289

Residual seasonality present in the last 3 years at the 10.0 per cent level: F=3,5213

FIGURE 8.55
Results of the Seasonality Test.

Sliding spans summary

Time spans

Span 1: from I-2007 to II-2015
Span 2: from I-2008 to II-2016
Span 3: from I-2009 to II-2017
Span 4: from I-2010 to II-2018

Tests for seasonality

	Span 1	Span 2	Span 3	Span 4
Stable seas.	14847,3	6002,3	2449,8	4911,3
Kruskal-Wallis	30,9	30,9	30,9	30,9
Moving seas.	7,0	3,5	1,7	5,7
Identifiable seas.	YES	YES	YES	YES

Means of seasonal factors

	Span 1	Span 2	Span 3	Span 4
Q1	0,9758	0,9753	0,9756	0,9748
Q2	1,0118	1,0125	1,0120	1,0131
Q3	0,9685	0,9685	0,9687	0,9685
Q4	1,0441	1,0446	1,0439	1,0436

FIGURE 8.56
Slip Span Analysis.

8.5.4.4 Revision History Analysis

The diagnostic series revision history analyses the types of revisions that result from adding new observations at the end of the series. It provides a SA chart of the trend cycle series and its growth rate. JD+ displays the revision history of two methods (TRAMO-SEATS and X-13). In Figure 8.57, each hollow circle on the graph depicts the initial adjustment when that point is the last observation. The line represents the current state of the results. The analysis begins with a model that estimates the entire period. Since then, the period has gradually shortened, and the decomposition has been re-estimated. For each period, a series of successive estimates is obtained. By default, only the model parameters are re-estimated. However, if the options change, the program allows for a complete re-estimation and re-identification of outliers. By clicking on a point in the chart, such as an observation for November 2006, a pop-up window appears. It shows continuous estimates for a particular date over different periods (eg, Q2 2015), and Figure 8.56 provides an estimate of the continuous period.

In the case of Belgian GDP, it doesn't matter if the pop-up confirms that the revision will be made after three years. This section also includes a revision history table called Relative differences, which shows the difference between the first and last estimates for the past four years. If the decomposition is additive, JD+ shows an absolute revision; Otherwise, it uses relative differences. For faster analysis, JD+ shows the biggest differences in red. Any red observations (expressed in absolute terms) are greater than twice the absolute or relative modified root mean square error. The Belgian series includes only one observation

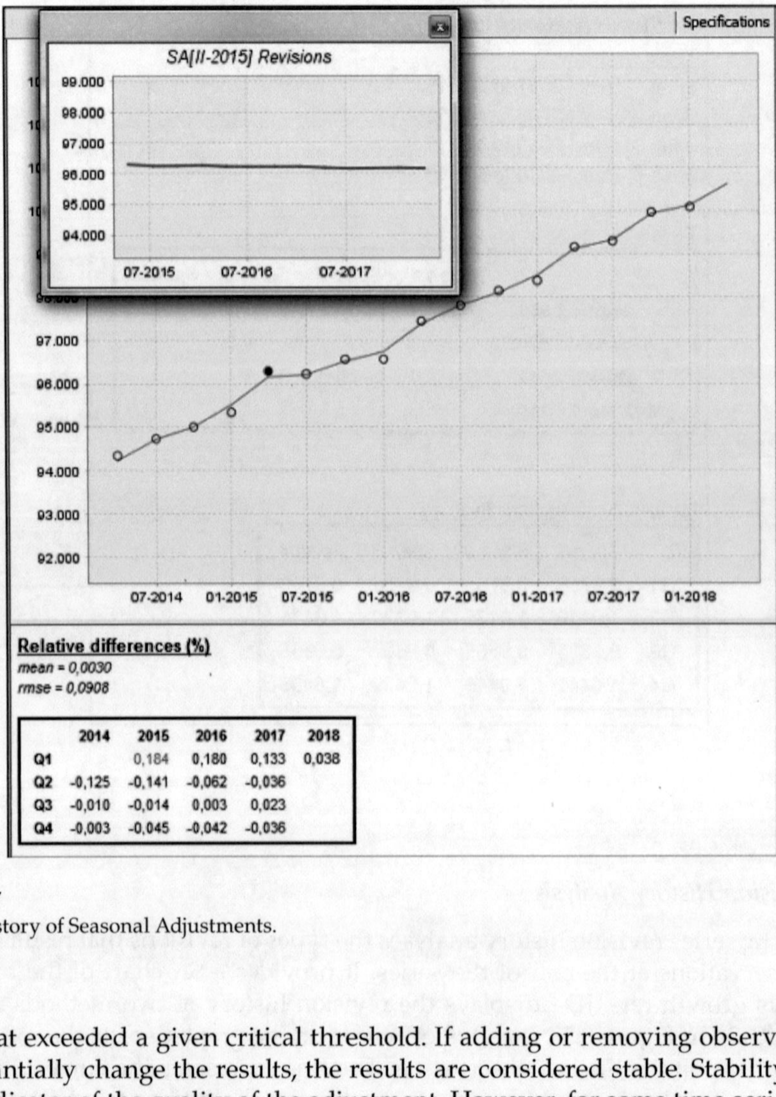

FIGURE 8.57
Revision History of Seasonal Adjustments.

in 2015 that exceeded a given critical threshold. If adding or removing observations does not substantially change the results, the results are considered stable. Stability is usually a good indicator of the quality of the adjustment. However, for some time series, even the best seasonal adjustments can be unstable. In this case, balancing quality with the number of revisions is challenging. If you detect an unstable seasonal component, you can try changing the model specifications.

8.5.4.5 Model Stability Analysis

The model stability analysis of the diagnostics assesses the stability of ARIMA parameters and regression coefficients across different periods. It displays the results in a visual form. The model stability analysis is based on an eight-year moving window that advances one year at a time. The points displayed correspond to continuous estimates of the parameter values.

Model stability analysis in Belgium includes 'time spans' from 2000 to 2008, 2001 to 2009, and 2002 to 2010 (Figure 8.58). In the airline model (left), the regular moving average

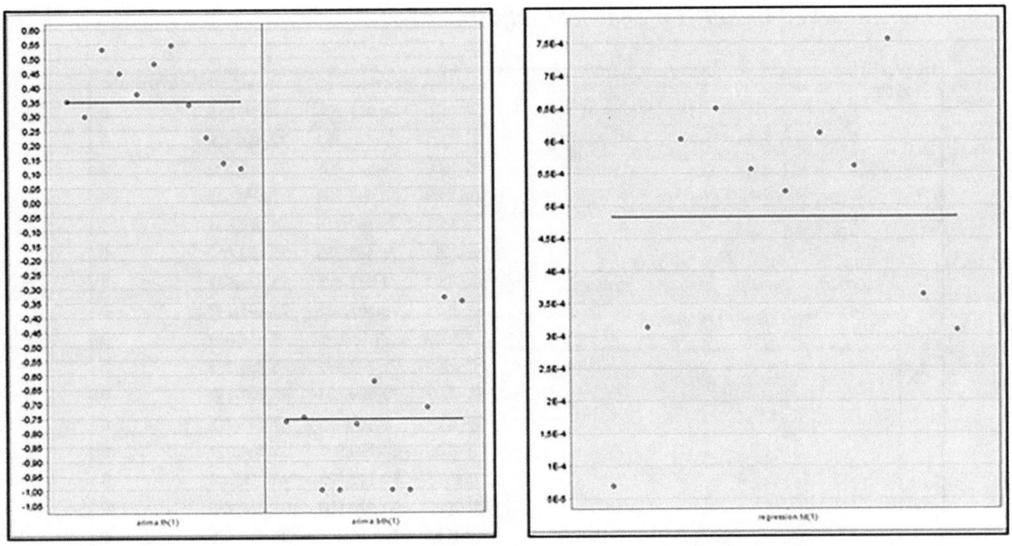

FIGURE 8.58
Model Stability Analysis.

parameter represents the trend component, whereas the seasonal moving average parameter reflects the seasonal component. The graph displays the steady evolution of the regular and seasonal moving average parameters, but also indicates that their behaviour has become unstable in recent years. The trading day parameters (figure on the right) also show the pattern of volatility over recent years. This implies that the results from the most recent year of the survey may be subject to revision. The results should be carefully evaluated.

If the quality diagnostics reveal that the year-on-year summary discrepancies are excessive, benchmarking can be employed to align the totals of the raw and adjusted data.

8.5.4.6 Benchmarking and Optimisation

Benchmarking provides limited information that is used to assess the quality of the process. The benchmark results are divided into four sections (Figure 8.59). The first chart shows the SA series before and after benchmarking. The graph below shows the difference in values for the SA series before and after benchmarking. This difference should be as low as possible. If the original time series contains outliers, this difference may be higher, but it does not necessarily indicate lower quality. The information for these two charts is summarised in the table next to it, including the following values: SA Series Before Benchmark (Raw), SA Series After Benchmark (Result), and Variance. The final section of the results presents descriptive statistics on the differences, including the maximum, minimum, mean, and standard deviation. For example, in the case of Belgian GDP, the differences are relatively close to zero. This indicates that the random structure of the SA series remains intact after the benchmarking process.

8.5.4.7 Benchmarked Results for Seasonal Adjustments

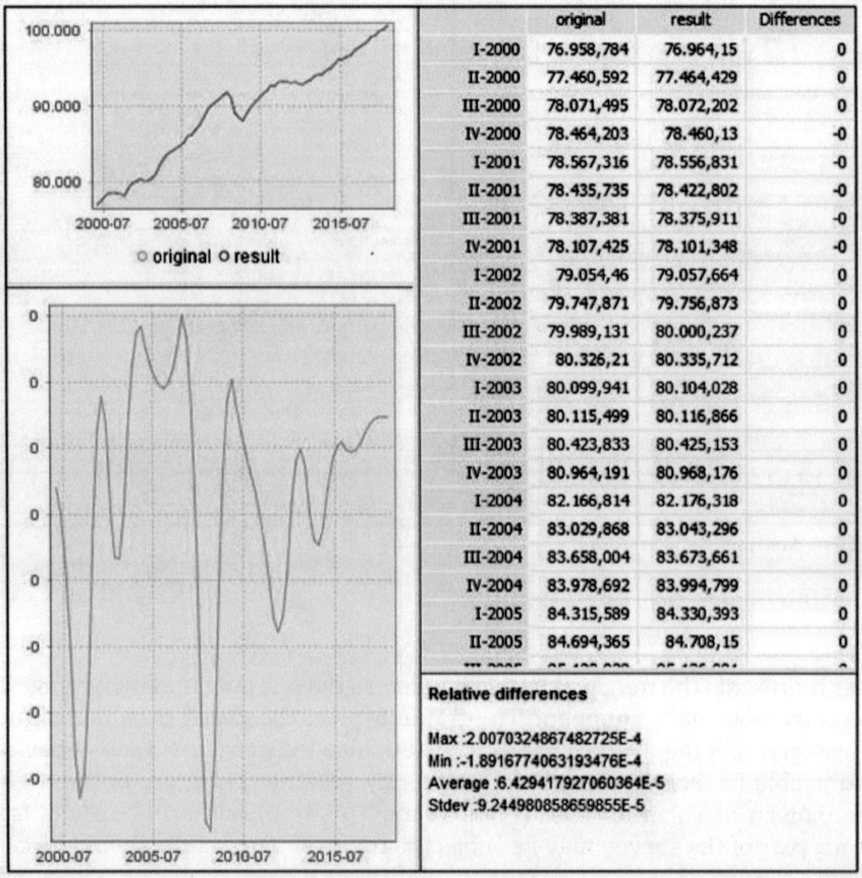

	original	result	Differences
I-2000	76.958,784	76.964,15	0
II-2000	77.460,592	77.464,429	0
III-2000	78.071,495	78.072,202	0
IV-2000	78.464,203	78.460,13	-0
I-2001	78.567,316	78.556,831	-0
II-2001	78.435,735	78.422,802	-0
III-2001	78.387,381	78.375,911	-0
IV-2001	78.107,425	78.101,348	-0
I-2002	79.054,46	79.057,664	0
II-2002	79.747,871	79.756,873	0
III-2002	79.989,131	80.000,237	0
IV-2002	80.326,21	80.335,712	0
I-2003	80.099,941	80.104,028	0
II-2003	80.115,499	80.116,866	0
III-2003	80.423,833	80.425,153	0
IV-2003	80.964,191	80.968,176	0
I-2004	82.166,814	82.176,318	0
II-2004	83.029,868	83.043,296	0
III-2004	83.658,004	83.673,661	0
IV-2004	83.978,692	83.994,799	0
I-2005	84.315,589	84.330,393	0
II-2005	84.694,365	84.708,15	0

FIGURE 8.59
Benchmarking Results for Seasonal Adjustments.

8.5.4.8 Optimise Your Results

Often, some diagnostic tests may fail after the first seasonal adjustment run. A key feature of JD+ is its capability to facilitate exploration of the seasonal adjustment process. For each round of seasonal adjustments, the user can modify any options in the specification and immediately observe the impact of the changes. Testing various options and their effects is valuable for gaining a deeper understanding of both the data and the software's capabilities. Figure 8.60 illustrates how to read the currently used spec and modify it using the Specification menu of the document. Pressing 'Apply' is sufficient to immediately observe the impact of the changes on the results.

8.5.5 Summary of Multi-Series Processing Results

8.5.5.1 Multi-Series Processing Pages

Figure 8.61 includes (1) the series name; (2) the method used (TS stands for user-defined spec); (3) the type of estimation (eg, current, parameter, last outlier, outlier, ARIMA (+

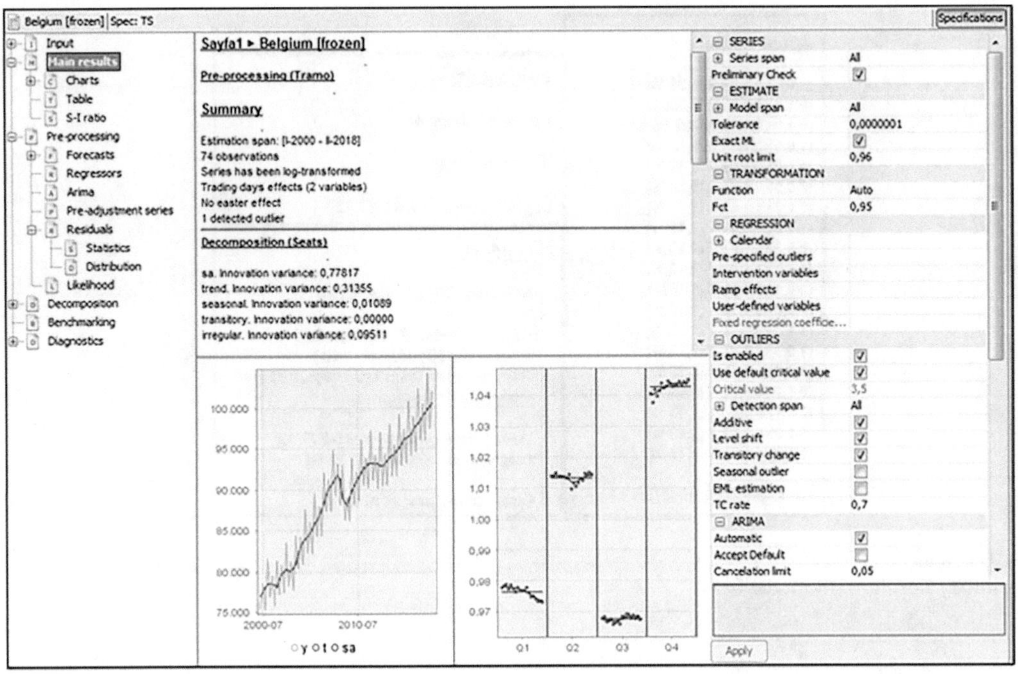

FIGURE 8.60
Specification Modifications.

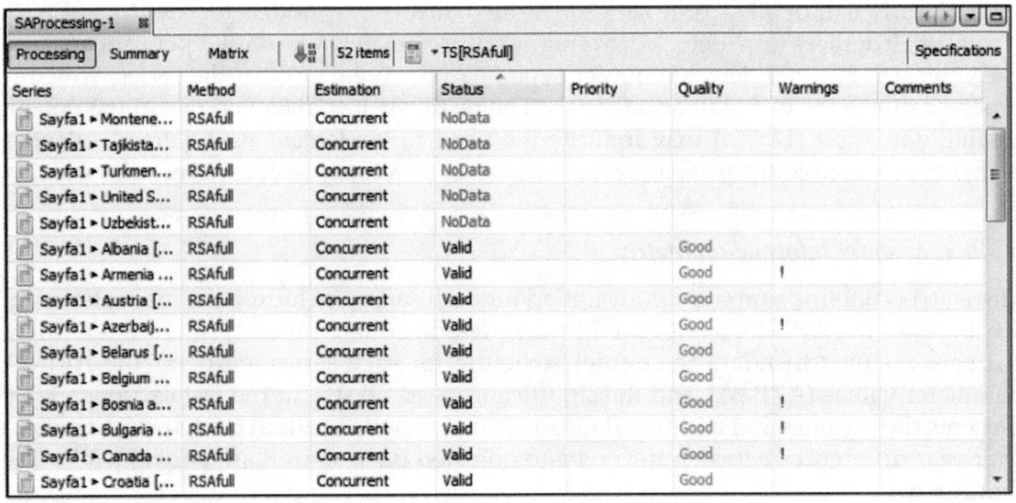

FIGURE 8.61
Multiprocessing Interface.

outlier), and concurrency); (4) the state of the multi-process series (eg, unprocessed, no data, and valid); (5) priority (a value of 10 indicates the priority of logarithmic/level selection); (6) the overall quality of seasonal adjustments; (7) possible warnings, eg, for short time series, non-decomposable models (SEATS), or undetermined seasonality. Clicking on the series name will open a results panel for each series, such as single-processing.

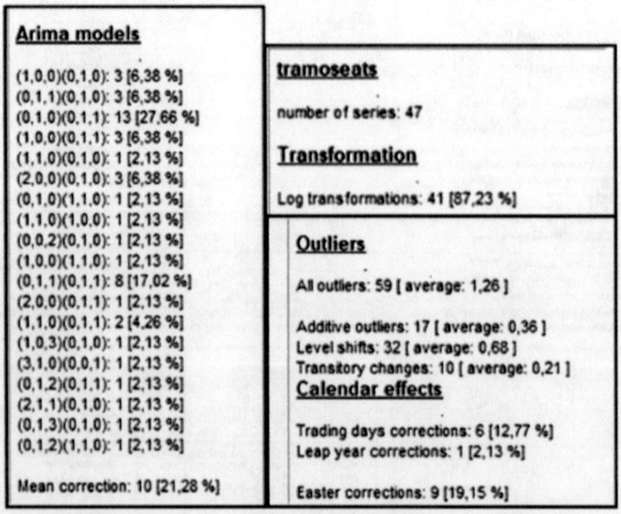

FIGURE 8.62
Summary Results of the Series Set.

8.5.5.2 Abstract Summary

The abstract gives general information about the model used for series sets (Figure 8.62). In this example, 47 of the 52 time series are analysed because 5 have no data. Forty-one of the analysed time series are modelled as logarithmic transformations of the original time series. Nearly half of all 47 time series identified airline-like models for TRAMO-SEATS. A total of 59 outliers were detected, averaging approximately one outlier per series.

The most frequently detected outlier type is a horizontal shift in the series, which is expected due to the global financial crisis. The relatively low detection rate of the default trading day effect (12.27%) may indicate the need to implement user-defined calendar regressors for each country's GDP series. The Easter effect was detected in nine time series.

8.5.5.3 Abstract Information Matrix

The matrix contains summary information for each series, including the main statistical attributes of the ARIMA model used in the pre-adjustment phase (primary), the calendar spec results (calendar), the outlier structure for each series (outliers), the ARIMA parameter values (ARIMA), and the significance tests (Tests). In the test section, a table presents the p-values and additional information from the residual tests, including year-over-year differences between the raw and adjusted data, as well as spectral peaks (see Figure 8.63).

You can copy (Ctrl+c) the Matrix part to the clipboard and then paste it (Ctrl+v) into another program. You can also export it as a CSV file, as described in Export Results and Metadata.

8.5.5.4 Quality Review

Based on the first results obtained from multiprocessing (Figure 8.64), TRAMO-SEATS can make appropriate seasonal adjustments to the quarterly GDP of all countries, except for

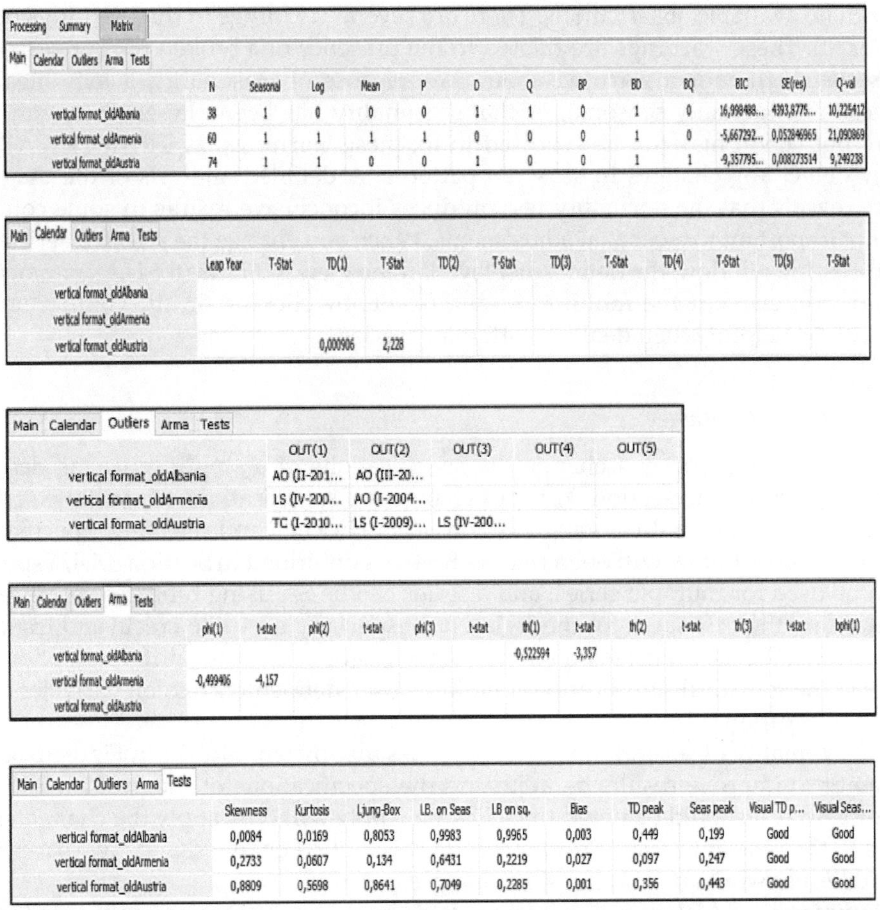

FIGURE 8.63
Matrix View in Multiple Processes.

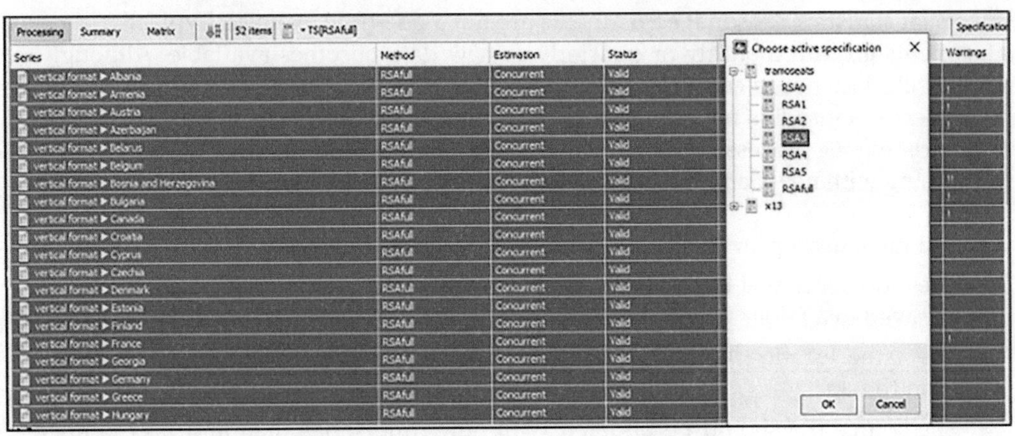

FIGURE 8.64
Modifying the Spec in Multiple Processes.

those with no available observations. There are several warnings in the Warnings column (!) and/or (!!). These warnings are related to the presence of a broken part or seasonality in the series. If there is a warning about the presence of seasonality, it may mean that the series is not actually seasonal, or that seasonality has been altered through journal creation. But action may not be required to fix these warnings, as TRAMO-SEATS will typically detect any changes in seasonal patterns. A detailed analysis of the diagnostic statistics reveals that the normality test produces inconclusive results in some countries. This is not important for seasonal adjustments. When you change the selected series in the multiprocessing window, the details displayed in the view in Figure 8.64 are automatically refreshed. You can sort the multiprocessing results by clicking on the column headers. Sorting can be helpful when dealing with many series.

8.5.5.5 Optimise Results

For multiprocessing, you can change the specifications of a single series, as described in the Single-Processing section. You can change the specifications of multiple series by selecting the series in the Processing window, right-clicking, and selecting 'Specification/Select'. A pop-up window will open (Figure 8.64). As described in Section 8.4.4, a specification can be used for multiple series, and a series can be SA using multiple specifications. This helps find the best specs for the series. It is especially useful to create and test different specs when the user regression volume of the calendar contains different options. JD+ cannot test the statistical significance of multiple user-defined regression variables within a single specification.

If the flow quality of several time series appears insufficient (unlike in Figure 8.60), you can attempt to improve results by adjusting the specifications of the problematic series (see Figure 8.64). It should be noted that clicking Apply will only apply the changed specifications to the relevant series. Restore reverts the applied changes, and Save saves them. Restore does not work after saving spec changes.

The Summary and Matrix sections are updated based on the changes only after the multiprocessing is rerun by clicking the green button (See Figure 8.65).

8.5.6 Refresh and Export Results

For official statistics, seasonal adjustments are not a one-time exercise. Typically, seasonal adjustments are run monthly or quarterly as new data becomes available. Although the model itself does not re-estimate monthly/quarterly, the SA process should be rerun and the SA series revised as new data becomes available for each period. In JD+, it's easy to revise existing SA processes and adjust new data. To enable successful refreshing of multiprocessing with new data, the following conditions must be met:

- You must use a provider to import the data source into JD+.
- The workspace that contains the multiprocessing of seasonal adjustments should be saved in a folder where the path or name should not be changed (Figure 8.66).
- The name, location, and data structure of the source file need to remain the same from the start.
- The source file should be updated with new time series data and user-defined regressions.

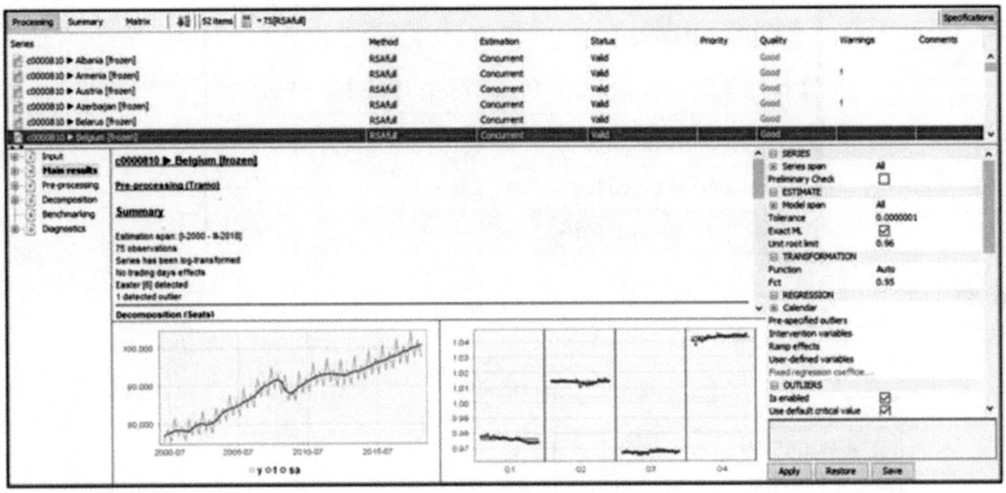

FIGURE 8.65
Optimising Series in Multiprocessing.

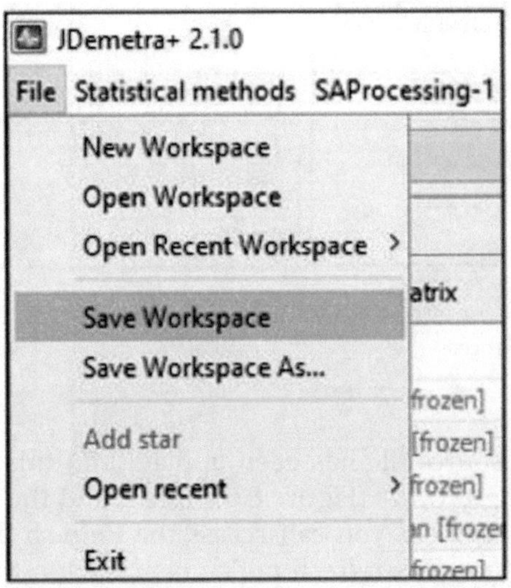

FIGURE 8.66
Save the Workspace for Future Processing.

8.5.6.1 Single-Series Processing

You can perform a refresh process for either single-series or multi-series processing. The first step is always to reload the previously saved workspace (Figure 8.67).

For a single process, double-click on the seasonal adjustment document (in this case, TramoSeatsDoc-1) that can be found under 'Workspace/documents' to open it (Figure 8.68, left). It contains the frozen results of the previous season's adjustment. You can now use the document item in the menu bar (in this case, TramoSeatsDoc-1) to refresh a document

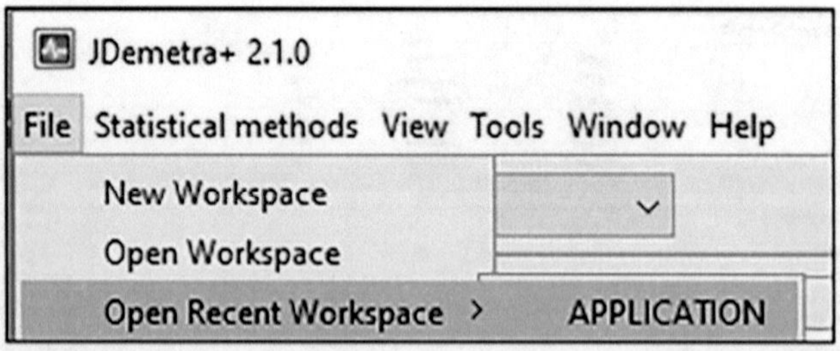

FIGURE 8.67
Reloading the Workspace.

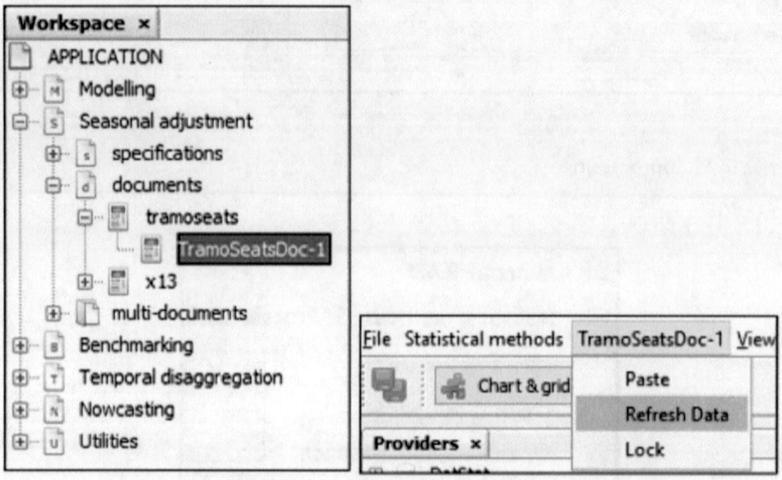

FIGURE 8.68
Invokes and Refreshes a Single Process.

with updated data (if the source file has been updated). In this menu, clicking 'Refresh Data' restarts the seasoning process (Figure 8.68, right), and the document then displays the updated results. Alternatively, you can access the Refresh data by right-clicking on a document in the workspace. The refresh policy is pinned to the spec used at the start. Immediately inspect the data source file and integrate the updated data with the SA process (Figure 8.69). The series results indicate that the estimated period was extended to Q2 2018, now comprising 74 observations. The time series retains a logarithmic transformation, includes an Easter effect, two detected outliers, and a trading day effect, with 73 observations yet to be identified.

8.5.6.2 Multiprocessing

Once the workspace is saved, if the above conditions are met (Figure 8.66), you can obtain the refresh seasonality results for an existing multi-process by following these steps (Figure 8.70):

Step 1: The previously saved workspace will open in JD+. You can also use Open Recent Workspace and select the Ask workspace for the question.

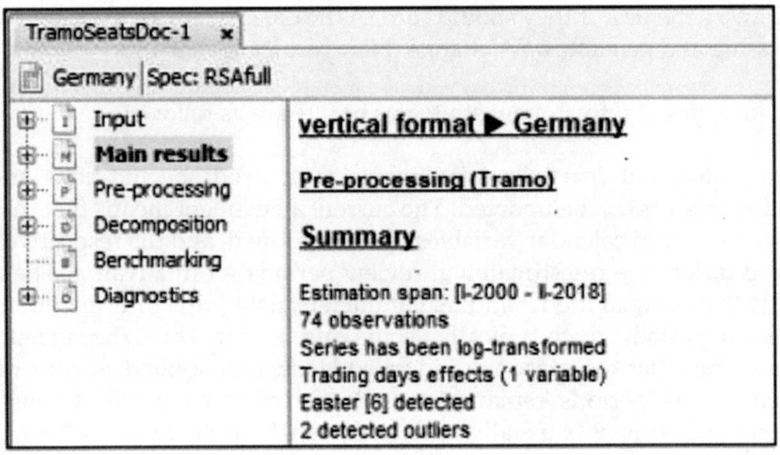

FIGURE 8.69
After the Update.

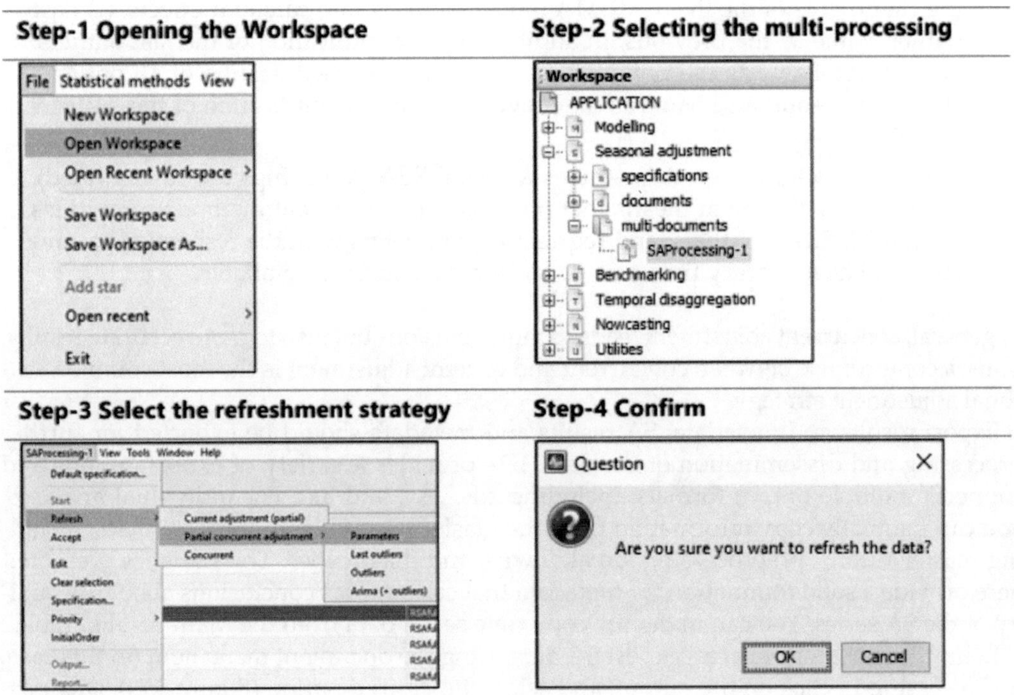

FIGURE 8.70
Steps for Refreshing the Multiprocessing.

Step 2: Double-click to open the problematic multiprocessing document under workspace/multi-documents (here SAProcessing-1).

Step 3: The project for multiprocessing documents (in this case, SAProcessing-1) appears in the menu bar.

In this menu, under Refresh, you can select a refresh policy.

Step 4: JD+ asks the user if they should refresh the data. Click OK to update the multiprocessing and generate a SA change (New results in step 4).

The three main types of refreshment strategies in JD+ are as follows:

- Current adjustment (partial): Adjustments with fixed specifications and user-defined regressors can be updated. The current adjustment means that the model, filters, outliers, and calendar variables are re-identified, and the respective parameters and factors are re-estimated at review periods set in advance. The method forecasts the seasonal model and its parameters, using this information until the next review period, which typically occurs once a year. Thus, the current adjustment assumes that the seasonal and calendar factors applied to new raw data between review periods remain fixed. Partial concurrent adjustment: Partial concurrent adjustment is usually used in cases where the model, filters, outliers, and calendar variables are re-identified once a year by the user. And, this refreshment strategy re-estimates the respective parameters and factors each time new or revised observations become available. It is possible to modify the partial concurrent adjustment with the following options: (1) Parameters: Re-estimation of the coefficients of the fixed ARIMA model, outliers, and calendar effects. (2) Last outlier: Same as the previous, in addition to re-identification of the last outliers. (3) Outliers: Same as the previous, in addition to re-identification of all outliers. (4) ARIMA (+outliers): Same as the previous, with re-identification of the ARIMA model.
- Concurrent adjustment: Adjustment with the RSAs (all things are re-identified). Concurrent adjustment means that the seasonal adjustment method re-identifies the model, filters, outliers, and regression parameters with the respective parameters and factors every time new or revised data becomes available.

In general, concurrent adjustment leads to more revisions but also to more accurate results. Thus, a compromise between concurrent and current adjustment is the most common seasonal adjustment strategy.

Export results and metadata: SA results and metadata should be exported for further processing and dissemination of the data. JD+ provides a variety of export options and supports multiple output formats, including .xls, .csv, and .txt. For individual analyses, you can manually copy information from the master page for summary statistics, including master results, preprocessing, breakdowns, and diagnostics. The statistics presented here provide a solid foundation for metadata that can support conclusions about the quality of the SA series. You can manually copy time series data from the Main results/Table.

In multiprocessing, you can export SA data using the document menu item (in this case, SAProcessing-1), click on the output, and select the desired option (Figure 8.71). Diagnostics and other results can be exported using the CSV matrix option. The batch output window (Figure 8.71, right) allows you to set the location, name, layout, and contents of the output file.[19] Under Content, it is recommended to include at least three series: y, y_c, and the benchmarking result, where y is the original time series, y_c is the calendar-adjusted series, and the benchmarking result is the seasonal adjustment and the benchmarking series.

Moreover, you can use the document item in the main menu (in this case, SAProcessing-1), then select Report to generate a report. This provides some useful diagnostics that can be used for SA statistical metadata (Figure 8.72).

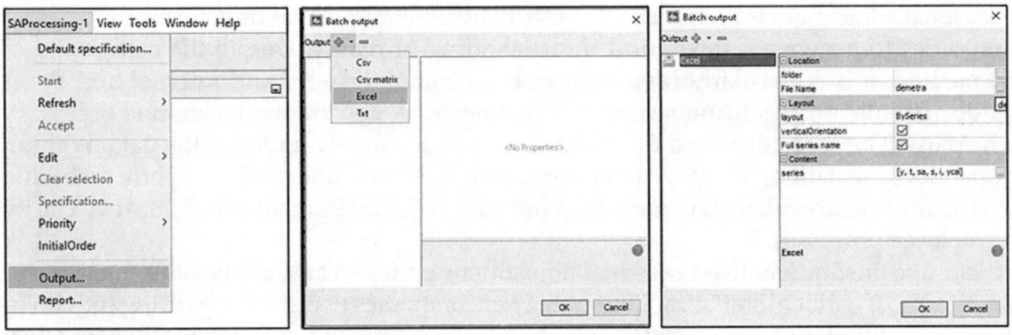

FIGURE 8.71
Export Results for Multiprocessing.

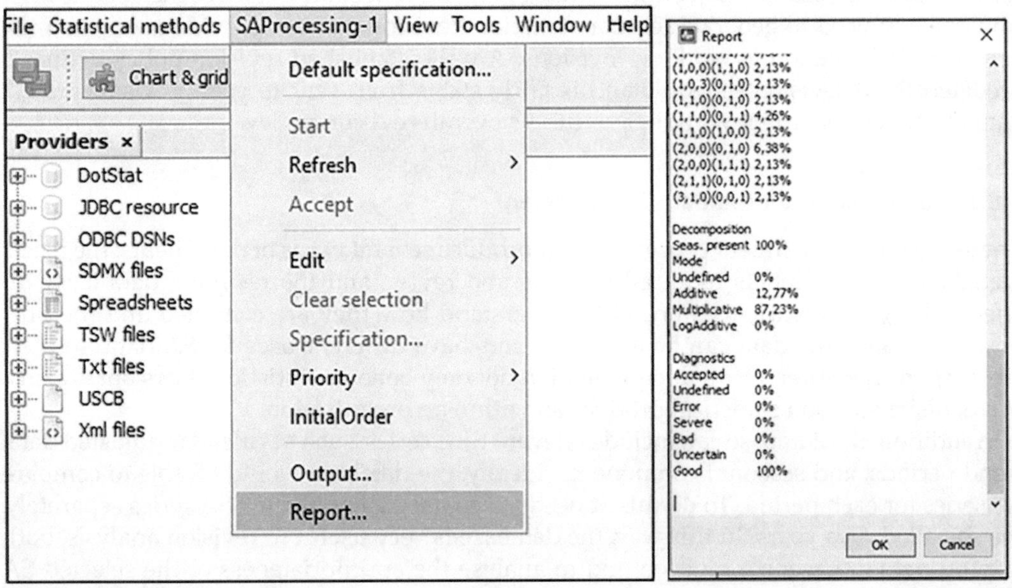

FIGURE 8.72
Seasonal Adjustment Report.

8.6 User Communication and Publishing Practices

Only the statistics used are useful, and this principle also applies to SA data. The seasonal adjustment process does not end with confirmation of the result quality. To assist statistics users in working with SA data the following is provided: (1) drafting internal and external documentation; (2) defining seasonal adjustment policies; (3) revising SA data; (4) preparing data for release; (5) providing the necessary support to the user. The seasonal adjustment of the economic time series is intended to support the users of statistical data in interpreting economic development.

Seasonal adjustment is a sophisticated estimation technique that transforms the original time series to generate a theoretical understanding of reality. Due to the complexity of the method, it is particularly important to have comprehensive and internal and external documentation regarding seasonal adjustments. A general explanation is necessary to help users better understand the technology and accurately interpret the data. Without proper documentation, SA data can be confusing for users. Internally, properly archiving SA data and clear work instructions help produce reliable SA results and improve clarity in employee processes.

Clear and institutionalised seasonal adjustment policies can enhance the quality and consistency of national statistics. The strategy encompasses decisions regarding the choice of common software and the methodology to be implemented. It takes a stance on when and how to modify seasonal models, how to handle outliers, and the content of documentation. As a result, a single statistician doesn't have to make all these tough decisions alone. Revisions are an important aspect of both the SA time series and the underlying original series. Thus, the seasonal adjustment strategy should reflect the choice of the generic revision strategy used to generate regular statistics. Revision policies provide users with the necessary information to work on revisions. A well-established revision policy defines a predetermined revision schedule and is fairly stable from year to year. Revisions to SA data should be consistent with the producer's overall revision policy.

8.6.1 Document and Metadata Management

The seasonality adjustment process should produce several types of documents: the user's metadata, the documents required to make and revise, and the resulting data itself. To effectively use statistics, users need to understand how they are compiled and the purposes for which the data can be applied. Users have different uses for SA data, so their metadata needs differ. Good documentation not only benefits statistical users but is also a prerequisite for good statistical production within an organisation.

In addition, the database can include relevant time series, such as calendar-adjusted data, trend periods, and seasonal components. Ideally, the database would be able to compare SA series for each period. To do this, it needs to contain all published SA series separately, the so-called data years. In this way, the database is very useful in revision analysis, both to propagate the average revision and to analyse the appropriateness of the selected SA models and specs. Using archived time series can enhance the quality of seasonal adjustments over the long term. It can analyse the behaviour of SA series, for example, during economic turning points.

The purpose of the internal documentation is to maintain the high quality of the seasonal adjustments. Internal documentation includes the following: (1) step-by-step work instructions for performing seasonal adjustments; (2) an internal quality report describing the quality and characteristics of SA data and raw data; (3) a regularly updated list of national holidays (if not included in a predefined calendar by JD+); and (4) a list of outlier causes for monitoring and enumeration. Internal documentation should allow any colleague familiar with seasonal adjustments to repeat the process. Adequate documentation helps retain the skills needed for seasonal adjustments within the organisation and supports briefing and training others.

JD+ provides extensive quality diagnostics based on the SA method chosen. Metadata can also be shared by using XML files generated by JD+. For key series, you can copy the summary statistics from the Results panel, including Main Results, Preprocessing,

Decomposition, and the first page of Diagnostics (such as Export Results and Metadata). Ability to go back to the old quality diagnosis and compare the quality of the new results with the previous diagnosis, especially after re-estimating seasonal models and parameters. More time should be devoted to preparing and reviewing documentation for key time series results.

A more comprehensive quality analysis is generally conducted once a year. This is because new data points available during the year may alter the time series trend, potentially affecting the models used in seasonal adjustments. For regular use, the internal quality report should be simple enough for statisticians to update easily contain the most relevant details. This should be understandable to colleagues unfamiliar with seasonal adjustment, although it may be biased, as charts and tables are generally easier to interpret. User documentation is often more concise than an internal quality report, but it could use the same content. User documentation is typically more static, allowing for quick updates as part of the monthly or quarterly production process.

Several international guidelines emphasise the need for greater transparency in statistical production. The United Nations (2014) emphasised the importance of transparency in statistical production, stating that, to promote accurate data interpretation, statistical agencies should disclose information on data sources, methods, and procedures in accordance with scientific standards. When publicly available information is abundant, the metadata for statistical data users must be carefully designed. It should contain only relevant information.

In addition, the OECD (2007) provides relevant data and metadata. Metadata describes the statistics (to some extent), the processes and tools involved in generating and using them, and explains how the results are interpreted and used. Many statistical offices maintain a description of their methods and practices in a dedicated section of their websites, for example, labelled 'Understanding Statistics'. As recommended by the OECD (2007), statistical offices should provide at least minimal information to enable assessment of the reliability of each SA series. Statistical offices should maintain comprehensive metadata to enable users to perform consistent seasonal adjustments on other series that may not have been SA. The U.S. Bureau of Economic Analysis (BEA) provides a website with frequently asked questions related to seasonal adjustments. The BEA applies X-12-ARIMA for seasonal adjustment and defines a set of minimum quality diagnostics, which are published alongside its statistical releases. Among other quality indicators, they publish overall quality assessment statistics and indicators that analyse the stability of estimates and the presence of moving seasonality in key time series.

In addition to revisions due to unadjusted raw data and changes in the regARIMA model, revisions are caused by two-sided filters and the use of forecasts in seasonal adjustments. In TRAMO-SEATS and X-13, the primary SA filters are bilateral, meaning that the estimate of the seasonal component depends on the observations before and after a particular period. Thus, estimates for the most recent period require observations that are not yet available. First, preliminary estimates are generated using ARIMA predictions. As new observations become available, these estimates are updated until the filter is complete and the historical estimates are finalised.

To anticipate revisions, JD+ includes a revision history test by Findley et al (1990) to indicate which series may have excessive revisions. SEATS also provides revised standard deviations in the seasonal and trend period components. When it's too high, the series may not warrant adjustment. The publication of any statistics should include metadata and direct readers to more detailed information.

It is a common practice to publish links to metadata in statistical publications. The link may lead to an archive of quality reports and historical quality descriptions for the relevant metric.

8.6.2 Seasonal Adjustment and Revision Policy

8.6.2.1 Seasonal Adjustment Revision Policy

Many statistical offices need to generate large amounts of SA data. As a result, the production units themselves carry out seasonal adjustments to the indicators. In other cases, responsibility for applying seasonal adjustment tests lies with the method group. Even with modern software, seasonal adjustments remain time-consuming. Key statistical aggregate data will be the main focus of attention for both domestic and international users. As a result, statisticians need to spend more time making seasonal adjustments to these aggregated time series. To make wider use of seasonal adjustments, organisations need sufficient computer resources for the dissemination of data and the storage of yearly data. Before embarking on a large-scale seasonal adjustment, producers should weigh its advantages and disadvantages.

Seasonal adjustment methods and options should be thoroughly tested before policies are developed. Defining the policies to be implemented across the organisation requires consideration and experience. The policy could be developed in stages, starting with basic options such as SA software and methods, revision times, and guidelines for publishing adjusted data and its metadata. As experience accumulates, the policy can be expanded to provide more detailed explanations for statisticians with problematic time series, time series disruptions, and the application of seasonal adjustments during periods of economic uncertainty.

Similarly, the range of SA data released can be increased gradually. SA policies should incorporate practical knowledge, empower statisticians within the organisation, adhere to international standards, and meet the expectations of statistical users. At a minimum, the policy should cover the following issues: (1) common seasonal adjustment methods to be applied; (2) software solutions for seasonal adjustment, dissemination and storage; (3) the method and timing of the reanalysis, ie, the revised policy of seasonal adjustment; (4) aggregation from lower-level classifications of industrial activities to higher-level aggregations, or from regional indicators to national levels and over time; (5) handling of outliers; (6) requirements for internal documentation and metadata of users; and (7) guidelines for the release of SA data as part of a regular release schedule.

Not all statistical offices have adopted a SA approach. For example, sometimes national accounts, industry, trade, and labour statistics can all be applied to their practices. However, for clarity and consistency, it is advisable to adopt a uniform approach across the organisation, while regularly comparing the results of alternative methods to ensure their usefulness. The international comparability of the statistics generated can be improved by using commonly used software and methods.

Publishing SA data may require changes to the propagation software. Users prefer to access time series data in a format that enables reprocessing with minimal manual effort. Seasonal adjustment multiplies the number of data units generated for each statistical indicator, making the propagation database a smart choice for publishing longer time series. A reliable database structure is also needed to store the results of seasonal adjustments, especially for revision analysis.

8.6.2.2 Revision of SA Data

One important issue in the development of SA policies is the definition of revision practices. Since seasonal adjustments are based on estimates, they are subject to adjustment. As BEA explains, SA data are modified, first, by corrections to the original data due to new observations becoming available; second, by refreshed and accumulated data that allow for better estimation of seasonal patterns and revisions to regARIMA models, filters, and forecasts. This also affects the recognition of seasonal patterns, which can lead to revisions to historical data.

Theoretically, the quality of the forecasts used in seasonal adjustment increases with the frequency of updates to the underlying model. As a result, there is a trade-off between the cost of frequent revisions and the quality of the released data. However, to complete the equations, frequent updates to the seasonal model can also decrease the stability of the results and lead to revisions in the opposite direction. The seasonal adjustment strategy should be selected from among alternative update policies. A balanced alternative between these two extremes may provide a better quality of adjustment.

Partial concurrency tuning is a widely used strategy. In its basic form, it keeps models, filters, outliers, and calendar variables fixed until they are re-identified annually or semi-annually. Every time new or revised data becomes available, it re-identifies the corresponding parameters and factors. As mentioned earlier, JD+ offers four different partial concurrency adjustment strategies for you to choose from. Eurostat (2015) recommends following one of these balancing approaches between the current and simultaneous adjusted extremes. The choice of refresh strategy also depends on the properties of the family.

According to Eurostat (2015), the optimal frequency to update a seasonal model is at least once a year. However, if any issues are detected between updates, they should be corrected immediately. There are more degrees of freedom in the frequency of parameter updates for the seasonal model. In addition, the impact on historical data should be evaluated, and the significance of the regression variables examined to determine whether the specification requires adjustment. The refresh policy needs to take into account the period for which the results were modified. Revisions resulting from a revised seasonal model can, but don't necessarily have to, be published in full. A comprehensive revision from the beginning of the series promotes methodical uniformity across all values and makes it easy to replicate SA results. However, some statisticians have questioned whether the newly added figures contain relevant information for significant revisions to historical seasonal patterns.

To balance information gain and revision, some offices have chosen to limit the revision period of SA data to a period of about four years longer than the revision period of the original data. In the early days, some offices chose to keep their seasonal factors unchanged. Information from the diagnostics, such as the sliding span and revision history, can help in selecting the appropriate revision range.

If there are any restrictions on the scope of revisions, they need to be communicated to statistical users. If the original data have been revised from the beginning of the series – for example, due to a change in definition or sampling scheme – the entire SA series should be updated accordingly. Consistency between the original series and the SA series should be maintained.

As part of the SA policy, the revision practice will be based on an analysis of the expected magnitude of the revision and the impact of the different choices made during the SA period. It provides information about revisions and their reasons to inform users about statistics. In addition, as suggested by Eurostat and the ECB (2002), the Global Revision Policy introduces a predictable framework for revision.

The time aggregation method should be included in the seasonal adjustment strategy. Seasonality is not neutral throughout the year. You can force the annual sum or average of the SA data to match that of the original data. Some users may find it difficult to understand why year-to-year variations differ between SA data and raw data. However, from a theoretical point of view, there is no reason to force the change to be equal from year to year.

To date, only mixed evidence has been presented regarding the superiority of direct versus indirect aggregation of SA data (Eurostat, 2015). The direct approach means that the aggregated time series and subcomponent series are SA independently. In the indirect method, a higher-level series is obtained by aggregating SA component series using a weighted scheme.

The direct method is generally preferred due to its transparency and accuracy, particularly when the component series exhibits similar seasonal patterns. When components of a series show significant differences in seasonal patterns, an indirect approach may be preferred. Regardless of whether the direct approach is simple, the indirect approach may help meet the user's strong demand for consistency between lower-level and higher-level aggregations. If indirect seasonal adjustment is used, residual seasonality must be carefully monitored.

8.6.3 Data Publishing Policies

If you don't make seasonal adjustments, you can calculate the change between one month (quarter) and the same month (quarter) of the previous year, ie, compare January to January, February to February, and so on. However, this comparison is not without a calendar effect, as the number and classification of different days in a month vary from one year to another. Thus, comparisons with calendar-adjusted series are more robust than the original time series. In year-over-year comparisons, this is the recommended series to use as the basis. One of the key advantages of SA data is that it enables meaningful comparisons between consecutive months or quarters. This may show changes in the economy more quickly because they don't have to be counted from 12 months ago. This only works if the underlying series is not too unstable.

In addition, users may be interested in learning about the changes compared to the same month a year ago. The adjusted calendar is a good source for calculating this measure. The cumulative growth rate may be available as additional information. Selecting the appropriate growth rate for statistical releases requires careful judgment by the statistics producer. Period-to-period growth rates are the rate of change in the previous period, month, or quarter. Such rates are expressed as (Q_{t-4}/M_{t-1})-1 or (Q_t/Q_{t-1})-1. Mt represents the value of the monthly time series in t months, and Qt represents the value of the quarterly time series in the t quarter. The year-over-year growth rate is the rate of change in the same period, month, or quarter of the previous year. They can be referred to as the year-over-year growth rate: the rate of change compared to the same period in the previous year, or the 12-month rate of change. These rates are expressed as (M_t/M_{t-12})-1 or (Q_t/Q_{t-4})-1.

The year-to-date growth rate is the data expressed as a cumulative value since the beginning of the year; this is sometimes referred to as cumulative data. For example, they might compare the sum of values from January 2019 to April 2019 to the same period in 2018.

The growth rates described earlier can be calculated for any component of a time series, including the raw series, the calendar-adjusted series, the SA series, or the trend-cycle component. Table 8.12 is based on the use of different growth rates for Belgium's quarterly GDP. There is no information on the growth rate of the original time series, as the

TABLE 8.12

The Explanation of the Growth Rates of Different Kinds

Type	Original time series	Original time series	Calendar-adjusted series	Calendar-adjusted series	Seasonal and calendar-adjusted series	Seasonal and calendar-adjusted series
JD+ name	y	y	ycal	ycal	result	result
Growth rate	q-to-q	y-to-y	q-to-q	y-to-y	q-to-q	y-to-y
Note	Not Informative	slowest	Not Informative	slow	fast	fast
Q1–2017	–6.10%	2.00%	–6.20%	1.70%	0.50%	1.70%
Q2–2017	4.50%	1.50%	4.80%	1.80%	0.60%	1.80%
Q3–2017	–4.40%	1.40%	–4.50%	1.40%	0.10%	1.40%
Q4–2017	8.60%	1.90%	8.60%	1.90%	0.70%	1.90%
Q1–2018	–6.40%	1.50%	–6.40%	1.70%	0.30%	1.70%
Q2–2018	4.50%	1.50%	4.40%	1.30%	0.20%	1.30%

year-on-year (year-to-year) growth rate includes changes in calendar effects. That's why the calendar-adjusted year-over-year growth rate gave a real signal (increase) in the first quarter of 2018, while the original time series implied a decline over the same period.

Seasonally and calendar-adjusted series provide accurate signals for both year-over-year and quarter-on-quarter growth rates, although they are most commonly used for quarter-on-quarter comparisons. The inter-period growth rates of the original series are affected by seasonality because they compare different months that are affected by it.

Implementing seasonal adjustments requires restructuring the website's content to facilitate easier access to time series data. Redesigning statistical news releases takes time to better support users. The challenge of statistical publishing is to present relevant information concisely while keeping it simple and easy to understand. At the same time, the release should contain enough information for users to draw meaningful conclusions.

When seasonality is present and recognisable, seasonal adjustments will improve the readability of statistics. In these cases, the statistical office should publish the data in a SA form. Producers can choose the level of detail for SA data based on user needs and available resources. Over time, they can increase the amount of SA data available. If the seasonal impact indicator is indicated, ideally, statistical releases should focus on SA data. The user should also have access to the original series.

Since press releases are intended to deliver timely news, SA data is the most appropriate for presentation. In cases of user demand, producers can also disseminate SA components, such as adjusted calendar days and/or trend cycle series. The statistical dissemination of data affected by seasonality should include, at a minimum, the most recent inter-period growth rate. If the statistical tradition includes the level of the published measured variable, and if space allows, the change or value of the level can also be published in the posting. The cumulative growth rate may be useful as additional information, but it is not the main focus. Unless the seasonal pattern is stable, the annual growth rate of the raw and SA data is not always the same.

For highly volatile SA series, changes between periods may frequently reverse direction. Thus, it is useful to analyse the behaviour of these changes to understand the type of information they will convey to the reader. The release of statistics focuses on key economic

Main visit and spending estimates for UK residents' visits abroad in September 2018

UK residents* visits abroad				
	Visits (thousands)	**% change from year earlier**	**Expenditure (£ million)**	**%change from year earlier**
Non seasonally adjusted				
Sep 2018	7,708	–3	5217	+1
Jul to Sep 2018	24,180	+1	16,410	+3
Year to date 2018	57,370	–1	36,310	+1
Latest 12 months	72,110	–1	45,230	+0
Seasonally adjusted				
Apr	6,070	+0	3,550	–6
May	6,070	+3	3,890	+17
Jun	6,040	–4	3,860	–6
Jul	6,090	+4	3,880	+7
Aug	6,170	+3	3,830	+2
Sep	6,040	–5	3,730	–0

FIGURE 8.73
Travel Industry Press Release (ONS).

indicators, which are usually aggregates at a higher level and less volatile than individual economic activities.

Another example is the International Passenger Survey press release from the UK Office for National Statistics (ONS, 2019). In this release, SA data is used for a monthly comparison of the number of visitors and their spending (Figure 8.73).

The rate of change compared to the same period of the previous year, ie, the year-on-year change, should be calculated based on the calendar day adjustment series or, if not, based on the original series. The calendar day adjustment series, ie, the business day or trading day adjustment data, makes the same month in different years more comparable by correcting for changes caused by the number of business days, etc. Any special events in the previous year will affect the year-over-year change.

Statistical offices often have unique access to large data sets and can inform users of a variety of important changes in subpopulations. UN (2013) recommends the development of a product group or industry that is primarily responsible for the monthly movement of the total index.

8.6.4 Social Media and User Support

In recent years, it has also become increasingly common to publish official statistics through social media platforms. The motivation for using social media is that it reaches a wide audience effectively. There are some very specific advantages and disadvantages of posting statistics on social media compared to traditional press releases: (1) the advantage is the ability to instantly deliver (visual) content to a wider audience; (2) the disadvantage is that the space is more limited, connects to shorter messages, and results in content that is often more superficial. Although there are several studies in the literature on posting statistics on social media (the most important being Everson, Gundlach, and Miller (2013)), none have shown particular interest in using SA data.

There is a growing demand from statistical offices for timely statistical information on economic development. At the same time, the increased volume of information can overwhelm statistical users, and the rapid development of new communication tools has transformed the way information is exchanged. Publishing statistics doesn't happen in isolation. There are significant differences in the technical ability and understanding of statistics among statistics users. This poses a challenge in helping users draw accurate conclusions. Regular users of statistics benefit from the availability of SA data. As mentioned earlier, it can compare the developments of different industries and countries and indicate of economic changes more quickly. Some users may further reprocess the results in their work. For example, economists' and analysts' forecasting models are based on SA data. Sometimes, users may perform seasonal adjustments according to their specifications. Therefore, the data should be provided in an easily accessible format, with additional information – such as different components of the original data – available upon request. As mentioned earlier, users value transparency through metadata, the availability of general explanations of the statistical methods applied, the publication of statistics, and well-designed websites.

The Internet offers increasing opportunities for disseminating statistics, but it also contains an overwhelming amount of data of varying quality. The Internet is an effective tool for data dissemination. Still, it requires data producers to work hard to track unknown users of statistics to understand their needs and problems in using statistics. Ideally, this release will provide statistical users with contact information for more information on concepts, definitions, and statistical methods. A contact can be a generic contact that directs the request to an expert. Experts responsible for published data can answer users' questions about the latest statistical releases to share their expertise. The responsibility of statistical offices does not stop at publishing; on the contrary, statistical agencies have the right to comment on the misinterpretation and misuse of statistics, in accordance with the basic principles of national official statistics (UN, 2014).

Statistically generated users can benefit from consulting statistical users. Many statistical offices conduct stakeholder analysis, conduct user surveys, and organise workshops and conferences to discuss statistics with key users. This interaction helps producers stay relevant by understanding user needs and helps users understand how producer apps work. Customer training services on the use of statistical data to analyse economic development may include tips for understanding and using SA data.

The purpose of seasonal adjustment is to meet the user's need for relevant and comparable short-term statistics that provide information on cyclical movements and turning points. The role of a statistician is to provide high-quality data along with comprehensive documentation and interpretation to ensure users can correctly utilise and understand the statistics.

8.7 Summary and Conclusions

This chapter systematically introduces the functions and operation process of JD+, a professional seasonal adjustment tool. Recommended by Eurostat, JD+ integrates the two leading methods, TRAMO-SEATS and X-13ARIMA-SEATS, delivering a cross-platform, modular seasonal adjustment solution via the Java platform. This significantly enhances the usability and extensibility compared to traditional FORTRAN programs.

1. Data preparation details the format requirements for input data (eg, Excel, CSV, TSW files), preprocessing steps (missing value processing, outlier detection), and the application of visual analysis tools (cycle charts, growth charts).
2. The core functions of JD+ are analysed: (1) calendar effect modelling (2) support for custom national/industry holiday regression variables to accurately correct the impact of trading days and moving holidays; (3) flexible seasonal adjustment process: provide single-series interactive adjustment and multi-series batch processing, allowing users to choose preset specifications (such as RSAfull) or custom ARIMA model parameters.
3. Comprehensive quality diagnostics: Stability adjustments are assessed through residual testing, slip span analysis, and revision history to ensure reliable results.
4. In the results analysis section, this chapter emphasises the following: (1) the decomposition validation of trend, seasonality, and irregular components; (2) the application of benchmarking; (3) aligning the adjusted data with the original annual sum; (4) according to the actual needs of statistical work, the release strategy of SA data (such as giving priority to the month-on-month growth rate) was discussed; (5) revision policies (eg, annual model updates), and (6) user communication methods (metadata recording, social media dissemination).
5. With its open Java architecture and comprehensive analysis tools, JD+ has become a key choice for modern seasonal adjustment, particularly for statistical agencies requiring large-scale, automated processing. This chapter offers statisticians a standardised operational framework – from data import to result export – and provides advanced users with an in-depth reference for parameter tuning and result validation, including diagnostic tools and optimisation recommendations to address complex adjustment challenges, such as short series and highly fluctuating data.
6. Based on Eurostat's case experience, the key role of transparent documentation and user communication in improving data credibility is emphasised.

Notes

1 This app works with the X-13ARIMA-SEATS as well. Since we have explained the use of the X-13ARIMA-SEATS in detail in Chapter 5, we will focus only on the use of the TRAMO-SEATS in this chapter.

2 See Maravall, A., López-Pavón, R., & Pérez-Cañete, D. (2015). *"Reliability of the Automatic Identification of ARIMA Models in Program TRAMO"*. Retrieved from https://link.springer.com/chapter/10.1007/978-3319-03122-4_7; Maravall, A., Caporello, G., Pérez, D., & López, R. (2014). "Program TSW: Revised Reference Manual". Series: Banco de España Occasiónal Paper No. 1404, p. 120. Banco de España (Bank of Spain). https://doi.org/10.2139/ssrn.2524063.

3 Documentation on Win X-X13 can be found on a dedicated U.S. Census Bureau webpage. Also available for: Linux and Unix platform X-13ARIMA-SEATS Releases. See U.S. Census Bureau. (2015). "X-13ARIMA-SEATS Quick Reference for Unix/Linux". 1.1 BETA, U.S. Census Bureau, Statistical Research Division.
4 European Statistical System (ESS). (2009). *ESS "Guidelines on Seasonal Adjustment"*. Eurostat. Seasonal Adjustment Guidelines.
5 The JDemetra+ software, as well as other documentation and guidelines, are available on the Eurostat website. Retrieved from https://ec.europa.eu/eurostat/cros/content/software-jdemetra_en.
6 For more details, see Grudkowska, S. (2015). "Seasonal Adjustment with JDemetra+ – A Step-by-Step Guide". Series: Materials and Studies No. 318. Narodowy Bank Polski (National Bank of Poland). https://doi.org/10.2139/ssrn.2652746. In addition, you can import predefined data in XML format into JDemetra+ via a web service. To do this, you need a plugin called "DotStat" (Charles, 2016). But this topic is beyond the scope of this book.
7 The data series used in this example is from the UNECE statistical database (www.unece.org/stats). The selected dataset contains quarterly gross domestic product (GDP) for 52 countries at 2010 average prices (ISIC Rev. 3.1). Not all countries have the same start date. The initial format of the data is shown in the top half of Figure 8.2. Users can arrange time series sets vertically or horizontally. Then, transpose and reformat the data based on the date (column A), as shown in the bottom half of Figure 8.2.
8 A more detailed explanation of the spectrogram can be found at: Grudkowska (2015) Section 3.4.2. Grudkowska, S. (2015). "Seasonal Adjustment with JDemetra+ – A Step-by-Step Guide". Materials and Studies No. 318. Narodowy Bank Polski (National Bank of Poland). https://doi.org/10.2139/ssrn.2652746.
9 Gomez & Maravall (2000) provided a comprehensive description of the software. Maravall, A., López-Pavón, R., & Pérez-Cañete, D. (2015). *"Reliability of the Automatic Identification of ARIMA Models in Program TRAMO"*. Retrieved from https://link.springer.com/chapter/10.1007/978-3 319-03122-4_7 provided Information on the most up-to-date facilities and interpretations of the SEATS results. See Gómez, V., & Maravall, A. (1997). "Programs TRAMO and SEATS: Instructions for the User". Beta Version, Banco de España Working Paper No. 9709. Banco de España, Madrid, June.
10 There are two other options under Add Calendar that are beyond the scope of this book (Figure 8.9).
11 The algorithm for the user-defined regressors in JD+ is the same as the regression matrix (.rmx table) we introduced in Chapter 6. In fact, in the appendix to Chapter 6, an exhaustive explanation has already been given.
12 For more information on the X-13, see Chapters 5 and 6 of this book.
13 see Maravall, A., López-Pavón, R., & Pérez-Cañete, D. (2015). *"Reliability of the Automatic Identification of ARIMA Models in Program TRAMO"*. Retrieved from https://link.springer.com/chapter/10.1007/978-3319-03122-4_7.
14 For more information on the M-statistic for X-13, see Grudkowska, S. (2015). "Seasonal Adjustment with JDemetra+ – A Step-by-Step Guide". Materials and Studies No. 318. Narodowy Bank Polski (National Bank of Poland). https://doi.org/10.2139/ssrn.2652746, and for SEATS diagnostics for TRAMO-SEATS, see Maravall, A., López-Pavón, R., & Pérez-Cañete, D. (2015). *"Reliability of the Automatic Identification of ARIMA Models in Program TRAMO"*. Retrieved from https://link.springer.com/chapter/10.1007/978-3319-03122-4_7.
15 See Grudkowska, S. (2015). "Seasonal Adjustment with JDemetra+ – A Step-by-Step Guide". Materials and Studies No. 318. Narodowy Bank Polski (National Bank of Poland). https://doi.org/10.2139/ssrn.2652746.
16 See Maravall, A., López-Pavón, R., & Pérez-Cañete, D. (2015). "Reliability of the Automatic Identification of ARIMA Models in Program TRAMO". Retrieved from https://link.springer.com/chapter/10.1007/978-3 319-03122-4_7.
17 For more details on diagnosis, see Grudkowska, S. (2015). "Seasonal Adjustment with JDemetra+ – A Step-by-Step Guide". Materials and Studies No. 318. Narodowy Bank Polski

(National Bank of Poland). https://doi.org/10.2139/ssrn.2652746. JDemetra+ reports all test statistics as "messages" using the classifications in Table 4.1. This allows for quick interpretation of test statistics.

18 Grudkowska, S. (2015). "The Development of the Seasonal Adjustment Procedure TRAMO-SEATS: R Package Implementation and Application". Doctoral Dissertation, Warsaw School of Economics (SGH), Poland.

19 A more detailed explanation of these characteristics can be found in Grudkowska, S. (2015). "The Development of the Seasonal Adjustment Procedure TRAMO- SEATS: R Package Implementation and Application". Doctoral Dissertation, Warsaw School of Economics (SGH), Poland.

9

Challenges and Future Directions of Seasonal Adjustment Methods

9.1 Introduction

In the field of time series decomposition, seasonal adjustment methods have evolved from simple moving averages to sophisticated, complex models. As two mainstream methods, X-13ARIMA-SEATS and TRAMO-SEATS have formed their own unique technical routes and application ecosystems. The birth and development of these two methods reflect the deepening of the statistical community's understanding of time series decomposition problems. Extensive practical experience has demonstrated that the flexibility and customisability of X-13ARIMA-SEATS enable analysts to adjust filter parameters or choose different decomposition components to address specific problems. This method is particularly suitable for dealing with economic data with complex calendar effects. On the other hand, due to its theoretical rigour and consistency, TRAMO-SEATS greatly lowers the barrier to entry via its automatic model selection function.

However, given the rapidly changing data landscape, both of these classical approaches face new challenges. The COVID-19 pandemic has exposed the inadequacies of traditional methods in responding to structural mutations. According to its 2021 report, Eurostat noted that the magnitude and frequency of pandemic-induced outliers went beyond what conventional methods could accommodate, forcing many statistical agencies to implement temporary manual interventions in seasonal adjustment.[1] The proliferation of high-frequency data has also placed new demands on traditional methods designed for monthly or quarterly data. However, existing methods make it difficult to achieve satisfactory results when processing daily or real-time data.

These challenges have spurred new research directions. On one hand, efforts have been made to combine traditional methods with modern techniques, such as the hybrid architecture proposed by Bandara et al. (2021),[2] which seeks to integrate the robustness of the STL method with the theoretical strengths of SEATS. On the other hand, the introduction of machine learning technologies has enabled breakthroughs beyond the limitations of traditional linear models. For example, Zeng et al. (2023) investigated the application of deep learning models to improve the prediction and adjustment of residuals.[3]

This chapter focuses on the challenges and future perspectives of the X-13ARIMA-SEATS and TRAMO-SEATS approaches. The structure of this chapter is organised as follows: following the introduction of Section 9.1, Section 9.2 reviews the traditional standard methods of seasonal adjustment – nonparametric and parametric methods; Section 9.3 discusses the contributions and challenges of X-13ARIMA-SEATS and TRAMO-SEATS; Section 9.4 discusses the prospects of combining tradition and modern methods; and finally, Section 9.5 gives a summary and concluding remarks.

DOI: 10.1201/9781003716655-9

9.2 Traditional Seasonal Adjustment Methods: Nonparametric and Parametric Methods

Seasonal adjustment methods were initially developed using linear smoothing filters, which decompose time series components through moving averages. They served as tools for analysing seasonal economic variables in the absence of suitable statistical models. By the late 1920s, new smoothing filters allowed wider dissemination of nonparametric or 'empirical' approaches. However, smoothing techniques based on linear filtering, particularly those using specific moving averages, proved unable to accurately track cyclical turning points in series longer than five years. Additionally, moving averages are highly sensitive to outliers, requiring special preprocessing to handle extreme values. Coupled with the primarily manual implementation of adjustments at the time, these methods saw limited practical use. These constraints spurred the development of more robust estimation methods for seasonal factors, such as moving medians.

9.2.1 Nonparametric Methods (Implicit Models)

Modern seasonal adjustment methodology began with the X-11 method developed by the U.S. Census Bureau. Its core approach involves iteratively applying a series of moving average filters to progressively separate trend, seasonal, and irregular components. The 1965 X-11 version was the first automated seasonal adjustment method, an empirically developed, nonparametric (implicit model) approach. It was based on various types of weighted moving averages, both symmetric and asymmetric. X-11's strengths lie in its flexibility and intuitiveness, enabling adaptation to diverse patterns in economic time series. However, its most significant limitation is the endpoint problem: lack sufficient past or future observations, resulting in substantial distortion in seasonally adjusted values near the series boundaries. This issue is particularly acute in economic forecasting and policy analysis, where decision-makers often focus most critically on the most recent data.

9.2.2 Semiparametric Methods (Mixed Models)

Following Box and Jenkins' (1970) work on the ARIMA model,[4] a new variant of the X-11 was developed. The X11-ARIMA uses ARIMA modelling to forecast/extend the series endpoints via backcasting, allowing the use of less asymmetric metric filters. The original X-11 method was modified to infer only missing values arbitrarily. As a result, the number of revisions is significantly reduced once the missing data becomes available. The introduction of X-11-ARIMA marked a significant milestone in the evolution of seasonal adjustment methods. By integrating the Box–Jenkins ARIMA model, the new method performs forward prediction and backward pushback of the series, thereby extending the length of the series and providing more data support for the moving average. This improvement significantly mitigates the endpoint problem and enhances the stability of the seasonal adjustment at the series' tails. However, without prior analysis of the ARIMA model, it cannot be applied directly to economic time series due to issues of stationarity and deterministic effects. X-11-ARIMA still relies on a fixed moving average filter and is unable to adaptively handle structural changes in the series, such as mutant outliers or moving holiday effects.

9.2.2.1 Improvements to the X-12-ARIMA and regARIMA Framework

X-12-ARIMA builds on the same approach as X-11-ARIMA, incorporating most of its improvements along with several recent advancements. X-12-ARIMA introduced a new approach, regARIMA, that includes an additional pre-adjustment step enabling the use of seasonal ARIMA models while simultaneously estimating outliers, trading day variations, and calendar effects. The latter is selected from an automated program based on five models. Similar to the X-11-ARIMA for inferring the original series. X-12-ARIMA provides the same diagnostics as X-11-ARIMA for assessing seasonal adjustments, while also introducing new quality tests, such as spectral tests for seasonal and residual trading day variations and diagnostics for the stability of seasonally adjusted series. X-12-ARIMA retains the X-11 nonparametric core but incorporates parametric pre-adjustment via regARIMA, so X-12-ARIMA and the later-developed X-13ARIMA-SEATS belong to the semiparametric (hybrid) method.

The development of X-12-ARIMA enhances the robustness of seasonal adjustments, and its core contribution is the introduction of the regARIMA framework. This framework allows for the direct inclusion of external regression variables such as outliers, trading day effects, and moving holidays (eg, Easter, Lunar New Year) in the modelling process. With regARIMA, analysts can more precisely identify and correct known data interferences, enabling them to pre-adjust the series before the seasonal breaks down.

9.2.2.2 Diagnostic Tools and Spectrum Analysis for X-12-ARIMA

The X-12-ARIMA improves the selection mechanism of the moving average filter and introduces more flexible spectrum diagnostic tools, such as spectrum analysis, to optimise the extraction of seasons and trends. Another important advancement in this phase is the systematic assessment of the seasonal adjustment quality, including quantifying the stability of adjustment results through sliding spans and historical revisions analysis. However, X-12-ARIMA still faces a fundamental limitation: its decomposition process is nonparametric, relying on empirical filters rather than explicit statistical models of the data generation process. This feature makes it difficult to rigorously determine the statistical properties of the decomposition results and also limits its applicability in more complex scenarios.9.2.2.3 Other Seasonal Adjustment Tools Similar to X-12-ARIMA

Nonetheless, the X-11 family (X-11-ARIMA, X-12-ARIMA, X-13ARIMA-SEATS) rapidly became the global standard. In the years following World War II, advancements in computer science greatly facilitated the popularisation and refinement of non-(semi-) parametric methods. Sharing the same fundamental principles as the X-11 series, a wide variety of seasonal adjustment methods emerged, including GLAS, STL, SABL, and SEASABS.

GLAS (General Linear Abstraction for Seasonality), developed by Young (1992) at the Bank of England,[5] was used to obtain seasonally adjusted monetary variables. It employs moving averages of the data with a triangular weighting pattern to estimate and smooth the trend and seasonal components. GLAS was designed to provide only additive linear adjustments. The method implements end-of-sample weighting, as evaluated by Lane (1972). The program also incorporates adjustments for weekdays (accounting for implied effects), but cannot handle holiday effects.

STL (Seasonal-Trend decomposition based on LOESS), proposed by Cleveland et al. (1990) at Bell Laboratories,[6] shares the same fundamental principle as GLAS: the concept of nonparametric regression based on locally weighted moving averages of the data. Using a tri-cube weight function, it fits polynomials via weighted least squares to obtain robust estimates and smoothing of the trend and seasonal components, making it more resilient to outliers. STL employs linear or quadratic adjustment for the trend component, while the seasonal component uses either a fixed (constant) or a linearly evolving adjustment.

SABL (Seasonal Adjustment at Bell Laboratories), developed by Cleveland et al. (1978) at Bell Laboratories,[7] shares a similar structure in principle to X-11 and consequently has no explicit underlying model.[8] SABL was specifically created to handle anomalous data (outliers). It employs a nonlinear filter based on moving median estimation before applying linear filters, achieving higher robustness. The principle of moving median estimation, proposed by Huber (1964),[9] uses a minimax estimator that is robust to outliers. Smoothing of the trend and seasonal components is based on a fairly complex system of moving averages and regression (built upon weighted moving medians). Additionally, SABL integrates the workday adjustment feature from X-11.

SEASABS (SEA System Analysis at the Australian Bureau of Statistics), developed by the Australian Bureau of Statistics (1987),[10] is a knowledge-based tool for seasonal adjustment. Its seasonal adjustment algorithm is based on the X-11 package. SEASABS maintains records of previous analyses of data series, allowing for comparison of X-11 diagnostics and identification of specific parameters that produced acceptable adjustments in earlier runs.

BV4 (Berliner Verfahren 4 – Berlin Procedure Version 4) is a seasonal adjustment method based on moving filter applications derived from approximation functions via regression techniques, proposed by Nourney (1983, 1984). It was originally developed by the Technical University of Berlin (Technische Universität Berlin) and the German Institute for Economic Research (Deutsches Institut für Wirtschaftsforschung, DIW).[11] This method serves as the official seasonal adjustment procedure for the Federal Statistical Office of Germany. Due to its use of local regression methods, BV4 is theoretically better grounded than ad hoc approaches. However, the technique can lead to identification problems and yields better results only when specific specifications are applied, such as using a cubic polynomial for trend estimation and trigonometric functions for the seasonal component.

9.2.3 Parametric Methods (Explicit Models)

Parametric seasonal adjustment methods only emerged in the 1950s. As early as the late 20s, some authors,[12] for example, Slutsky (1927) and [13] Yule (1927), criticised nonparametric seasonally adjusted methods based on linear filters or moving averages. They argued that moving averages induce spurious cycles in the data and that regression (parameter) models should be used as a method of seasonal adjustment. Two types of methods have been developed: (1) deterministic model-based methods, and (2) stochastic model-based methods.

9.2.3.1 Deterministic Models: Early Applications of Regression Techniques

Deterministic regression models were the first to offer a parameter-based (explicit model) approach to seasonal adjustment. As early as the mid-19th century, Buys-Ballot (1847)[14] pioneered the deterministic seasonal adjustment method based on global regression. In the mid-20th century, Fisher (1937)[15] and Mendershausen (1939)[16] proposed adjusting polynomials by least squares to remove seasonal components. By the 1960s, the use of multiple regression techniques to adjust economic time series was widely available. This is mainly due to (1) the construction of econometric models for monthly and quarterly time series, and (2) the development of computers. The deterministic regression model method is based on modelling the original series and each component using a simple parameter function, and on estimating the parameters by the ordinary least squares method.[17] However, methods based on deterministic models do not account for the stochastic nature of economic time series, and therefore they are rarely used for seasonal adjustment.[18]

9.2.3.2 Stochastic Models: UCARIMA and Structural Time Series

The advancement of stochastic process theory after World War II, coupled with the development of computers, greatly accelerated the progress of parametric methods and, consequently, the enhancement of seasonal adjustment tools. Modern model-based methods (such as TRAMO/SEATS and X-12-ARIMA) are fundamentally grounded in the Unobserved Components (UC) framework, often formulated using ARIMA models (UCARIMA). Within this framework, seasonal adjustment is treated as a signal extraction problem. The core objective is to estimate and remove the unobserved seasonal component (S_t) from the observed series (Z_t), which is modelled as the sum of unobserved components, typically including a trend (T_t), a seasonal component (S_t), and an irregular component (I_t), ie, $Z_t = T_t + S_t + I_t$ Optimal signal extraction filters derived from the time series model are applied to estimate S_t to obtain the seasonally adjusted series $(T_t + I_t)$.

Stochastic model-based methods are the most commonly used models for parametric methods. There are two types of stochastic models: (1) ARIMA model-based (AMB) method, and (2) Structural Time Series (STS)-based method.

9.2.3.2.1 AMB (ARIMA Model-Based) Method

The purpose of the AMB method is to model the observed time series using a Seasonal ARIMA (SARIMA) framework and to infer components using spectral estimation based on the model structure. Since these components are unobservable, a unique decomposition of the general ARIMA fit to the original time series is obtained by maximising the variance of the irregular component and minimising the variance of the seasonal component, combining the regression model with dummy variables and ARIMA errors. Because the ARIMA model is highly sensitive to outliers and cannot accurately estimate deterministic components, the TRAMO module is employed to remove these deterministic effects and smooth the series. A mixed model combining regression variables with ARIMA errors is then used for estimation, forecasting, and interpolation. SEATS then first fits the ARIMA model to a steady-state linear series produced by TRAMO. It is determined by a model-identification algorithm that applies constraints on the seasonal and non-seasonal polynomial orders and selects among candidate models using the Schwarz Information Criterion (SIC). Finally, SEATS uses the AMB method to break down the time series into trend periods, seasonality, and irregular components.

9.2.3.2.2 STS (Structural Time Series-Based) Method

The STS method is a structured time series approach that directly specifying the ARIMA model for each unobserved component. It follows the principles of a regression approach; instead of using a global deterministic model or a local random regression model to estimate each component, it uses a simple random walk model. The structural time series decomposition method begins with a measurement equation that incorporates each unobservable component, namely the trend, seasonal, and irregular components. For each unobservable variable defined by state equations, we assume a fairly simple ARIMA process. The structural model is cast in State Space form and typically estimated via the Kalman filter. STAMP serves as the standard software implementing integrated STS methods.

9.3 Contributions and Limitations of X-13ARIMA-SEATS and TRAMO-SEATS

Seasonal adjustment of time series is a core task in modern economic statistics, aiming to separate trend cycles, seasonal components, and irregular fluctuations from raw data to reveal the underlying patterns of economic variables.

9.3.1 X-13ARIMA-SEATS: Fusion of Nonparametric and Parametric Parameters

The fusion of the X-13ARIMA-SEATS represents a combination of nonparametric and parametric methods and is the final form of the U.S. Census Bureau method. The most significant innovation of the X-13ARIMA-SEATS is the integration of the SEATS, a model-based decomposition method, into the X-12-ARIMA framework.

9.3.1.1 Theoretical Advantages of the SEATS Module

The core idea of SEATS, derived from the TRAMO-SEATS system, is to treat the time series as a stochastic process and directly derive the optimal filter for trend, seasonal, and irregular components.

The integration of the X-13ARIMA-SEATS allows users to choose between nonparametric (X-11 class) and parametric (SEATS) decomposition methods on the same platform, providing both flexibility and statistical rigour. An advantage of SEATS is that its filter design is entirely based on the series's ARIMA model structure, allowing it to capture the data's dynamic characteristics more precisely, particularly when the series is short or the seasonal pattern is unstable.

9.3.1.2 Model Misdesign and Stationarity Limitations

However, the limitation of SEATS is its strong dependence on ARIMA model settings: if the model is mis-designed (eg, ignoring higher-order seasonality or nonlinear effects), the decomposition results can be significantly biased. In addition, SEATS imposes strict stationarity requirements for Real-time processing, which may be difficult to meet in some economic series, such as GDP data with structural breakpoints.

9.3.2 TRAMO-SEATS: Fully Parameterised Paths

Unlike the incremental improvements of the X-13ARIMA-SEATS, the TRAMO-SEATS has been firmly committed to a fully parametric modelling path since its inception. The TRAMO (Time Series Regression with ARIMA Noise, Missing Observations, and Outliers) module handles the preprocessing of the series, including outlier detection, trading day effect correction, and imputation of missing values, while the SEATS module focuses on signal extraction based on the ARIMA model.

9.3.2.1 Rigour of Spectral Decomposition

TRAMO-SEATS has a stronger theoretical foundation, enabling precise spectral decomposition where the trend, seasonal, and irregular components correspond to distinct frequency bands. Theoretically, this method offers improved statistical properties, including unbiasedness and minimum variance.

9.3.2.2 Computational Complexity and Seasonal Unit Root Problems

However, its practical application faces two major challenges. Firstly, the computational complexity is high, especially for processing long series or multivariate data; secondly, the sensitivity to model settings, for example, misjudging the unit root of the seasonality, may lead to systematic distortion of the decomposition results.

The difference between X-13ARIMA-SEATS and TRAMO-SEATS is essentially the difference between nonparametric empiricism and parametric modelling. The former strengthens its theoretical foundation by continuously incorporating the advantages of the latter (eg, the introduction of SEATS in X-13), while the latter enhances its practical utility through engineering improvements (eg, TRAMO's heuristic for outlier detection). In the field of time series decomposition and seasonal adjustment, X-13ARIMA-SEATS and TRAMO-SEATS represent the two main approaches for seasonal adjustment. The technological advances in these two methods reflect not only advances in statistical technology but also a gradual conquest of the complexity of real-world data. This is, without a doubt, the most impressive innovation in the seasonally adjusted approach over the past few decades.

9.4 Modern Challenges: The Pandemic and the Impact of High-Frequency Data

Although over the past 100 years or so, the theory and methods of seasonal adjustment have been more developed and many more timely tools have become available, although the X-13ARIMA-SEATS and TRAMO-SEATS have performed well on traditional monthly or quarterly economic data (eg, GDP, unemployment, industrial production index), the COVID-19 pandemic has had an unprecedented impact on global economic and social activity; complex seasonal patterns of high-frequency data. Real-time processing amid revision volatility poses challenges, particularly for real-time adjustments. These recent developments have introduced numerous new challenges for seasonal adjustment.

9.4.1 Structural Fractures and Abrupt Changes

COVID-19 undermines traditional assumptions. The core assumption of seasonal adjustments is that 'seasonal patterns are relatively stable', but during COVID-19 the following were observed: (1) Structural breaks were caused, such as the global lockdown in March–April 2020 that led to a sudden change in data (eg, the U.S. unemployment rate soared from 3.5% to 14.8%); (2) The traditional holiday peak disappeared due to lockdown (eg, Christmas retail), while the e-commerce data showed a new seasonal peak (new e-commerce promotion cycle); and (3) The magnitude/frequency of outliers exceeded conventional methods' robustness thresholds (eg, IQR adjustment for STL). Seasonal patterns in traditional time series (eg, retail sales, unemployment, energy consumption) are heavily distorted. Lockdown policies, supply chain disruptions, and abrupt changes in consumer behaviour have rendered standard seasonally adjusted methods (eg, X-13ARIMA-SEATS, TRAMO-SEATS) ineffective.

9.4.2 Complex Seasonality of High-Frequency Data

High-frequency data (ie, data collected on an hourly, daily, or weekly basis) can provide an immediate indication of market volatility (the impact of the pandemic on consumption) and identify short-term seasonal patterns (the peaks on 'Black Friday'), while traditional monthly data can lag by weeks. However, multiple seasonal patterns, non-integer periodicity, and noise interference in high-frequency data present another significant challenge for current seasonal adjustment methods.

Multiple seasonality (weekly, yearly nested structure): Traditional seasonality adjustment methods (such as X-11 or SEATS) assume that the data contains only a single seasonality (eg, 12 months for monthly data or 4 quarters for quarterly data). However, high-frequency data often has multiple seasonality, ie, 'nested seasonal structures', eg, 'daily data' may be included at the same time as 'weekly seasonality' (7-day cycles) and 'annual seasonality' (365-day cycles); 'hourly data' (eg, power load or stock volume) may have 'intraday mode' (24-hour period), 'weekly mode' (7-day period), and 'annual mode' (365-day cycle). As an example using daily retail sales, suppose that the daily sales of an e-commerce platform indicate the higher sales volume on the weekend (Saturday and Sunday) than on weekdays (Monday, Tuesday, . . .) showing a weekly seasonal. Sales have surged during promotional seasons such as 'annual seasonality', Christmas, and 'Singles Day'.

Moving holiday effect: This refers to the impact of holidays with irregular dates, such as Easter, Chinese New Year, etc. The default settings for X-13ARIMA-SEATS or TRAMO-SEATS can only handle a single seasonality, so this 'multiple seasonality' cannot be directly modelled. While this can be partially addressed by pre-decomposition (eg, adjusting for weekly seasonality followed by annual seasonality) or external regression variables (eg, virtual 'day of week' sub-variables), this increases model complexity and may introduce the risk of overfitting.

Non-integer periodicity and noise interference: Traditional seasonal adjustment methods assume that the seasonal cycle is fixed (eg, 12 months, 4 quarters), but the period of some high-frequency data is not an integer, for example, the number of trading days in each year is different due to holidays (eg, about 252 days/year for U.S. stocks and 240 days/year for A-shares). As a result, the annual seasonal cycle is not fixed. Since the 'weekly data' is not exactly 52 weeks (about 52.14 weeks) in a year, the yearly seasonality cannot be accurately aligned. Take stock market volatility (daily data) as an example: the volatility of the stock market usually exhibits a 'weekly effect' (eg, higher volatility on Monday and lower on Friday), but due to the different number of days the market is closed each year (eg, the U.S. stock market is closed for Thanksgiving, Christmas), the direct application of X-13 or TRAMO-SEATS may lead to seasonal estimation bias.

Increased interference from noise and outliers: High-frequency data often contains stronger 'noise' and 'transient outliers' (eg, news shocks, market flash crashes), which places higher demands on seasonally adjusted approaches. Moving average filters in X-13ARIMA-SEATS (eg, the Henderson filter) may 'over-smooth' short-term fluctuations when reducing high-frequency noise, leading to the loss of genuine signals. Although the regARIMA module can detect outliers, such as additive outliers (AO), temporary changes (TC), and level shifts (LS), the high frequency of outliers in daily data (eg, sharp market fluctuations during a financial crisis) can make stable model estimation difficult. In certain scenarios, such as minute-level cryptocurrency price data, the value of assets like Bitcoin can fluctuate dramatically during extreme events, such as regulatory announcements or exchange hacks. The outlier detection mechanism of the X-13ARIMA-SEATS may not adapt quickly to this mutation, leading to substantial residual noise in the seasonally adjusted series. With TRAMO-SEATS, TRAMO relies on the ARIMA model, and the 'short-term autocorrelation' of high-frequency data can be extremely strong (eg, the GARCH effect of financial data), making standard ARIMA a poor fit. Signal extraction for SEATS is based on 'spectral decomposition', but the spectrum of high-frequency data may appear 'continuum' (rather than discrete seasonal peaks), making component separation difficult. For example, in the foreign exchange market, the EUR/USD exchange rate (hourly data) can be affected by global trading sessions (Asian, European, and American market openings and closings), leading to complex intraday seasonality. TRAMO-SEATS may mistakenly interpret certain types of noise arising from market microstructure, such as liquidity variations, as seasonal components.

The moving average filter of the X-13ARIMA-SEATS and the spectral decomposition of the SEATS are extremely computationally intensive for long series (eg, daily data over ten years). TRAMO's outlier detection algorithms may not function effectively in real-time when applied to ultra-high-frequency data, such as tick-level transaction data. For example, with social media sentiment data (updated every minute), the batch mode of X-13ARIMA-SEATS or TRAMO-SEATS may not be able to meet the low-latency requirements. For example, if the Twitter Sentiment Index is seasonally adjusted in real-time, the computational burden increases dramatically, and simplified methods such as scrolling window adjustments may introduce lag bias.

9.4.3 Real-Time Processing and Correction of Volatility

The seasonal adjustment of economic data is usually accompanied by 'revisions', ie, historical adjustments may change when new data is available. For high-frequency data, such revisions can occur frequently, affecting decision consistency. For example, the U.S. Department of Labour's weekly jobless claims data may be revised after its initial release, and the seasonal adjustment factors used in X-13ARIMA-SEATS and TRAMO-SEATS may be updated in subsequent months, potentially confusing market interpretation.

In response to the challenges posed by COVID-19, national statistical offices have adopted some temporary adjustment schemes, such as (1) the Direct Intervention Approach, which marks the pandemic period as an 'abnormal interval' to exclude its impact on seasonal estimates; (2) the U.S. BLS introduced 'pandemic dummy variables' in the 2020–2021 unemployment data to prevent outliers from distorting moving average filtering;[19] (3) the ECB uses manual adjustment of seasonal factors in X-13ARIMA-SEATS, and the 2020 Q2 data adopts interpolation instead of standard filtering (Eurostat, 2021).[20] These real-time decision-making needs require timely data for policymaking, but standard methods require long-term observations to calibrate seasonality. Early adjustments may be subject to revision risk due to subsequent data updates (eg, the U.S. Census Bureau's multiple revisions to retail sales data in 2020). Relying too heavily on experts' subjective judgments may introduce human bias.

9.5 Methodological Innovations: Adaptive and Intelligent

Recent research on seasonally adjusted methods has emphasised the use of more advanced statistical techniques and the use of more powerful statistical tools, including (1) Hybrid methods: machine learning combined with traditional seasonal methods, such as the use of LSTM or Transformer models to capture complex seasonality, combined with X-13ARIMA-SEATS for post-processing, eg, Facebook's Prophet model[21] allows for custom multi-seasonality, partially complementing traditional methods; (2) State space models (SSMs) offer a flexible framework for capturing multiple seasonal patterns in high-frequency time series data, for example, the Bayesian approach employs MCMC sampling to quantify uncertainty in the seasonally adjusted series; (3) Online learning and incremental calculation, such as recursive filtering (such as Kalman filtering), achieve real-time seasonal adjustment and avoid batch processing delays.

9.5.1 Hybrid Architecture of Machine Learning and Traditional Methods

The seasonal adjustment of high-frequency time series (such as daily, weekly, and intraday data) faces challenges such as multiple seasonality, noise interference, and nonlinear patterns, and it is difficult for traditional methods (X-13ARIMA-SEATS, TRAMO-SEATS) to be directly applicable. Deep learning models (such as LSTM and Transformer) can automatically learn complex time series patterns but lack interpretability. Therefore, it is a feasible solution to combine the feature extraction capabilities of deep learning with the interpretability of traditional seasonal adjustment.

9.5.1.1 LSTM/Transformer Feature Extraction Capability

Hochreiter & Schmidhuber (1997) proposed the LSTM Network.[22] LSTM is a variant of RNN (Recurrent Neural Network)[23] that uses additional features to help memorise time series in data. It is used for time series forecasting, especially when long-term dependencies are present. It has a memory unit that can store state in time, which is essential for identifying or remembering relevant information over a fairly wide window. The core structure of an LSTM relies on a gating mechanism – comprising the input gate, forget gate, and output gate – to control the flow of information, making it well-suited for modelling long-term dependencies. The formula for the LSTM gating mechanism is as follows:

$$\begin{aligned}
f_t &= \sigma\left(W_f \cdot [h_{t-1}, x_t] + b_f\right) && \text{forgetting gate} \\
i_t &= \sigma\left(W_i \cdot [h_{t-1}, x_t] + b_i\right) && \text{input gate} \\
\hat{C}_t &= tanh\left(W_C \cdot [h_{t-1}, x_t] + b_C\right) && \text{candidate memory} \\
C_t &= f_t \odot C_{t-1} + i_t \odot \hat{C}_t && \text{memory update} \qquad (9.1) \\
o_t &= \sigma\left(W_o \cdot [h_{t-1}, x_t] + b_o\right) && \text{output gates} \\
h_t &= o_t \odot \tanh(C_t) && \text{hidden status}
\end{aligned}$$

The forgetting gate (f_t) determines what to forget from the cell state (C_t), the input gate (i_t) helps us decide what new data to put into the cell state (C_t), and the output gate (o_t) calculates the elements that ultimately remain in each dimension at a given moment. $\hat{\boldsymbol{C}}_t$ stands for candidate memory, C_t. represents a new memory update vector; h_t and h_{t-1} denote the current and previous hidden states, respectively; weight matrix W_f, W_i, W_C, and W_o corresponds to the forgetting gate (f_t), the input gate $(\boldsymbol{i}_t)$, the cell state (C_t), and the output gate (o_t), respectively; the biases associated with these gates are represented by b_f, b_i, b_C, and b_o; x_t represents input at time t; the hyperbolic tangent function (tanh) maps values to the range [−1,1], which is used to activate the function; σ represents the sigmoid activation function.

LSTMs autonomously capture long-term seasonality (annual trends) and short-term dependencies (intra-week effects). It also applies to 'non-fixed-period seasonality' (eg, trading day data). However, due to the large parameter scale, LSTM requires at least 10,000 pieces of historical data to be stably trained, which is computationally intensive, and it is difficult to directly decompose the trend, seasonal, and residual components, resulting in a 'black box' problem.

9.5.1.2 Transformer – Self-attention Timing Model

Vaswani et al. (2017)[24] proposed a Transformer (self-attention timing) model. Transformer relies on the self-attention mechanism to capture global dependencies and is suitable for high-frequency data modelling. The key calculations are as follows:

$$Attention(Q,K,V) = softmax\left(\frac{QK^T}{\sqrt{d_k}}\right)\mathrm{V} \tag{9.2}$$

where (Q, K, V) represent the Query, Key, and Value matrices, respectively. (d_k) is the scaling factor that prevents the gradient from disappearing.

Improved variants of time series Transformers, such as Informer[25] and Autoformer[26], have been developed. Informer introduces ProbSparse attention to reduce computational complexity, while Autoformer proposes an autocorrelation mechanism that replaces traditional attention and more effectively captures periodic patterns.

An improved version of the time series Transformers can model 'very long series dependence' (eg, multi-year daily data) and partially explain seasonal patterns (eg, which dates have the most impact on forecasts) through attention weight visualisation. However, large-scale training data is required; otherwise, it is easy to overfit. It is still not possible to directly export seasonally adjusted economic indicators (such as trend-cycle terms).

9.5.1.3 A Hybrid Approach to Deep Learning + X-13ARIMA-SEATS

(1) The deep learning model (LSTM/Transformer) is used as the pre-processor: the raw high-frequency time series data $(\boldsymbol{y}_t)$ is input (eg, daily sales); output prediction, $\hat{\boldsymbol{y}}_t$, and calculate the residuals $e_t = y_{t-}\hat{y}_t$.

(2) X-13ARIMA-SEATS post-processing residuals: first, apply a conventional seasonal adjustment to $\boldsymbol{e}_t$, decomposing it as follows:

$$e_t = Trend_t + Seasonal_t + Irregular_t$$

Finally, adjust the series:

$$y_t^{adjusted} = \hat{y}_t + Trend_t + Irregular_t$$

9.5.1.4 *Balance of Interpretability and Statistical Rigour*

The advantage of the hybrid approach combining deep learning with X-13ARIMA-SEATS lies in its complementary strengths: deep learning captures complex seasonality (eg, multiple cycles, moving holidays), while X-13ARIMA-SEATS offers explainable decomposition into trend, seasonal, and irregular components. This hybrid LSTM/Transformer + X-13ARIMA-SEATS method combines 'nonlinear modelling capabilities' and 'explainable decomposition', which is suitable for seasonal adjustment of high-frequency data. It is the future direction of automatic machine learning (AutoML) optimisation model selection, and causal inference (eg, dealing with seasonal breaks under policy interventions). The LSTM + X-13ARIMA-SEATS hybrid model can be practically implemented using Python and R.[27]

9.5.2 Flexibility of the SSM

State space models (SSMs) offer a universal and flexible framework for time series analysis, enabling effective decomposition and forecasting of high-frequency data with trend, seasonal, and irregular components. Their core lies in dynamically characterising a system's evolution through state and observation equations, while allowing structural extensions tailored to specific needs. SSMs can also be readily extended to incorporate calendar effects, sampling errors, and outlier components. Outliers and holiday effects can be specified as needed to address diverse impacts, such as missing values. Users may select from six different maximum likelihood estimation optimisers, with automatic detection of level shifts and additive outliers, offering extensive options for time series processing. Trends can be modelled in multiple ways, and SSMs handle high-frequency data effectively, [28] functionally operating as regression models with time-varying coefficients and smoothing capabilities. Furthermore, SSMs perform robustly in both Linux and PC environments. For weekly data, seasonal components can be constructed using trigonometric seasonality, dummy variables, or time-varying periodic splines.

9.5.2.1 *Dynamic Decomposition of High-Frequency Data*

When analysing daily or weekly data, SSM decomposes the observation series $\{y_t\}$ in the following form:

$$y_t = \mu_t + \gamma_t + e_t \tag{9.3}$$

where, y_t represents observations (eg, sales, temperature), μ_t is the trend component, γ_t. is the seasonal component, and e_t is the irregular noise. The state space form consists of a state equation and an observation equation:

$$\text{observation equation: } y_t = Z\alpha_t + e_t \tag{9.4}$$

$$\text{state equation: } \alpha_t = T\alpha_{t-1} + R\eta_t, \quad \eta_t \sim N(0,Q) \tag{9.5}$$

Where, α_t is a state vector (contains μ_t, v_t γ_t); *T*, *R*, and *Z* are the transition, selection, and observation matrices, respectively. Trend (μ_t) typically employs local linear trend models, and the dynamic process is described by the following equations:

$$\mu_t = \mu_{t-1} + v_{t-1} + \xi_t \tag{9.6}$$

$$v_t = v_{t-1} + \zeta_t \tag{9.7}$$

where v_t is the trend slope ξ_t and ζ_t are independent noise terms. The model allows the trend slope to change randomly over time, adapting to nonlinear features in the data, such as bursts or dips. The modelling of the seasonal component (γ_t) requires different methods depending on the cycle length.

(1) Short periods (eg, weekly seasonality with frequency = 7): A time-varying cycle model can be used to directly estimate the latent state associated with each cyclical component:

$$\gamma_t = -\sum_{j=1}^{6} \gamma_{t-j} + \omega_t \tag{9.8}$$

(2) Long period (eg, annual seasonality, frequency = 365): To avoid high-dimensional state vectors, a Fourier series approximation is used to capture seasonal patterns by finite harmonics (usually $K \leq 5$):

$$\gamma_t = \sum_{k=1}^{k} \left[a_{k,t} \cos\left(\frac{2\pi kt}{365}\right) + b_{k,t} \sin\left(\frac{2\pi kt}{365}\right) \right] \tag{9.9}$$

Where, $a_{k,t}$ and $b_{k,t}$ are time-varying coefficients.

State space models provide a powerful tool for high-frequency time series analysis through dynamic decomposition and flexible modelling techniques such as Fourier approximation and local linear trends. Its core strengths lie in its customisable model structure and its ability to adapt to complex data patterns. In practice, the above model can be implemented with the 'KFAS' or 'DLM' packages in R.[29] Because 'KFAS' has efficient Fourier seasonality support, It offers fast computational performance and is well suited to high-frequency data analysis; DLM, on the other hand, offers more flexibility but is computationally slower. Noise variance is usually optimised automatically by MLE to avoid bias introduced by subjective settings.

9.5.2.2 Fourier Series and Local Linear Tendencies

Combining the Fourier series with a local linear trend enhances the functionality of SSMs, enabling them to effectively capture both periodic and evolving components of a time series. (1) A Fourier series approximates long-period seasonality through harmonic superposition, which significantly reduces the dimension of the state space. It has a strong advantage: only a few harmonics, such as (K = 5), are needed to capture major seasonal patterns. This is computationally efficient, suitable for daily or higher frequency data. At the same time, it is dynamically adaptable. The time-varying coefficients, $a_{k,t}$ and $b_{k,t}$ allow seasonal patterns to change slowly over time, adapting to the non-fixed periodic nature of the actual data. (2) The local linear trend model enhances the dynamic adaptability of the trend component by introducing the random walk slope (v_t): The nonstationary trend processing makes the slope noise term, ζ_t respond to sudden changes in the data (such as policy shocks or market fluctuations) in the model, and the superposition of the trend and the Fourier seasonal component can form a complete decomposition framework of 'trend-seasonal-noise', similar to the 'd11 + d13' adjustment series in the classical X-11 method.

This framework can be directly applied to high-frequency scenarios such as daily sales and financial volatility.

SSM is a powerful and flexible time series framework, but users must manually construct certain seasonal adjustment diagnostics. For those accustomed to using seasonally adjusted programs such as X13ARIMA-SEATS, SSM has a low degree of automation: it does not automatically select additive or multiplicative models, and it lacks built-in identification of outliers or trading day effects. Diagnostic tools rely on external programs: SAS, R, or Python drawing tools are required to generate seasonal diagnostic charts. The filter output needs to be manually extended: the forward-filtering step of the Kalman filter must be implemented through iterative prediction, which increases the complexity of the implementation. Despite these limitations, SSM requires a single-step prediction error that can be generated by iterating over the series, which takes much less time than expected. Users can easily generate diagnostic graphs using SAS/GRAPH, PROC SGPLOT, or a variety of other statistical software tools. Users must complement the diagnostic steps with domain knowledge and balance computational efficiency against model complexity. In practice, it is recommended to choose a toolkit such as 'KFAS' (efficient) or 'DLM' (flexible), depending on the characteristics of the data, to realise the full potential of SSM.

9.5.3 Bayesian Methods and Uncertainty Quantification

A prevailing trend is the growing popularity of Bayesian methods for time series decomposition. These methods provide a probabilistic framework that integrates prior knowledge with observed data to achieve comprehensive posterior inference on trend, seasonality, and noise components. At their core, Bayesian approaches recast decomposition as a joint estimation of latent states/parameters via probability distributions, utilising MCMC sampling for numerical solutions. This offers a flexible framework for seasonal adjustment and time series decomposition. Posterior distributions yield credible intervals for the trend and seasonal components, enabling effective quantification of uncertainty. By incorporating prior information, such as domain-specific constraints on seasonal amplitude or flexible structures like random walks or Gaussian processes (GPs) for modelling complex trends, the framework enhances both interpretability and robustness.

9.5.3.1. Bayesian Time Series Decomposition Model

Assuming the series of observations (y_t) can be decomposed into additive components,

$$y_t = \mu_t + \gamma_t + e_t \tag{9.10}$$

where μ_t is the trend component (such as a random walk or local linear trend); γ_t is a seasonal component (eg, Fourier series or periodic autoregressive process); e_t is the observed noise (usually assumed to be Gaussian white noise). The state space model is defined as a Dynamic Linear Model (DLM).

$$\text{Observation equation: } y_t = Z_t\alpha_t + e_t \;\; e_t \sim N(0,V) \tag{9.12}$$

$$\text{State equation: } \alpha_t = T_t\alpha_{t-1} + R_t\eta_t,\; \eta_t \sim N(0,W) \tag{9.13}$$

where α_t is the state vector (containing μ_t, γ_t etc.); Z_t and T_t are observation matrix and state transition matrix, respectively; and V and W are the covariance matrices of the observed noise and the state noise.

9.5.3.2. Bayesian Inference (Decomposition) Process

9.5.3.2.1 A Priori Distribution Setting

Assuming trends and seasonality obey a priori Gaussian process state parameters, eg, trend (μ_t) and seasonal components (γ_t):

$$\mu_t \sim \mathcal{GP}(0, K_{trend}), \qquad \gamma_t \sim \mathcal{GP}(0, K_{seasonal})$$

where μ represents the trend component at time *t* and is a state parameter that represents the long-term trend portion of the time series. It is assumed to obey the Gaussian process a priori. γ_t denotes the seasonal component at time *t* and is also a state parameter, and $\mathcal{GP}$ denotes the Gaussian process (a nonparametric prior commonly used to model functions such as trends or seasonality). In the formula, $\mu_t \sim \mathcal{GP}(0, K_{trend})$ and $\gamma_t \sim \mathcal{GP}(0, K_{seasonal})$, 0 indicates that both the trend and the seasonal components follow a Gaussian process with a mean value of 0. K_{trend} represents the covariance kernel function for the trend. This function defines the covariance structure of the trend Gaussian process and can be constructed based on a random walk. Similarly, $K_{seasonal}$ denotes the covariance kernel function used to model seasonality. This function that defines the covariance structure of seasonal Gaussian processes can be based on periodic kernels such as exponential, sinusoidal square kernels, ie, $K(t,t') = \exp\left(-\frac{2sin^2(\frac{\pi|t-t'|}{p}}{\ell^2}\right)$, where p is the period and ℓ is the length scale to capture the periodic pattern.

Noise variance: An inverse Gamma conjugate prior is used for V and W.

$$V^{-1} \sim Gamma(a_0, b_0),\ W^{-1} \sim Gamma(c_0, d_0)$$

where V^{-1} denotes the inverse of observation noise variance and obeys the Gamma prior distribution for conjugate Bayesian inference. This is equivalent to V obeying the inverse Gamma distribution. W^{-1} denotes the inverse of state noise variance and obeys the Gamma prior distribution for conjugate Bayesian inference. This is equivalent to W obeying an inverse Gamma distribution. a_0 and b_0 are the hyperparameters of the Gamma prior distribution for V^{-1}. a_0 is the shape parameter of the Gamma distribution, which controls the skewness of the distribution. b_0 denotes the scale parameter or rate parameter of the Gamma distribution. c_0 and d_0 are the hyperparameters of the Gamma prior distribution for W^{-1}. $\mathbf{c}_0$ denotes the shape parameter of the Gamma distribution. d_0 denotes the scale parameter or rate parameter of the Gamma distribution.

9.5.3.2.2 The Derivation of the Posterior Distribution (Joint Posterior Distribution) Is

$$p(\alpha_{1:T}, V, W \mid y_{1:T}) \propto p(y_{1:T} \mid \alpha_{1:T}, V) \cdot p(\alpha_{1:T} \mid W) \cdot p(V) \cdot p(W) \tag{9.14}$$

where the likelihood function is based on the Gaussian distribution of the observed equation

$$p(y_{1:T} \mid \alpha_{1:T}, V) = \Pi_{t=1}^{T} \mathcal{N}(y_t \mid Z_t\alpha_t, V) \tag{9.15}$$

The probability of a Gaussian transition is defined a priori in the state equation:

$$p(\alpha_{1:T} \mid W) = p(\alpha_0)\Pi_{t=1}^{T}\, \mathcal{N}(\alpha_t \mid T_t\alpha_{t-1}, W) \tag{9.16}$$

9.5.3.2.3 MCMC Sampling and Posterior Samples Are Generated Iteratively through Gibbs Sampling

1. State sampling uses forward filtering, and backwards sampling (FFBS) is drawn from $p(\alpha_{1:T} \mid y_{1:T}, V, W)$. Fetch the state series.
2. Variance sampling updates the noise variance from the conjugate posterior distribution:

$$V^{-1} \mid \cdot \sim Gamma\left(a_0 + \frac{T}{2}, b_0 + \frac{1}{2}\sum_{t=1}^{T}(y_t - Z_t\alpha_t)^2\right) \tag{9.17}$$

3. Hyperparameter adjustment: If hyperparameters (such as seasonal cycles) are introduced, they can be updated through the random walk Metropolis–Hastings step.

By combining MCMC sampling and Bayesian modelling, the Bayesian method can transform the seasonal adjustment problem into a probabilistic inference problem, well-suited to handling missing data or irregularly spaced time series. Bayesian methods can be implemented with the R(DLM) package.[30] Appendix 9.4 provides procedures for Bayesian modelling using the 'DLM' package, including (1) generating simulation data; (2) setting the DLM model; (3) MCMC parameter initialisation; (4) MCMC cycle: constructing the current DLM model; FFBS sampling status θ; Sampling Observation Noise V; Sampling state noise W; (5) updating parameters (random walk); (6) a series of posterior analyses. MCMC sampling provides a complete characterisation of the posterior distribution. This method is suitable for nonstationary series and quantifies the uncertainty.

9.5.4 Online Learning and Real-Time Adjustments

The concept of online learning can be traced back to the stochastic approximation theory proposed by Robbins and Monro[31] in the 1950s, and later systematically developed by scholars such as Ljung, Widrow, and Hoff[32] in adaptive filtering research. The mathematical basis of incremental computation is derived from the recursive estimation algorithm proposed by Kolmogorov in the 1960s.[33] Both approaches are essentially solutions for processing data streams. Ideal for scenarios where observations arrive sequentially in chronological order and need to be processed in real-time.

9.5.4.1 Recursive Update of the Kalman Filter

In the field of time series analysis, Cleveland and Tiao first applied the state space model to seasonal adjustment in 1976,[34] and the recursive nature of the Kalman filter[35] naturally became an ideal tool for online seasonal adjustment. Unlike the traditional batch processing X-11 ARIMA method, the e-learning approach does not require storing the entire historical dataset; instead, it maintains a fixed-size state vector updated recursively.

9.5.4.2 Recursive Algorithm for Kalman Filtering

The Kalman filter was proposed by Rudolf Kalman in 1960.[36] Its recursive characteristics are reflected in two aspects. (1) Time recursion: the current state estimation only depends on the state estimation of the previous moment and the current observations, and does not need to reprocess all historical data; (2) Computational recursiveness: The error covariance matrix is recursively updated via the Riccati equation, avoiding matrix inversion operations

9.5.4.3 State Space Models and Seasonal Adjustments

Taking seasonal adjustment as an example, let the time series $\{y_t\}$ follow the additive decomposition model:

$$y_t = T_t + S_t + I_t \tag{9.18}$$

Where T_t is the trend term, S_t is the seasonal term, and I_t is the irregular term. After the state space model is established, the Kalman filter implements recursive estimation through the following steps:

Prediction Step:

$$X_t|_{t-1} = F_t X_{t-1}|_{t-1} \tag{9.19}$$

$$P_t|_{t-1} = F_t P_{t-1}|_{t-1} F_t T + Q_t \tag{9.20}$$

Where x_t is the state vector (including trend and seasonal components), F_t is the state transition matrix, and Q_t is the process noise covariance.

Update Step:

$$K_t = P_t|_{t-1} H_t^T (H_t P_t|_{t-1} H_t^T + R_t)^{-1} \tag{9.21}$$

$$X_t|_t = X_t|_{t-1} + K_t (Y_t - H_t X_t|_{t-1}) \tag{9.22}$$

$$P_t|_t = (I - K_t H_t) P_t|_{t-1} \tag{9.23}$$

where H_t is the observation matrix, R_t is the covariance of the observed noise, and K_t is the Kalman gain.

The 'one-step ahead forecasting' mechanism allows the model to update the estimates of each component in real-time. Harvey's Structural Time Series Model (1989)[37] is based on this framework to achieve online seasonal adjustment.

9.5.4.4 Incremental X-11/X-13 Algorithm

The traditional X-11 algorithm, developed by Shiskin et al.[38] in 1967, employed a batch processing approach using moving average filters. In contrast, the incremental X-11/X-13 algorithm incorporates the recursive principles of Kalman filtering. In the refinement scheme proposed by Dagum (1988),[39] seasonal component estimation was transformed into an exponentially weighted recursive form, ie,

$$S_t = \alpha(L)(y_t - T_t) + (1 - \alpha(L) S_{t-s} \tag{9.24}$$

where $\alpha(L)$ is the weight function that depends on the lag operator L, and s is the seasonal period. This form essentially models the seasonal components through a 'random walk' model:

$$S_t = S_{t-s} + \omega_t \quad \omega_t \sim N(0, Q_s) \tag{9.25}$$

Within the state space framework, seasonal components are represented either as linear combinations of trigonometric functions or as dummy variables, both of which are recursively updated using the Kalman filter.

9.5.4.5 Adaptive Kalman Filtering

The choice of process noise covariance Q and observed noise covariance R directly affects the filtering performance. The 'Adaptive Kalman Filter' proposed by Anderson and Moore in 1979[40] uses the following method:

(1) Covariance matching based on the innovation series $\epsilon_t = y_t - H_t X_t |_{t-1}$

(2) MLE is implemented via Prediction Error Decomposition:

$$logL = -1/2\sum[log\,|F_t| + \varepsilon_t^T F_t^{(-1)}\varepsilon_t], \quad F_t = H_t P_t |_{t-1} H_t^T + R_t \tag{9.26}$$

9.5.4.6 Non-Gaussian State Space Model – An Approach to Outlier Handling

The 'non-Gaussian state space model' proposed by Kitagawa and Gersch in 1984[41] deals with outliers through the following mechanisms:

(1) Replacing the normal distribution assumption with a t-distribution;

(2) Introducing intervention analysis:

$$y_t = H_t X_t + \sum \beta_i I_t^{(i)} + \varepsilon_t \tag{9.27}$$

where $I_t^{(i)}$ is the outlier indicator variable.

9.5.4.7 Multi-Frequency Data Modelling – Hybrid Processing of Multi-Frequency Data

Harvey and Pierse's (1984)[42] 'unobserved components' framework allows for the integration of data at different frequencies:

(1) Establish a continuous-time state space model for high-frequency data;

(2) A set of time series with different sampling frequencies can be linked through observation equations:

$$y_t^{(k)} = Z^{(k)} X(t_k) + \varepsilon_t^{(k)}, k = 1, ..., K \tag{9.28}$$

where X(t) is a continuous-time state process.

The fundamental differences between the traditional seasonal adjustment method and the online method are as follows: (1) the batch processing method (X-11-ARIMA) requires a complete sample interval, and needs to use a symmetric moving average filter, the endpoint processing relies on backward prediction, and the seasonal component is fixed or extrapolated through the ARIMA model; the online learning method (state space model) can be recursively processed point by point, asymmetric filters can be used, state estimates can be updated in real-time, and more importantly, the seasonal components can obey random processes. This paradigm shift has enabled real-time monitoring of economic indicators, exemplified by the EUROLIB system adopted by the European Central Bank, which leverages this technology to assess the state of the economy daily. Subsequent developments, such as the 'Small Area Estimation' method proposed by Pfeffermann in 2013,[43] have further expanded the use of e-learning in official statistics.

9.5.5 Moving Regression (MovReg)

MovReg (Moving Regression) is another notable new method in seasonal adjustment, showing promising potential for real-time economic monitoring and early warning of financial risks. MovReg is a time series analysis approach based on local regression principles that captures time-varying characteristics through dynamic modelling within sliding windows. Its core premise assumes approximate stationarity of the data-generating process (DGP) over short windows, permitting regression coefficients to evolve gradually over time, thereby enabling dynamic estimation of trends and seasonality. Evolving from locally weighted regression (LOESS) proposed by Cleveland and Devlin (1988),[44] MovReg was systematically implemented via computational procedures by Cleveland, Evans, and Scott (2018),[45] establishing itself as a vital tool for high-frequency seasonal adjustment.

9.5.5.1 MovReg Model Building

The mathematical model of moving regression is built on a sliding window regression framework. Given the time series y_t to fit the linear model within the window width (w):

$$y_t = \beta_{0,k} + \beta_{1,k} x_t + \epsilon_t \quad t \in [k, k+\omega-1] \tag{9.29}$$

where $\beta_{0,k}$ and $\beta_{1,k}$ are the intercept and slope coefficients of the window (k), respectively, and x_t can represent point-in-time or seasonal dummy variables. Parameter estimation is done by least squares (OLS):

$$\hat{\beta}_k = \left(X_k^T X_k\right)^{-1} X_k^T Y_k \tag{9.30}$$

Here X_k is the design matrix within the window, and Y_k is the vector of observations. For greater flexibility, MovReg extends the weighted version to implement locally weighted least squares estimation by introducing diagonal weight matrices W_k (eg, exponential decay weights):

$$\hat{\beta}_k = \left(X_k^T W_k X_k\right)^{-1} X_k^T W_k Y_k \tag{9.31}$$

9.5.5.2 Advantages of MovReg

The advantage of MovReg over traditional methods such as X-11/X-13 lies in its ability to model explicit and time-varying parameters. The X-13ARIMA-SEATS relies on a fixed moving average filter, in which the time-varying nature of seasonality and trend is implicit; MovReg, on the other hand, directly estimates the dynamic coefficient through sliding window regression, which is more suitable for dealing with high-frequency data or abrupt trends. For example, in a minute-by-minute stock price analysis, MovReg can quickly adapt to market volatility with a short window (eg, 30 minutes), while X-13's fixed filter may lag real-time changes.

9.5.5.3 Local Adaptability of MovReg

In high-frequency data scenarios (eg, finance, IoT sensors), MovReg's core value lies in is its real-time adaptability and flexibility. Taking the hourly sales data of an e-commerce platform as an example, a typical workflow includes trend extraction: using MovReg with a window width w = 24 × 7 (7 days) to fit local trends and capture intraday and intra-week patterns; seasonality decomposition: STL or Fourier analysis is applied to residuals, diurnal (24 hours) and weekly (168 hours) seasonality is separated, and outlier handling is combined to reduce the impact of extreme values,[46] in combination with Robust regression,[47] such as Huber's loss.

9.5.5.4 MovReg Hybrid Modelling

However, several challenges arise when applying MovReg: (1) window selection sensitivity, the window width (w) needs to be optimised by rolling cross-validation, and high-frequency data usually requires a smaller w; (2) computational efficiency, the use of recursive least squares (RLS) can avoid repeated matrix inversion, improve real time performance, and (3) multiple seasonality, and add Fourier functions or dummy variables to the regression (such as modelling 24-hour and 168-hour periods at the same time).

To further improve the performance, the MovReg method can be extended to be combined with other techniques for hybrid modelling, such as the following. (1) MovReg + STL: Multiple seasonality of STL[48] decomposition residuals after MovReg extracts trends. (2) MovReg + LSTM: The trend is preprocessed with MovReg, and the LSTM predicts the nonlinear mode of the residuals. (3) Adaptive weights: Dynamically adjust the window weights based on data volatility (with weight allocation guided by models such as GARCH).

Moving regression enables dynamic estimation of time series parameters within a local regression framework, providing an explicit and flexible approach to seasonal adjustment for high-frequency data. Future research directions include automated window selection; dynamic optimisation integrated with reinforcement learning; distributed computing using frameworks such as Spark to accelerate large-scale high-frequency data modelling; and improved interpretability through the development of visualisation tools to illustrate the economic implications of time-varying coefficients. It remains necessary to further examine the trade-off between computational complexity and model stability.

9.5.6 Future Perspectives: The Harmony of Tradition and Modernity

Despite the many challenges associated with traditional seasonal adjustment methods, and regardless of how the seasonal adjustment tools evolve, the foundations laid by

X-13ARIMA-SEATS and TRAMO-SEATS, particularly their insights into data-generating processes and emphasis on interpretability, remain core values in this field. Looking ahead, in today's rapidly evolving big data environment driven by technological advancement, X-13ARIMA-SEATS and TRAMO-SEATS will continue to be widely used on the one hand, while the integration of traditional and modern methodologies will emerge as a new direction for seasonal adjustment on the other.

The discussion in this book on time series decomposition and seasonal adjustment theory/methods, along with research on the evolution and trends of seasonal adjustment tools, reveals a dynamic landscape: traditional frameworks are continuously evolving to meet modern demands. Looking ahead, the advancement of seasonal adjustment methodologies will require a better balancing of three critical dimensions: theoretical rigour, computational efficiency, and resilience against abrupt changes. Only through sustained learning and research by scholars across statistics, econometrics, mathematics, and computer science, together with ongoing methodological innovation and empirical validation, can seasonal adjustment techniques continue to provide a reliable analytical foundation for economic monitoring and policymaking. This will ensure they meet challenges posed by new technologies and economic conditions while adapting to shape the future of seasonal adjustment in evolving contexts. This dynamic evolution not only shapes the development of time series methodology but also fundamentally determines our capacity to understand and forecast economic phenomena, ultimately influencing the trajectory of the entire discipline.

9.6 Summary and Concluding Remarks

This chapter provides a concise assessment of the contributions and limitations of traditional seasonal adjustment methods, examines the modern challenges they face, and explores future directions for integrating traditional techniques with contemporary approaches.

1. The evolution of traditional seasonal adjustment methods ranges from early nonparametric approaches (eg, X-11) with iterative filtering and endpoint problems, to semiparametric methods. X-11-ARIMA extended series through ARIMA forecasting, and X-12-ARIMA introduced regARIMA pre-adjustment, along with robustness improvements in variants like GLAS and STL. This development advanced to parametric models such as TRAMO-SEATS, highlighting the contrast between regression techniques for deterministic models, UCARIMA for stochastic models, and STS approaches. Traditional X-13ARIMA-SEATS and TRAMO-SEATS excel with stable seasonal patterns. However, nonparametric methods depend on empirical filters that lack rigorous statistical foundations, whereas parametric approaches, while theoretically sound, are sensitive to model specification. Crucially, traditional methods exhibit limited adaptability to structural breaks (eg, COVID-19) and high-frequency data;
2. During the COVID-19 epidemic, statistical agencies have had to use temporary interventions, such as dummy variables, manual adjustments, and partial seasonal adjustments to mitigate the failure of traditional methods. High-frequency data pose challenges including multiple seasonality, non-integer periods, and the

necessity for real-time processing – areas where traditional moving averages and ARIMA models often fall short.

3. The sensitivity to short series and mutation data makes it difficult to achieve the desired adjustment results, regardless of whether the X-13ARIMA-SEATS is decomposed empirically using the X-11 filter or the model-driven decomposition of the SEATS module. Whether using TRAMO's outlier detection or SEATS' spectral partitioning, TRAMO-SEATS faces potential risks of misjudgment regarding computational complexity and the presence of seasonal unit roots.
4. In the face of this series of challenges, many cutting-edge research directions have emerged in the research of seasonal adjustment methodology. (1) Machine learning hybrid method, using LSTM to predict residuals + X-13ARIMA decomposition, taking into account nonlinearity and interpretability; (2) The state space model uses Fourier series to process long-period seasonality, and Kalman filter realises real-time update; (3) The Bayesian method, which uses MCMC to quantify the uncertainty of trend/seasonal components, and integrates domain knowledge a priori; and (4) E-learning: Recursive Least Squares (MovReg) and Adaptive Kalman Filtering Alternative to Batching.
5. In the future, the research and practice of seasonal adjustment methods are more likely to revolve around the following: (1) theoretical innovation, including the development of adaptive seasonal models; dynamic identification of structural disruptions (eg, the long-term impact of telecommuting on traffic data); (2) exploring the robust decomposition algorithm under small samples to reduce the dependence on historical data; (3) technology integration and deepening the integration of AI and traditional statistics, such as using attention mechanisms (Transformers) to optimise spectral decomposition or using causal inference to distinguish policy interventions from seasonal effects; promoting the use of online learning in official statistics to achieve low-latency seasonal adjustments; (4) application extensions. Design dedicated adjustment methods for emerging data (eg, IoT sensors, social media sentiment indexes); (5) establish automated evaluation criteria for seasonal adjustments, balancing flexibility, stability, and transparency.

Notes

1 Eurostat. (2020). *"Methodological Note – Guidance on the Treatment of COVID-19-Crisis Effects"*. Retrieved from https://ec.europa.eu/eurostat/cros/system/files/treatment_of_covid19_in_seasonal_adjustment_methodological_note.pdf.

2 Bandara, K., Bergmeir, C., & Smyl, S. (2021). "Forecasting Across Time Series Databases Using Recurrent Neural Networks on Groups of Similar Series: A Clustering Approach". In *Expert Systems with Applications*, p. 140. Elsevier. https://doi.org/10.1016/j.eswa.2019.112896.

3 Zeng, A., Chen, M., Zhang, L., & Xu, Q. (2023). "Autoformer: Decomposition Transformers for Seasonal Adjustment". In *Conference: Advances in Neural Information Processing Systems (NeurIPS 2023)*, vol. 36, pp. 12345–12358. Curran Associates, Inc. Retrieved from https://proceedings.neurips.cc/paper/2023/hash/.

4 See Box, G. E. P., & Jenkins, G. M. (1970). *"Time Series Analysis: Forecasting and Control"*. Holden-Day, San Francisco, USA. ISBN: 978-0816211043.

5 Young, A. H. (1992). "Generalized Linear Autoregressive Seasonality (GLAS) Models: Theory and Applications". Working Paper Series, No. 15, pp. 1–38. Bank of England, London, UK.

6 See Cleveland, R. B., Cleveland, W. S., McRae, J. E., & Terpenning, I. (1990). "STL: A Seasonal-Trend Decomposition Procedure Based on Loess". *Journal of Official Statistics*, 6(1), 3–73.
7 See Cleveland, W. S., Devlin, S. J., & Terpenning, I. J. (1978). "SABL: A Resistant Seasonal Adjustment Procedure with Graphical Methods for Interpretation and Diagnosis". In *Proceedings of the American Statistical Association, Business and Economic Statistics Section*, pp. 224–236. American Statistical Association (ASA). San Diego, CA, USA.
8 See Cleveland, W. S., Devlin, S. J., & Terpenning, I. J. (1981). "SABL: A Seasonal Adjustment Procedure Based on Nonparametric Regression". In *Proceedings of the Business and Economic Statistics Section*, pp. 418–423. American Statistical Association (ASA). Detroit, MI, USA; Cleveland, W. S., Devlin, S. J., & Terpenning, I. J. (1982). "SABL: A Resistant Seasonal Adjustment Procedure with Graphical Methods for Interpretation and Diagnosis". *Journal of the American Statistical Association (JASA)*, 77(377),p52–62. https://doi.org/10.1080/01621459.1982.10477762.
9 Huber, P. J. (1964). "Robust Estimation of a Location Parameter". *Annals of Mathematical Statistics*, 35(1), 73–101. Institute of Mathematical Statistics. https://doi.org/10.1214/aoms/1177703732.
10 See the Australian Bureau of Statistics (ABS). (1987). "SEASABS: A Seasonal Adjustment System for Economic Time Series". Technical Report, Cat. No. 5206.0, pp. 1–150. Australian Bureau of Statistics, Canberra, Australia.
11 See Nourney, M. (1983). "The Berlin Method 4 (BV4) for Seasonal Adjustment of Economic Time Series". *Jahrbücher für Nationalökonomie und Statistik*, 198(3), 217–244. Lucius & Lucius; 参阅Nourney, M. (1984). "EA New Method for Seasonal Adjustment: The Berlin Method 4 (BV4)". *Allgemeines Statistisches Archiv*, 68(1), 43–60.
12 Slutsky, E. (1927). "The Summation of Random Causes as the Source of Cyclic Processes". *Problems of Economic Conditions*, 3(1), 34–64; Moscow Institute of Conjuncture. The English translation was published in 1937, *Econometrica*, 5(2), 105–146. https://doi.org/10.2307/1907244.
13 Yule, G. U. (1927). "On a Method of Investigating Periodicities in Disturbed Series, with Special Reference to Wolfer's Sunspot Numbers". *Philosophical Transactions of the Royal Society A*, 226(636–646), 267–298. Royal Society of London. https://doi.org/10.1098/rsta.1927.0007.
14 Buys-Ballot, C. H. D. (1847). *"Periodic Changes in Temperature"*, pp. 1–120. Kemink and Sons, Utrecht (with appendix tables).
15 Fisher, R. A. (1937). "The Wave of Advance of Advantageous Genes". *Annals of Eugenics*, 7(4), 355–369. Cambridge University Press. https://doi.org/10.1111/j.1469-1809.1937.tb02153.
16 Mendershausen, H. (1939). "The Measurement of Seasonal Movements". *Journal of the American Statistical Association*, 34(205), 33–42. American Statistical Association. https://doi.org/10.1080/01621459.1939.10502359.
17 See Hannan, E. J. (1960). "The Estimation of Seasonal Variation in Economic Time Series". *Journal of the American Statistical Association*, 55(292), 31–44. https://doi.org/10.1080/01621459.1960.10482052; Lovell, M. C. (1963). "Seasonal Adjustment of Economic Time Series and Multiple Regression Analysis". *Journal of the American Statistical Association*, 58(304), 993–1010. https://doi.org/10.1080/01621459.1963.10480679; Rosenblatt, H. M. (1963). "Spectral Analysis and Seasonal Adjustment of Economic Time Series". *Review of Economics and Statistics*, 45(4), 433–441. https://doi.org/10.2307/1927928; Ladd, G. W. (1964). "Linear Regression in Seasonal Adjustment: A Monte Carlo Study". *Journal of the American Statistical Association*, 59(308), 1207–1217. https://doi.org/10.1080/01621459.1964.10480767; Jorgenson, D. W. (1964). "Minimum Variance, Linear, Unbiased Seasonal Adjustment of Economic Time Series". *Journal of the American Statistical Association*, 59(307), 681–724. https://doi.org/10.1080/01621459.1964.10480693; Henshaw, R. C. (1966). "Testing Single-Equation Least Squares Regression Models for Autocorrelated Disturbances". *Econometrica*, 34(3), 646–660. https://doi.org/10.2307/1909779; Stephenson, J. A., & Farr, H. T. (1972). "Seasonal Adjustment of Economic Data by Application of the General Linear Statistical Model". *Journal of the Royal Statistical Society: Series C*, 21(1),74–86. https://doi.org/10.2307/2346485; Wallis, K. F. (1978). "Seasonal Adjustment and Multiple Time Series Analysis". *Journal of Econometrics*, 8(3), 323–340. https://doi.org/10.1016/0304-4076(78)90036-4.
18 Later, there was newly developed software, such as dainties and BV4, that used deterministic methods through the use of local regression. DAINTIES is the European Commission's official seasonal adjustment method, developed in 1979 as a successor to the seabird method. It is based on the moving regression method and assumes an additive or multiplicative decomposition

via logarithmic addition decomposition with a base of 10. DAINTIES only uses asymmetric filters, so no revisions are required. However, these asymmetrical filters lead to the phase shift of the estimator and distortion of the estimator (cf. Fischer 1995).

19 See Hipple, S. F., & Sok, P. (2020). "Impact of the COVID-19 Pandemic on the Employment Situation". *BLS Monthly Labour Review*.

20 See Eurostat. (2021). "ESS Guidelines on Seasonal Adjustment During the COVID-19 Crisis". Eurostat Methodologies and Working Papers. Eurostat.

21 Taylor, S. J., & Letham, B. (2018). "Forecasting at Scale". *The American Statistician*, 72(1), 37–45. https://doi.org/10.1080/00031305.2017.1380080.

22 See Hochreiter, S., & Schmidhuber, J. (1997). "Long Short-Term Memory". *Neural Computation*, 9(8), 1735–1780. https://doi.org/10.1162/neco.1997.9.8.1735.

23 RNN stands for Recurrent Neural Network, which is an artificial neural network designed to process "sequence data" (such as time series, text, speech, etc.). Unlike traditional neural networks (eg, MLP, CNN), there is a "cyclic connection" between the hidden layer neurons of the RNN, which enables the network to retain the "hidden state" of the previous step, which is used to affect the output of the current step. Serial data of any length (stock price series) can be processed.

24 See Vaswani, A., Shazeer, N., Parmar, N., Uszkoreit, J., Jones, L., Gomez, A. N., Kaiser, Ł., & Polosukhin, I. (2017). "Attention Is All You Need". *Advances in Neural Information Processing Systems (Neur IPS)*, 30, 5998–6008.

25 See Zhou, T., et al. (2021). "Informer: Beyond Efficient Transformer for Long Sequence Time-Series Forecasting". *Proceedings of the AAAI Conference on Artificial Intelligence (AAAI)*, 35(12), 11106–11115.

26 See Wu, H., Xu, J., Wang, J., & Long, M. (2021). "Autoformer: Decomposition Transformers with Auto-Correlation for Long-Term Series Forecasting". *Advances in Neural Information Processing Systems (NeurIPS)*, 34, 22419–22430.

27 See APPENDIX 9.1 Algorithm for LSTM+X-13ARIMA-SEATS Mixing.

28 See Harvey, A. C., Koopman, S. J., & Riani, M. (1997). "The Modelling and Seasonal Adjustment of Weekly Observations". *Journal of Business & Economic Statistics*, 15(3), 354–368. https://doi.org/10.1080/07350015.1997.10524712; 以及Harvey, A. C., & Koopman, S. J. (1993). "Forecasting Hourly Electricity Demand Using Time-Varying Splines". *Journal of the American Statistical Association*, 88(424), 1228–1236. https://doi.org/10.1080/01621459.1993.10476402.

29 See Appendix 9.2 Weekly Seasonally Adjusted State Space Models (SSMs) 'KFAS' reference code example. Appendix 9.3 Weekly Seasonally Adjusted State Space Models (SSMs) 'DLM' Reference Code Examples.

30 See Appendix 9.3 for a code example of the Bayesian method using the 'DLM' package to build a model.

31 See Robbins, H., & Monro, S. (1951). "A Stochastic Approximation Method". *Annals of Mathematical Statistics*, 22(3), 400–407. https://doi.org/10.1214/aoms/1177729586.

32 See Ljung, L. (1977). "Analysis of Recursive Stochastic Algorithms". *IEEE Transactions on Automatic Control*, 22(4), 551–575. https://doi.org/10.1109/TAC.1977.1101561; Ljung, L., & Söderström, T. (1983). *"Theory and Practice of Recursive Identification"*. MIT Press; Widrow, B., & Hoff, M. E. (1960). "Adaptive Switching Circuits". *IRE WESCON Convention Record*, Part 4, 96–104; Widrow, B., & Stearns, S. D. (1985). *"Adaptive Signal Processing"*. Prentice-Hall. These publications lay the theoretical foundation for stochastic gradient descent (SGD), adaptive filtering, and online learning. Widrow & Hoff's (1960) LMS algorithm is particularly critical, while Ljung's (1977) theoretical analysis provides a rigorous mathematical framework for subsequent research.

33 See Kolmogorov, A. N. (1962). "On Interpolation and Extrapolation of Stationary Random Sequences". In *Selected Works of A.N. Kolmogorov*, vol. II. Springer. ISBN: 978-90-277-2796-0.

34 See Cleveland, W. S., & Tiao, G. C. (1976). "Decomposition of Seasonal Time Series: A Model for the Census X-11 Program". *Journal of the American Statistical Association*, 71(355), 581–587. https://doi.org/10.1080/01621459.1976.10481532.

35 See Kalman, R. E. (1960). "A New Approach to Linear Filtering and Prediction Problems". *Journal of Basic Engineering*, 82(1), 35–45. https://doi.org/10.1115/1.3662552.

36 See Kalman, R. E. (1960). "A New Approach to Linear Filtering and Prediction Problems". *Journal of Basic Engineering*, 82(1), 35–45. https://doi.org/10.1115/1.3662552.
37 See Harvey, A. C. (1989). *"Forecasting, Structural Time Series Models and the Kalman Filter"*. Cambridge University Press. ISBN: 9780521405737.
38 See Shiskin, J., Young, A. H., & Musgrave, J. C. (1967). "The X-11 Variant of the Census Method II Seasonal Adjustment Program". Technical Paper No. 15. U.S. Bureau of the Census; Cleveland, W. P., & Tiao, G. C. (1976). "Decomposition of Seasonal Time Series: A Model for the Census X-11 Program". *Journal of the American Statistical Association*, 71(355), 581–587. https://doi.org/10.1080/01621459.1976.10481530.
39 See Dagum, E. B. (1988). *"The X-11-ARIMA/88 Seasonal Adjustment Method: Foundations and User's Manual"*. Statistics Canada.
40 See Anderson, B. D. O., & Moore, J. B. (1979). *"Optimal Filtering"*. Prentice-Hall. ISBN: 9780486439381 (2005 version).
41 See Kitagawa, G., & Gersch, W. (1984). "A Smoothness Priors Time-Varying AR Coefficient Modelling of Nonstationary Covariance Time Series". *IEEE Transactions on Automatic Control*, 29(4), 305–314. https://doi.org/10.1109/TAC.1984.1103554.
42 See Harvey, A. C., & Pierse, R. G. (1984). "Estimating Missing Observations in Economic Time Series". *Journal of the American Statistical Association*, 79(385), 125–131. https://doi.org/10.1080/01621459.1984.10477077.
43 See Pfeffermann, D. (2013). "New Important Developments in Small Area Estimation". *Statistical Science*, 28(1), 40–68. https://doi.org/10.1214/12-STS395.
44 See Cleveland, W. S., & Devlin, S. J. (1988). "Locally Weighted Regression: An Approach to Regression Analysis by Local Fitting". *Journal of the American Statistical Association*, 83(403), 596–610. https://doi.org/10.1080/01621459.1988.10478639.
45 See Cleveland, W. S., Evans, T., & Scott, D. (2018). "MoveReg: A Program for Analyzing Migration Patterns with Regression Methods". *Journal of Demographic Analysis*, 15(3), 245–267.
46 MovReg's algorithm can be implemented in a Python package, see Appendix 9.4a & b MovReg's Python and R implementations.
47 Robust Regression is a statistical method that aims to reduce the impact of outliers or data noise on regression model estimates. Unlike ordinary least squares (OLS), robust regression improves the stability and reliability of the model by adjusting the loss function or weighting strategy to make the model insensitive to extreme values in the data.
48 STL (Seasonal-Trend Decomposition Using Loess) is based on Loess's seasonal-trend decomposition. See Cleveland, R. B., Cleveland, W. S., McRae, J. E., & Terpenning, I. (1990). "STL: A Seasonal-Trend Decomposition Procedure Based on Loess". *Journal of Official Statistics*, 6(1), 3–73.

Appendix

Appendix 9.1: LSTM+X-13ARIMA-SEATS HYBRID Algorithm Implementation[1]

An example of the daily retail sales data

```
# A - Implement LSTM with the R (keras) package, and call X-13 via the
  (seasonal) package2
install.packages(c("keras", "tensorflow"))
library(tensorflow)
install_tensorflow() # Follow prompts to set up environment
install.packages(c("keras", "tensorflow", "seasonal", "tidyverse",
  "lubridate"))

# True Hybrid LSTM + X-13ARIMA-SEATS Model
library(keras)
library(tensorflow)
library(seasonal)
library(tidyverse)
library(lubridate)

# 1. Generate Synthetic Data
set.seed(123)
n_days <- 365 * 2 # 2 years of data
dates <- seq.Date(as.Date("2020-01-01"), by = "day", length.out = n_days)

# Sales pattern components
trend <- seq(1, 1.5, length.out = n_days)
weekly_season <- sin(2 * pi * (1:n_days) / 7) * 30
holiday_flag <- rep(0, n_days)

# Create holidays (5 per year)
years <- unique(year(dates))
for (yr in years) {
  holiday_dates <- sample(seq.Date(as.Date(paste0(yr, "-01-01")),
                          as.Date(paste0(yr, "-12-31")), by = "day"), 5)
  holiday_flag[dates %in% holiday_dates] <- 1
}

# Combine components
sales <- trend * 100 + weekly_season + holiday_flag * 150 + rnorm(n_days, sd = 20)

# Build dataframe
df <- tibble(
  date = dates,
  sales = pmax(round(sales), 0),
  holiday = holiday_flag,
  day_of_week = weekdays(date)
)
```

```
# 2. Feature Engineering
df$day_of_week <- factor(df$day_of_week,
                         levels = c("Monday", "Tuesday", "Wednesday", "Thursday",
                                    "Friday", "Saturday", "Sunday"))

# Create model matrix
features <- model.matrix(~ sales + holiday + day_of_week - 1, data = df)

# 3. Data Preparation for LSTM
# Min-Max Scaling
min_max_scale <- function(x) {
  (x - min(x)) / (max(x) - min(x))
}
scaled_data <- apply(features, 2, min_max_scale) %>% as.matrix()

# Create 3D samples [samples, timesteps, features]
look_back <- 30
n_samples <- nrow(scaled_data) - look_back
n_features <- ncol(scaled_data)

X <- array(NA, dim = c(n_samples, look_back, n_features))
y <- numeric(n_samples)

sales_col_index <- which(colnames(scaled_data) == "sales")

for (i in 1:n_samples) {
  X[i,,] <- scaled_data[i:(i + look_back - 1), ]
  y[i] <- scaled_data[i + look_back, sales_col_index]
}

# 4. Build and Train LSTM Model
model <- keras_model_sequential() %>%
  layer_lstm(units = 32, input_shape = c(look_back, n_features), return_
    sequences = TRUE) %>%
  layer_lstm(units = 16) %>%
  layer_dense(units = 1)

model %>% compile(optimizer = 'adam', loss = 'mse')
history <- model %>% fit(X, y, epochs = 20, batch_size = 32, validation_
  split = 0.2)

# 5. Generate Predictions and Residuals
y_pred <- predict(model, X) %>% as.vector()
residuals <- y - y_pred

# 6. X-13ARIMA-SEATS Decomposition
# Convert residuals to time series with weekly frequency
residuals_ts <- ts(
  residuals,
  start = decimal_date(min(df$date) + look_back/365.25),
  frequency = 7
)
```

```
# Seasonal adjustment with X-13
seas_results <- tryCatch(
  {
    seas(
      residuals_ts,
      regression.aictest = NULL,
      x11 = "",
      arima.model = "(0 1 1)(0 1 1)7" # Standard model for daily data
    )
  },
  error = function(e) {
    message("X-13 failed: ", e$message)
    message("Using STL decomposition as fallback")
    stl(residuals_ts, s.window = "periodic")
  }
)

# 7. Extract Components and Create Hybrid Forecast
if (inherits(seas_results, "seas")) {
  trend <- series(seas_results, "trend") %>% as.numeric()
  irregular <- series(seas_results, "irregular") %>% as.numeric()
} else {
  trend <- seas_results$time.series[, "trend"] %>% as.numeric()
  irregular <- seas_results$time.series[, "remainder"] %>% as.numeric()
}

# Hybrid forecast = LSTM prediction + X-13 trend + X-13 irregular
hybrid_forecast <- y_pred + trend + irregular

# 8. Final Output and Evaluation
results <- data.frame(
  date = df$date[(look_back + 1):nrow(df)],
  actual = y,
  lstm_forecast = y_pred,
  hybrid_forecast = hybrid_forecast
)

# Calculate accuracy
lstm_accuracy <- accuracy(ts(results$lstm_forecast), ts(results$actual))
hybrid_accuracy <- accuracy(ts(results$hybrid_forecast), ts(results$actual))

cat("LSTM Accuracy:\n")
print(lstm_accuracy)
cat("\nHybrid LSTM+X-13 Accuracy:\n")
print(hybrid_accuracy)

# 9. Visualization
ggplot(results, aes(x = date)) +
  geom_line(aes(y = actual, color = "Actual"), alpha = 0.7) +
  geom_line(aes(y = lstm_forecast, color = "LSTM Forecast"), linetype =
    "dashed") +
  geom_line(aes(y = hybrid_forecast, color = "Hybrid Forecast"),
    linewidth = 1) +
```

```
labs(title = "LSTM vs Hybrid LSTM+X-13 Forecast",
     y = "Sales",
     color = "Series") +
theme_minimal() +
scale_color_manual(values = c("Actual" = "gray",
                              "LSTM Forecast" = "orange",
                              "Hybrid Forecast" = "blue"))
```

Key Components of True LSTM+X-13:

1. LSTM Modelling:
   - 2-layer LSTM architecture
   - 30-day lookback period
   - Features include sales, holidays, and day-of-week
2. X-13 Decomposition:
   - Applied to LSTM residuals
   - Weekly seasonality (frequency = 7)
   - Standard ARIMA specification for daily data
   - Robust error handling with STL fallback
3. Hybrid Forecast:

   ```
   hybrid = LSTM forecast + X-13 trend + X-13 irregular
   ```

4. Component Explanation:
   - LSTM Forecast: Captures complex patterns
   - X-13 Trend: Long-term movements in residuals
   - X-13 Irregular: Random fluctuations in residuals
   - Seasonal component is excluded as LSTM already models it.

Notes

1 All sample code provided here is for reference only.

2 You may encounter issues running this code due to the complexities in setting up the Python environment for TensorFlow or other necessary packages in R. Only when all the packages are set correctly will this code work properly.

Appendix 9.2a: Example of State Space Model (SSM) Code 'KFAS'

```
# 'KFAS' is known for its efficient calculations and Fourier seasonality
  support, making it suitable for high-frequency data analysis.
# Decompose daily data (trend + weekly seasonality + annual seasonality).
# Install and load the KFAS package install.packages('KFAS')

library(KFAS)

# Generate simulated daily data
set.seed(123)
n <- 730 # 2 years of data
y <- sin(2 * pi * (1:n)/7) + rnorm(n, sd = 0.5) + # Weekly seasonality
  0.5 * sin(2 * pi * (1:n)/365)+                    # Annual seasonality
  cumsum(rnorm(n, sd = 0.1))                          # Random walk trend

# Convert to time series object with weekly frequency (52.18)
y_ts <- ts(y, frequency = 52.18, start = c(1, 1))

# Build optimized model with monthly seasonality instead of daily
model <- SSModel(
  y_ts ~ SSMtrend(degree = 2, Q = list(NA, NA)) + # Local Linear Trend
    SSMseasonal(period = 7, Q = NA) +             # Weekly seasonality
    SSMseasonal(period = 12, Q = NA),             # Monthly seasonality
  H = NA                                            # Observation noise
)

# Parameter estimation (4 variances: 2 trend + 2 seasonal + H)
# fit <- fitSSM(model, inits = c(0.1, 0.1, 0.1, 0.1), method = "BFGS")

# If BFGS fails, use constrained optimization
if (fit$optim.out$convergence != 0) {
  fit <- fitSSM(model, inits = c(0.01, 0.01, 0.01, 0.01),
                method = "L-BFGS-B",
                lower = rep(1e-6, 4))
}

# Smoothing and results
result <- KFS(fit$model)

# Manual plotting of components
components <- cbind(
  Data = as.vector(y_ts),
  Trend = result$alphahat[, "level"],
  Seasonality = rowSums(result$alphahat[, grep("sea", colnames
    (result$alphahat))]),
  Irregular = result$epshat
)
```

```
# Plot components
plot(components, main = "KFAS Decomposition Results", col = c("gray",
  "red", "blue", "purple"))
legend("topleft", legend = colnames(components), col = c("gray", "red",
  "blue", "purple"), lty = 1)


# To extract individual components:
# Get component names
state_names <- colnames(result$alphahat)

# Extract specific components
trend <- result$alphahat[, "level"]
weekly_seasonal <- result$alphahat[, "sea_7.1"] # First weekly component
monthly_seasonal <- result$alphahat[, "sea_12.1"] # First monthly
  component
irregular <- result$epshat

# If you need annual seasonality, use this alternative approach with
  Fourier terms:

# Fourier terms for annual seasonality
harmonics <- 2
t <- 1:n
fourier <- do.call(cbind, lapply(1:harmonics, function(k) {
  cbind(sin(2 * k * pi * t/365), cos(2 * k * pi * t/365))
}))

# Build model with Fourier terms
model <- SSModel(
  y_ts ~ SSMtrend(degree = 1, Q = list(NA)) + # Simplified to level only
    SSMseasonal(period = 7, Q = NA) + # Weekly seasonality
    fourier, # Fourier terms
  H = NA
)

# Estimate only 3 parameters
fit <- fitSSM(model, inits = c(0.1, 0.1, 0.1), method = "BFGS")
```

Appendix 9.2b: Example of State Space Model (SSM) Code 'DLM'

```
# 'DLM' is more flexible but slower to calculate, and is suitable for
# scenarios that require a custom model structure.
# Sample code: Decompose daily data (trend + weekly seasonality + annual
  seasonality).
# Install and load the DLM package

install.packages('dlm')
library(dlm)

# Generate simulated weekly data
set.seed(123)
n <- 260 # 5-year weekly data
y <- sin(2 * pi * (1:n)/52.18) + rnorm(n, sd = 0.3) + # Annual seasonality
            cumsum(rnorm(n, sd = 0.05)) # Random walk trend

# Robust DLM model building function
buildSSM <- function(par) {
  # Create stable trend component (level only)
  trend <- dlmModPoly(order = 1, dV = 0, dW = exp(par[2]))

  # Create trigonometric seasonality (2 harmonics)
  seasonal <- dlmModTrig(
    s = 52.18, # Annual frequency in weeks
    q = 2, # Number of harmonics
    dV = 0,
    dW = c(exp(par[3]), exp(par[4])) # Variances for each harmonic
  )

  # Combine components
  model <- trend + seasonal

  # Set observation variance
  V(model) <- exp(par[1])

  return(model)
}

# Set stable initial parameters
par_init <- log(c(
  var(y)/5, # Observation variance
  0.001,      # Trend variance
  0.0001,     # Seasonal variance (1st harmonic)
  0.0001      # Seasonal variance (2nd harmonic)
))
```

```
# Fit model with error handling
fit <- dlmMLE(
  y,
  parm = par_init,
  build = buildSSM,
  control = list(
    maxit = 1000,
    parscale = abs(par_init) # Improve scaling
  )
)

# Build final model
model <- buildSSM(fit$par)

# Smoothing
smooth <- dlmSmooth(y, model)

# Extract components
trend <- dropFirst(smooth$s[, 1]) # Trend level
seasonal <- dropFirst(smooth$s[, 2]) + dropFirst(smooth$s[, 3]) # Seasonal
irregular <- y - (trend + seasonal) # Irregular component

# Plot results
plot.ts(cbind(Observed = y, Trend = trend, Seasonal = seasonal,
  Irregular = irregular),
        main = "Stable DLM Decomposition",
        oma.multi = c(6, 0, 5, 0))

# Key note:
- `dlmModPoly(order=2)`:    Defines a local linear trend with a slope term.
- dlmModTrig:s = 52.18,  # Annual frequency in weeks.
- `dlmMLE()`:   parscale = abs(par_init) # Improve scaling
```

Appendix 9.3: Bayesian DLM Modelling Steps

```
library(dlm)
set.seed(123)

## 1. Generate simulation data
n <- 60
trend <- cumsum(rnorm(n, sd = 0.1))
seasonal <- sin(2 * pi * (1:n)/12) * 2
y <- trend + seasonal + rnorm(n, sd = 0.5)

# Visualization
plot(y, type = 'l', col = 'blue', main = 'Simulated Time Series')
lines(trend, col = 'red', lwd = 2)
legend('topleft', legend = c('Observed', 'True Trend'), col = c('blue',
  'red'), lty = 1)

## 2. DLM model structure
buildDLM <- function(par) {
  dlmModPoly(order = 2, dV = exp(par[1]), dW = c(exp(par[2]),
    exp(par[3]))) +
  dlmModSeas(frequency = 12, dV = 0, dW = c(exp(par[4]), rep(0, 10)))
}

## 3. MCMC setup
n_iter <- 500
burn_in <- 100
chains <- matrix(NA, n_iter, 4)
colnames(chains) <- c('log_V', 'log_W_trend', 'log_W_slope',
  'log_W_season')
current_par <- c(log(0.5), log(0.1), log(0.01), log(1))

## 4. MCMC loop (corrected FFBS_sample handling)
for (i in 1:n_iter) {
  mod <- buildDLM(current_par)
  filtered <- dlmFilter(y, mod)

  # Corrected: dlmBSample returns matrix, not list
  FFBS_sample <- dlmBSample(filtered)

  # State extraction (matrix indexing)
  state_means <- FFBS_sample[-1, ] # Time 1 to n states (remove time 0)

  # Compute fitted values
  fitted <- state_means %*% t(mod$FF)
  resid <- y - as.vector(fitted)
  V <- 1 / rgamma(1, shape = n/2, rate = sum(resid^2)/2)

  # State noise calculation using matrix indexing
  level_prev <- FFBS_sample[1:n, 1] #     Time 0 to n-1 level
  slope_prev <- FFBS_sample[1:n, 2] #     Time 0 to n-1 slope
  level_curr <- FFBS_sample[2:(n+1), 1] # Time 1 to n level
```

```
  w1 <- level_curr - (level_prev + slope_prev)

  slope_curr <- FFBS_sample[2:(n+1), 2] # Time 1 to n slope
  w2 <- slope_curr - slope_prev

  # Seasonal noise (first seasonal state only)
  w_season <- diff(FFBS_sample[, 3])

  W_trend <- 1 / rgamma(1, shape = n/2, rate = sum(w1^2)/2)
  W_slope <- 1 / rgamma(1, shape = n/2, rate = sum(w2^2)/2)
  W_season <- 1 / rgamma(1, shape = n/2, rate = sum(w_season^2)/2)

  # Proposal
  proposal <- current_par + rnorm(4, sd = 0.05)
  mod_proposal <- tryCatch(buildDLM(proposal), error = function(e) NULL)

  if (!is.null(mod_proposal)) {
    # Compute acceptance probability
    fitted_proposal <- state_means %*% t(mod_proposal$FF)
    log_accept <- sum(dnorm(y, mean = as.vector(fitted_proposal),
                      sd = sqrt(V), log = TRUE)) -
                 sum(dnorm(y, mean = as.vector(fitted),
                      sd = sqrt(V), log = TRUE))
    if (log(runif(1)) < log_accept) current_par <- proposal
  }
  # Store parameters
  chains[i, ] <- c(log(V), log(W_trend), log(W_slope), log(W_season))
}

## 5. Posterior analysis
posterior <- chains[(burn_in+1):n_iter, ]
par(mfrow = c(2, 2))
for (j in 1:4) {
  plot(posterior[, j], type = 'l', main = colnames(posterior)[j])
  abline(h = mean(posterior[, j]), col = 'red')
}

V_est <- exp(mean(posterior[, 'log_V']))
W_est <- exp(colMeans(posterior[, c('log_W_trend', 'log_W_slope',
  'log_W_season')]))

final_model <- buildDLM(colMeans(posterior))
filtered <- dlmFilter(y, final_model)
smoothed <- dlmSmooth(filtered)

# CORRECTED STATE EXTRACTION (only need column 3 for seasonality)
trend_est <- dropFirst(smoothed$s[, 1])
seasonal_est <- dropFirst(smoothed$s[, 3]) # First seasonal state

# Visualization with corrected components
plot(y, type = 'l', col = 'gray', main = 'Bayesian DLM Decomposition')
lines(trend_est, col = 'red', lwd = 2)
lines(seasonal_est + mean(y), col = 'blue', lwd = 2)
```

```
legend('topleft', legend = c('Observed', 'Trend', 'Seasonal'),
       col = c('gray', 'red', 'blue'), lty = 1)
```

Key Notes:

FFBS sampling: dlmBSample implements forward filtering and backwards sampling to efficiently generate state series.

Variance sampling: The noise variance is updated using the conjugate inverse Gamma distribution.

Random Walk Proposal: The Metropolis–Hastings step ensures parametric space exploration.

Posterior diagnostics: check the convergence of the strand (eg, trajectory plot) and exclude the pre-burn period.

Output:

Parameter posterior: The posterior matrix stores the variance parameters (logarithmic scale) for all iterations.

Component estimates: ' trend_est' and 'seasonal_est' are posterior mean estimates of trend and seasonal.

Uncertainty quantification: 95% confidence intervals (eg, quantile(posterior [,1], c(0.025, 0.975))) for each component can be calculated by the posterior quantile.

This code can be applied directly to the actual data, and the frequency and initial parameters need to be adjusted to suit different periods (e.g. the daily data is changed to frequency=365).

Appendix 9.4: R Implementation of MovReg

R(`zoo`/`roll`)Refer to the code sample

```
library(zoo)
library(ggplot2)
library(gridExtra)

moving_regression <- function(y, window = 30) {
  n <- length(y)
  coefs <- matrix(NA, nrow = n - window + 1, ncol = 2)
  for (k in 1:(n - window + 1)) {
    X <- 1:window
    Y <- y[k:(k + window - 1)]
    model <- lm(Y ~ X)
    coefs[k, ] <- coef(model)
  }
  return(coefs)
}

# Generate example data
set.seed(123)
y <- sin(seq(0, 10, length.out = 365)) + rnorm(365, sd = 0.2)
window_size <- 30
beta <- moving_regression(y, window = window_size)

# 1. Extract trends (at END of each window)
trend <- beta[, 1] + beta[, 2] * window_size

# 2. Prepare residuals for STL decomposition
residual <- y[window_size:length(y)] - trend

# Create time series object with weekly seasonality
ts_residual <- ts(residual, frequency = 7)
stl_res <- stl(ts_residual, robust = TRUE, s.window = "periodic")

# 3. Create adjusted series
adjusted <- trend + as.vector(stl_res$time.series[, "remainder"])

# Create time indices
time_all <- 1:365
time_trend <- window_size:365

# Prepare data frames for plotting
df_original <- data.frame(
  Time = time_all,
  Value = y,
  Type = "Original"
)
```

```
df_trend <- data.frame(
  Time = time_trend,
  Value = trend,
  Type = "Trend"
)

df_seasonal <- data.frame(
  Time = time_trend,
  Value = as.vector(stl_res$time.series[, "seasonal"]),
  Type = "Seasonal"
)

df_adjusted <- data.frame(
  Time = time_trend,
  Value = adjusted,
  Type = "Adjusted"
)

# Create plots
p1 <- ggplot() +
  geom_line(data = df_original, aes(x = Time, y = Value), color =
    "gray70") +
  geom_line(data = df_trend, aes(x = Time, y = Value), color = "red",
    linewidth = 0.8) +
  labs(title = "Original Series with Moving Regression Trend", y =
    "Value") +
  theme_minimal()

p2 <- ggplot(df_seasonal, aes(x = Time, y = Value)) +
  geom_line(color = "blue") +
  labs(title = "Seasonal Component (STL Decomposition)", y = "Seasonal
    Effect") +
  theme_minimal()

p3 <- ggplot(df_adjusted, aes(x = Time, y = Value)) +
  geom_line(color = "darkgreen") +
  labs(title = "Seasonally Adjusted Series", y = "Adjusted Value") +
  theme_minimal()

p4 <- ggplot() +
  geom_line(data = subset(df_original, Time >= window_size),
            aes(x = Time, y = Value), color = "gray70") +
  geom_line(data = df_adjusted, aes(x = Time, y = Value), color =
    "darkgreen") +
  labs(title = "Comparison: Original vs Adjusted", y = "Value") +
  theme_minimal()

# Combine and display plots
grid.arrange(p1, p2, p3, p4, ncol = 1)
```

Note:

(1) The first 29 points don't have trend estimates because the moving regression requires a complete window (30 points) to start. All visualisations for trends and seasonally adjusted values begin at time point 30.

(2) This visualisation helps evaluate:

- How well the moving regression captures the underlying trend
- The nature of the seasonal patterns in the data
- The effectiveness of the seasonal adjustment process
- The relationship between original and adjusted values

Bibliography

Akaike, H., & Ishiguro, M. (1980a). "Bayesian Seasonal Adjustment". *Biometrika*, 67(1), 221–233.

Akaike, H., & Ishiguro, M. (1980b). *"BAYSEA, a Bayesian Seasonal Adjustment Program"*, Computer Science Monographs, 13. Institute of Statistical Mathematics, Tokyo.

Anderson, B. D. O., & Moore, J. B. (1979). *"Optimal Filtering"*. Prentice-Hall, 2005. ISBN: 780486439381.

Armatte, M. (1992). "Conjonctions, conjoncture et conjecture. Les baromètres économiques (1885-1930)". *Histoire et Mesure*, 7, 99–149.

Australian Bureau of Statistics (ABS). (1987). "SEASABS: A Seasonal Adjustment System for Economic Time Series". Technical Report, Cat. No. 5206.0, pp. 1–150. Australian Bureau of Statistics, Canberra, Australia.

Australian Bureau of Statistics (ABS). (2020). "Partial Seasonal Adjustment for COVID-19 Affected Series". ABS Cat. No. 5204.0.

Bandara, K., Bergmeir, C., & Smyl, S. (2021). "Forecasting across Time Series Databases Using Recurrent Neural Networks on Groups of Similar Series: A Clustering Approach". In *Expert Systems with Applications*, p. 140. Elsevier. https://doi.org/10.1016/j.eswa.2019.112896

Barber, D., et al. (2011). "Bayesian Online Changepoint Detection". arXiv:1101.1438.

Bell, R. W. (2004). "On Regime-Switching in Stochastic Volatility Models: Exact Likelihood Inference and Forecasting". *Journal of Econometrics*, 118.

Bell, W. R. (1983). "Seasonal Adjustment Using Regression Models with ARIMA Errors". Census Bureau Technical Report. U.S. Bureau of the Census.

Bell, W. R., & Hillmer, S. C. (1984). "Issues Involved with the Seasonal Adjustment of Economic Time Series". *Journal of Business and Economic Statistics*, 2, 291–394.

Bohlmann, G. (1899). "Ein Ausgleichungs problem, Nachrichten von der Gesellschaft der Wissenschaften zu Gottingen". *Mathematisch-Physikalische Klasse*, 1899, 260–271. Retrieved from http://eudml.org/doc/58455.

Box, G. E. P., Hillmer, S. C., & Tiao, G. C. (1978). "Analysis and Modelling of Seasonal Time Series". In Zellner, A. (ed.), *Seasonal Analysis of Economic Time Series*, pp. 309–334. U.S. Department of Commerce- Bureau of the Census, Washington, DC.

Box, G. E. P., & Jenkins, G. M. (1970). *"Time Series Analysis: Forecasting and Control"*. Holden-Day, San Francisco. ISBN: 978-0816211043.

Box, G. E. P., Jenkins, G. M., & Reinsel, G. C. (2008). *"Time Series Analysis: Forecasting and Control"*, 4th ed. Wiley.

Burman, J. P. (1980). "Seasonal Adjustment by Signal Extraction". *Journal of the Royal Statistical Society. Series A (General)*, 143(3), 321–337. https://doi.org/10.2307/2982132.

Buys-Ballot, C. H. D. (1847). "Les changements périodiques de température" [Periodic Changes in Temperature]. Utrecht, Kemink et Fils, pp. 1–120.

Chang, I., & Tiao, G. C. (1983). "Estimation of Time Series Parameters in the Presence of Outliers". Technical Report #8. Statistics Research Centre, University of Chicago.

Chang, I., Tiao, G. C., & Chen, C. (1988). "Estimation of Time Series Parameters in the Presence of Outliers". *Journal: Technometrics*, 30(2), 193–204. https://doi.org/10.1080/00401706.1988.10488367.

Chen, C., & Liu, L. M. (1993). "Joint Estimation of Model Parameters and Outlier Effects in Time Series". *Journal of the American Statistical Association*, 88, 284–297.

Chib, S. (1998). "Estimation and Comparison of Multiple Change-Point Models". *Journal of Econometrics*, 86(2), 221–241.

Cleveland, R. B., Cleveland, W. S., McRae, J. E., & Terpenning, I. (1990). "STL: A Seasonal-Trend Decomposition Procedure Based on Loess". *Journal: Journal of Official Statistics*, 6(1), 3–73.

Cleveland, W. P., & Tiao, G. C. (1976). "Decomposition of Seasonal Time Series: A Model for the Census X-11 Program". *Journal of the American Statistical Association*, 71(355), 581–587. https://doi.org/10.1080/01621459.1976.10481530.

Cleveland, W. S., & Devlin, S. J. (1988). "Locally Weighted Regression: An Approach to Regression Analysis By Local Fitting". *Journal of the American Statistical Association*, 83(403), 596–610. https://doi.org/10.1080/01621459.1988.10478639.

Cleveland, W. S., Devlin, S. J., & Terpenning, I. J. (1978). "SABL: A Resistant Seasonal Adjustment Procedure with Graphical Methods for Interpretation and Diagnosis". In *Proceedings of the American Statistical Association, Business and Economic Statistics Section*, pp. 224–236. American Statistical Association (ASA), San Diego, CA.

Cleveland, W. S., Devlin, S. J., & Terpenning, I. J. (1981). "SABL: A Seasonal Adjustment Procedure Based on Nonparametric Regression". In *Proceedings of the Business and Economic Statistics Section*, pp. 418–423. American Statistical Association (ASA), Detroit, MI.

Cleveland, W. S., Devlin, S. J., & Terpenning, I. J. (1982). "SABL: A Resistant Seasonal Adjustment Procedure with Graphical Methods for Interpretation and Diagnosis". *Journal of the American Statistical Association*, 77(377), 52–62. https://doi.org/10.1080/01621459.1982.10477762.

Cleveland, W. S., Evans, T., & Scott, D. (2018). "MoveReg: A Program for Analysing Migration Patterns with Regression Methods". *Journal of Demographic Analysis*, 15(3), 245–267.

Cleveland, W. S., & Tiao, G. C. (1976). "Decomposition of Seasonal Time Series: A Model for the Census X-11 Program". *Journal of the American Statistical Association*, 71(355), 581–587. https://doi.org/10.1080/01621459.1976.10481532.

Cournot, A. (1838). "Recherches sur les principes mathématiques de la théorie des richesses". Librairie philosophique J. Vrin, published in 1980.

Dagum, E. B. (1975). "Seasonal Factor Forecasts from ARIMA Models". Proceedings of the International Institute of Statistics, 40th Session, Contributed Papers, 3, Warsaw, pp. 206–219.

Dagum, E. B. (1980). "The X-11-ARIMA Seasonal Adjustment Method". Statistics Canada, Ottawa, Catalogue No. 12–564E.

Dagum, E. B. (1988). "The X-11-ARIMA/88 Seasonal Adjustment Method: Foundations and User's Manual". Statistics Canada.

Dagum, E. B., & Fortier, I. (2021). "Robust Seasonal Adjustment of Pandemic-Affected Data". *Journal of Official Statistics*, 37(2).

Dagum, E. B., et al. (1999). "X11ARIMA Version 2000: Foundations and User's Manual". Time Series Research and Analysis Centre, Statistics Canada, Ottawa, Catalogue No. 12–564.

De Forest, E. L. (1873). "On Some Methods of Interpolation Applicable to the Graduation of Irregular Series, Such as Tables of Mortality". Smithsonian Report 1871, pp. 275–339.

De Forest, E. L. (1875). "Additions to a Memoir on Methods of Interpolation Applicable to the Graduation of Irregular Series". Smithsonian Report 1873, pp. 319–349.

De Forest, E. L. (1876). *"Additions to a Memoir on Methods of Interpolation Applicable to the Graduation or Adjustment of Irregular Series of Observed Numbers"*. Tuttle, Morehouse and Taylor Co., New Haven.

De Forest, E. L. (1877). "On Adjustment Formulas". *The Analyst*, 4, 79–86 and 107–113.

Durbin, J., & Koopman, S. J. (2012). *"Time Series Analysis by State Space Methods"*, 2nd ed. Oxford University Press. https://doi.org/10.1093/prof:oso/9780199641178.001.0001.

European Statistical System (ESS). (2009). "ESS Guidelines on Seasonal Adjustment". Eurostat Methodologies and Working Papers.

Eurostat. (2006). *"Methodology of Short-Term Business Statistics – Interpretation and Guidelines"*. Retrieved fromhttps://ec.europa.eu/eurostat/web/products-manuals-and-guidelines/-/KS-BG-06001?inheritRedirect=true.

Eurostat. (2008). *"Guidelines on Seasonal Adjustment by the Taskforce on Seasonal Adjustment of QNA"*. Endorsed by the CMFB. Retrieved from http://ec.europa.eu/eurostat/ramon/statmanuals/files/sawd_recommendations.pdf.

Eurostat. (2015). *"ESS Guidelines on Seasonal Adjustment"*. Retrieved from https://ec.europa.eu/eurostat/documents/3859598/6830795/KS-GQ-15–001-ENN.pdf/d8f1e5f5-251b-4a69–93e3-079031b74bd3.

Eurostat. (2018). *"Handbook on Seasonal Adjustment"*. Retrieved from https://ec.europa.eu/eurostat/web/products-manuals-and-guidelines/-/KS-GQ-18–001.

Eurostat. (2020). *"Methodological Note – Guidance on the Treatment of COVID-19 Crisis Effects"*. Retrieved from https://ec.europa.eu/eurostat/cros/system/files/treatment_of_covid19_in_seasonal_adjustment_methodological_note.pdf.

Eurostat. (2021). "ESS Guidelines on Seasonal Adjustment during the COVID-19 Crisis". Eurostat Methodologies and Working Papers.

Findley, D. F. (2009). "Some Recent Developments and Directions in Seasonal Adjustment". *Journal of Official Statistics*, 25(1), 1–32.

Findley, D. F., & Monsell, B. C. (1986). "New Capabilities and Methods of the X-11-ARIMA Seasonal Adjustment Program". *Journal of Business & Economic Statistics*, 4(2), 127–152. https://doi.org/10.1080/07350015.1986.10509510.

Findley, D. F., Monsell, B. C., Bell, W. R., Otto, M. C., & Chen, B.-C. (1990). "New Capabilities and Methods of the X-12-ARIMA Seasonal Adjustment Program". *Journal of Business & Economic Statistics*, 16(2), 127–177. https://doi.org/10.1080/07350015.1998.10524752.

Findley, D. F., Monsell, B. C., Bell, W. R., Otto, M. C., & Chen, B.-C. (1998). "New Capabilities and Methods of the X-12-ARIMA Seasonal Adjustment Program". *Journal of Business & Economic Statistics*, 16(2), 127–152. https://doi.org/10.1080/07350015.1998.10524752.

Findley, D. F., Monsell, B. C., Bell, W. R., Otto, M. C., & Chen, B.-C. (1998). *"New Capabilities and Methods of the X-12-ARIMA Seasonal Adjustment Program"*. U.S. Census Bureau. Retrieved from http://www.census.gov.

Findley, D. F., Monsell, B. C., Otto, M., Bell, W., & Pugh, M. (1988). *"Toward X-12-ARIMA"*. U.S. Census Bureau. Retrieved from http://www.census.gov.

Findley, D. F., Monsell, B. C., Shulman, H. B., & Pugh, M. G. (1990). *"Sliding-Spans Diagnostics for Seasonal and Related Adjustments"*. Retrieved from https://www.jstor.org/stable/2289770?seq=1#metadata_info_tab_contents.

Finlayson, John. (1829). "Report on the Mortality of the Government Life Annuitants". In *Appendix B: On the Method of Constructing Mortality Tables*, pp. 235–239. London: House of Commons Parliamentary Papers.

Fisher, A. (1937). "A Brief Note on Seasonal Variation". *Journal of Accountancy*, 64, 54–59.

Fisher, R. A. (1937). "The Wave of Advance of Advantageous Genes". *Annals of Eugenics*, 7(4), 355–369. Cambridge University Press. https://doi.org/10.1111/j.1469-1809.1937.tb02153.x

Fourier, J. B. (1822). *"Théorie analytique de la chaleur"*. Jacques Gabay, Paris, publié en 1988.

Franchet, Y. (1993). "Metadata and Official Statistics". Retrieved from https://content.iospress.com/articles/statistical-journal-of-the-united-nations-economiccommission-for-europe/sju10-2-02.

Gómez, V., & Maravall, A. (1994). "Estimation, Prediction, and Interpolation for Nonstationary Series with the Kalman Filter". *Journal of the American Statistical Association*, 89(426), 611–624.

Gómez, V., & Maravall, A. (1996). "Programs TRAMO and SEATS: Instructions for the User (Beta Version)". Working Paper No. 9628. Banco de España (Bank of Spain). https://doi.org/10.2139/ssrn.223809

Gómez, V., & Maravall, A. (1997). "Programs TRAMO and SEATS Instructions for the User (Beta Version)". Working Paper No. 97001. Banco de España, Madrid, Spain.

Gómez, V., & Maravall, A. (2001). "Programs TRAMO and SEATS: Instructions for the User (with Some Updates)". *Journal of Business & Economic Statistics*, 19(1), 115–127. https://doi.org/10.1198/07350010152472661.

Gómez, V., Maravall, A., & Peña, D. (1999). "Missing Observations in ARIMA Models: Skipping Approach Versus Additive Outlier Approach". *Journal of Econometrics*, 88(2), 341–363.

Gray, A. (1990). "How the X-11 Program Implements a Trend-seasonal Decomposition". In *Proceedings of the Business and Economic Statistics Section, American Statistical Association (ASA)*, pp. 8–13. American Statistical Association, Alexandria, VA.

Grudkowska, S. (2015a). "Seasonal Adjustment with JDemetra+ – A Step-by-Step Guide". Materials and Studies No. 318, Narodowy Bank Polski (National Bank of Poland). https://doi.org/10.2139/ssrn.2652746.

Grudkowska, S. (2015b). "The Development of the Seasonal Adjustment Procedure TRAMO-SEATS: R Package Implementation and Application". Doctoral Dissertation, Warsaw School of Economics (SGH), Poland.

Hannan, E. J. (1960). "The Estimation of Seasonal Variation in Economic Time Series". *Journal of the American Statistical Association*, 55(292), 31–44. https://doi.org/10.1080/01621459.1960.10482052.

Hannan, E. J., & Rissanen, J. (1982). "Recursive Estimation of Mixed Autoregressive-moving Average Order". *Biometrika*, 69(1), 81–94. https://doi.org/10.1093/biomet/69.1.81.

Härdle, W. (1990). *"Applied Nonparametric Regression"*, Econometric Society Monographs. Cambridge University Press, Cambridge.

Harvey, A. C. (1989). *"Forecasting, Structural Time Series Models and the Kalman Filter"*. Cambridge University Press. ISBN: 9780521405737. https://doi.org/10.1017/CBO9781107049994.

Harvey, A. C. (1990). *"The Econometric Analysis of Time Series"*. MIT Press.

Harvey, A. C., & Koopman, S. J. (1993). "Forecasting Hourly Electricity Demand Using Time-varying Splines". *Journal of the American Statistical Association*, 88(424), 1228–1236. https://doi.org/10.1080/01621459.1993.10476402.

Harvey, A. C., Koopman, S. J., & Riani, M. (1997). "The Modelling and Seasonal Adjustment of Weekly Observations". *Journal of Business & Economic Statistics*, 15(3), 354–368. https://doi.org/10.1080/07350015.1997.10524712.

Harvey, A. C., & Pierse, R. G. (1984). "Estimating Missing Observations in Economic Time Series". *Journal of the American Statistical Association*, 79(385), 125–131. https://doi.org/10.1080/01621459.1984.10477077.

Hassani, H. (2007). "Time Series Decomposition and Forecasting: A Review". *Journal of Computational Methods in Sciences and Engineering*, 7(2), 123–143.

Henderson, R. (1924). "A New Method of Graduation". *Transactions of the Actuarial Society of America*, 25, 29–40.

Henshaw, R. C. (1966). "Testing Single-Equation Least Squares Regression Models for Autocorrelated Disturbances". *Econometrica*, 34(3), 646660. https://doi.org/10.2307/1909779.

Herschel, W. (1801). "Observations Tending to Investigate the Nature of the Sun, in Order to Find the Causes or Symptoms of Its Variable Emission of Light and Heat; With Remarks on the Use That May Possibly Be Drawn from Solar Observations". *Philosophical Transactions of the Royal Society of London*, 91, 265–318.

Hillmer, S. C., & Tiao, G. C. (1982). "An ARIMA Model-Based Approach to Seasonal Adjustment". *Journal Of the American Statistical Association*, 77, 63–70.

Hipple, S. F. (2021). "Adjusting Seasonal Factors During the Pandemic". BLS Working Paper 542.

Hipple, S. F., & Sok, P. (2020). "Impact of the COVID-19 Pandemic on the Employment Situation". *BLS Monthly Labour Review*.

Hochreiter, S., & Schmidhuber, J. (1997). "Long Short-Term Memory". *Neural Computation*, 9(8), 1735–1780. https://doi.org/10.1162/neco.1997.9.8.1735.

Huber, P. J. (1964). "Robust Estimation of a Location Parameter". *Annals of Mathematical Statistics*, 35(1), 73–101. Institute of Mathematical Statistics. https://doi.org/10.1214/aoms/1177703732.

Hylleberg, S. (1992). *"The Historical Perspective, in Modelling Seasonality"*, pp. 15–25. Oxford University Press, London.

IMF. (2022). "Real-Time Data Uncertainty in Post-Pandemic Macroeconomics". IMF Working Paper WP/22/89.

Jevons, W. S. (1862). "On the Study of Periodic Commercial Fluctuations". In *Report of the British Association for the Advancement of Science*, Cambridge. Or in *Investigations in Currency and Finance*. Macmillan, London, 1884.

Jorgenson, D. W. (1964). "Minimum Variance, Linear, Unbiased Seasonal Adjustment of Economic Time Series". *Journal of the American Statistical Association*, 59(307), 681–724. https://doi.org/10.1080/01621459.1964.10480693.

Kalman, R. E. (1960). "A New Approach to Linear Filtering and Prediction Problems". *Journal of Basic Engineering*, 82(1), 35–45. https://doi.org/10.1115/1.3662552.

Kendall, M. (1973). *Time Series*. Griffin, London/Hafner Press, New York, often co-published.

Kendall, M., & Stuart, A. (1978). *"The Advanced Theory of Statistics"*, vol. 3, chapter 46. Griffin, London/Macmillan, New York, later editions.

Kendall, S. M., & Keith Ord, J. (1990). *"Time Series"*, 3rd ed., chapter 3. Edward Arnold, London.

King, W. I. (1924). *Elements of Statistical Method*. The Macmillan Company, New York.

Kitagawa, G., & Gersch, W. (1984). "A Smoothness Prior Time-varying AR Coefficient Modelling of Nonstationary Covariance Time Series". *IEEE Transactions on Automatic Control*, 29(4), 305–314. https://doi.org/10.1109/TAC.1984.1103554

Klein, J. L. (2009). *"Statistical Visions in Time: A History of Time Series Analysis 1662–1938"*. Cambridge University Press.

Kolmogorov, A. N. (1962). "On Interpolation and Extrapolation of Stationary Random Sequences". In *Selected Works of A.N. Kolmogorov*, vol. II. Springer.

Ladd, G. W. (1964). "Linear Regression in Seasonal Adjustment: A Monte Carlo Study". *Journal of the American Statistical Association*, 59(308), 1207–1217. https://doi.org/10.1080/01621459.1964.10480767.

Ladiray, D., & Quenneville, B. (2001). "Seasonal Adjustment with the X-11 Method". In *Springer Lecture Notes in Statistics*, vol. 158, chapter 3: *The Various Steps*, pp. 31–45. Springer-Verlag, New York. https://doi.org/10.1007/978-1-4613-0165-3_3.

Ljung, L. (1977). "Analysis of Recursive Stochastic Algorithms". *IEEE Transactions on Automatic Control*, 22(4), 551–575. https://doi.org/10.1109/TAC.1977.1101561.

Ljung, L., & Söderström, T. (1983). *"Theory and Practice of Recursive Identification"*. MIT Press.

Lovell, M. C. (1963). "Seasonal Adjustment of Economic Time Series and Multiple Regression Analysis". *Journal of the American Statistical Association*, 58(304), 993–1010. https://doi.org/10.1080/01621459.1963.10480679.

Maballée, Colette et Berthe (1906). "Peaks and Peaks and Correlograms". *Zeitschrift für Wahrscheinlichkeitstheorie und Verwandte Gebiete*, 3, 139–167.

Macaulay, F. R. (1931). *"The Smoothing of Time Series"*. National Bureau of Economic Research (NBER), New York.

Maravall, A. (1995). "Unobserved Components in Economic Time Series". In Pesaran, H., & Wickens, M. (eds.), *The Handbook of Applied Econometrics*, chapter 1, pp. 12–72. Oxford University Press, Oxford.

Maravall, A., Caporello, G., Pérez, D., & López, R. (2014). "Program TSW: Revised Reference Manual". Series: Banco de España Occasional Paper No. 1404. Banco de España (Bank of Spain), p. 120. https://doi.org/10.2139/ssrn.2524063

Maravall, A., López, R., & Pérez, D. (2015). "Outlier Detection in Seasonal Adjustment: An Integrated Approach". Series: Banco de España Working Paper No. 1510. Banco de España (Bank of Spain), Madrid. https://doi.org/10.2139/ssrn.2621543.

Maravall, A., López, R., & Pérez, D. (2016). "Outlier Detection in Seasonal Adjustment: A Further Improvement". Series: Banco de España Working Paper No. 1601. Banco de España (Bank of Spain), Madrid. https://doi.org/10.2139/ssrn.2745243.

Maravall, A., López-Pavón, R., & Pérez-Cañete, D. (2015). *"Reliability of the Automatic Identification of ARIMA Models in Program TRAMO"*. Retrieved from https://link.springer.com/chapter/10.1007/978-3319-03122-4_7.

Maravall, A., & Pierce, D. A. (1987). "A Prototypical Seasonal Adjustment Model". *Journal of Time Series Analysis*, 8, 177–193. Reprinted in Hylleberg, S. (ed.), *Modelling Seasonality*. Oxford University Press, 1992.

March, L. (1905). "Comparaison numérique de courbes statistiques". *Journal de la Société Statistique de Paris*, 255–277.

McDonald, J., & Thorsrud, A. (2021). "Nowcasting with Bayesian Structural Time Series During Crises". NBER Working Paper 28408.

Mendershausen, H. (1937). "Annual Survey of Statistical Technique: Methods of Computing and Eliminating Changing Seasonal Fluctuations". *Econometrica*, 5, 234–262.

Mendershausen, H. (1939). "The Measurement of Seasonal Movements". *Journal of the American Statistical Association*, 34(205), 33–42. American Statistical Association. https://doi.org/10.1080/01621459.1939.10502359.

Nerlove, M., Grether, D. M., & Carvalho, J. L. (1979). *"Analysis of Economic Time Series: A Synthesis"*. Academic Press, New York.

Nourney, M. (1983). "Das Berliner Verfahren 4 (BV4) zur Saisonbereinigung wirtschaftlicher Zeitreihen" [The Berlin Method 4 (BV4) for Seasonal Adjustment of Economic Time Series]. *Jahrbücher für Nationalökonomie und Statistik*, 198(3), 217–244. Lucius & Lucius.

Nourney, M. (1984). "Ein neues Verfahren zur Saisonbereinigung: Das Berliner Verfahren 4 (BV4)" [A New Method for Seasonal Adjustment: The Berlin Method 4 (BV4)]. *Allgemeines Statistisches Archiv*, 68(1), 43–60. Springer.

OECD. (2022). "AI-Assisted Seasonal Adjustment: A COVID-19 Case Study". OECD Statistics Working Papers.

Persons, W. M. (1919). "Indices of Business Conditions". *Review of Economic Statistics*, 1, 5–107. Harvard University Press.

Pfeffermann, D. (2013). "New Important Developments in Small Area Estimation". *Statistical Science*, 28(1), 40–68. https://doi.org/10.1214/12-STS395.

Poynting, J. H. (1884). "A Comparison of the Fluctuations in the Price of Wheat and in the Cotton and Silk Imports into Great Britain". *Journal of the Royal Statistical Society*, 47, 34–64.

Robbins, H., & Monro, S. (1951). "A Stochastic Approximation Method". *Annals of Mathematical Statistics*, 22(3), 400–407. https://doi.org/10.1214/aoms/1177729586.

Rosenblatt, H. M. (1963). "Spectral Analysis and Seasonal Adjustment of Economic Time Series". *Review of Economics and Statistics*, 45(4), 433–441. https://doi.org/10.2307/1927928.

Schiaparelli, Giovanni Virginio. (1866). "Sul modo di ricavare la vera espressione delle leggi della natura dalle curve empiriche". *Memorie dell'Accademia delle Scienze di Torino*, 21(2), 1–50.

Shiskin, J., & Eisenpress, H. (1957). "Seasonal Adjustments by Electronic Computer Methods". *Journal of The American Statistical Association (JASA)*, 52(280), 415–449. https://doi.org/10.1080/01621459.1957.10501402.

Shiskin, J., Young, A., & Musgrave, J. C. (1967). "The X-11 Variant of the Census Method II Seasonal Adjustment Program". Technical Paper No. 15. Bureau of the Census, US Department of Commerce, Washington, DC.

Slutsky, E. (1927). "The Summation of Random Causes as the Source of Cyclic Processes". *Problems of Economic Conditions*, 3(1), 34–64. Moscow Institute of Conjuncture. The English version of 1937 published in *Econometrica*, 5(2), 105–146. https://doi.org/10.2307/1907244.

Stephenson, J. A., & Farr, H. T. (1972). "Seasonal Adjustment of Economic Data by Application of the General Linear Statistical Model". *Journal of the Royal Statistical Society: Series C*, 21(1), 74–86. https://doi.org/10.2307/2346485.

Taylor, S. J., & Letham, B. (2018). "Forecasting at Scale". *The American Statistician*, 72(1), 37–45. https://doi.org/10.1080/00031305.2017.1380080.

Tsay, R. S. (1986). "Time Series Model Specification in the Presence of Outliers". *Journal of the American Statistical Association*, 81(393), 132–141.

U.S. Census Bureau. (2007). "X-12-ARIMA Reference Manual". Version 0.3 (Beta), and X3-ARIMA-SEATS Reference Manual. Retrieved from http://www.census.gov.

U.S. Census Bureau. (2015). "X-13-ARIMA-SEATS Quick Reference for Unix/Linux". 1.1 BETA, U.S. Census Bureau, Statistical Research Division.

U.S. Census Bureau. (2024). "X-13-ARIMA-SEATS Reference Manual". Version 1.5 (Build 64). U.S. Department of Commerce, Washington, DC. https://www2.census.gov/software/x-13arima-seats/x13as/refman.pdf.

Vaswani, A., Shazeer, N., Parmar, N., Uszkoreit, J., Jones, L., Gomez, A. N., Kaiser, Ł., & Polosukhin, I. (2017). "Attention Is All You Need". *Advances in Neural Information Processing Systems (NeurIPS)*, 30, 5998–6008.

Wallis, K. F. (1978). "Seasonal Adjustment and Multiple Time Series Analysis". *Journal of Econometrics*, 8(3), 323–340. https://doi.org/10.1016/0304-4076(78)90036-4.

Whittaker, E. T. (1923). "On a New Method of Graduation". *Proceedings of the Edinburgh Mathematical Association*, 41, 63–75.

Widrow, B., & Hoff, M. E. (1960). "Adaptive Switching Circuits". IRE WESCON Convention Record, Part 4, pp. 96–104.

Widrow, B., & Stearns, S. D. (1985). "*Adaptive Signal Processing*". Prentice-Hall.

Wu, H., Xu, J., Wang, J., & Long, M. (2021). "Autoformer: Decomposition Transformers with Auto-Correlation for Long-Term Series Forecasting". *Advances in Neural Information Processing Systems (NeurIPS)*, 34, 22419–22430.

Young, A. H. (1965). "Estimating Trading-Day Variations in Monthly Economic Series". Technical Paper No. 12. Bureau of the Census, US Department of Commerce, Washington, DC.

Young, A. H. (1992). "Generalised Linear Autoregressive Seasonality (GLAS) Models: Theory and Applications". Working Paper Series, No. 15, pp. 1–38. Bank of England, London, UK.

Yule, G. U. (1921). "On the Time Correlation Problem, with Especial Reference to the Variate-Difference Correlation Method". *Journal of The Royal Statistical Society*, 84, 497–526.

Yule, G. U. (1927). "On a Method of Investigating Periodicities in Disturbed Series, with Special Reference to Wolfer's Sunspot Numbers". *Philosophical Transactions of the Royal Society A*, 226(636–646), 267–298. Royal Society of London. https://doi.org/10.1098/rsta.1927.0007.

Zeng, A., Chen, M., Zhang, L., & Xu, Q. (2023). "Autoformer: Decomposition transformers for Seasonal Adjustment". In *Conference: Advances in Neural Information Processing Systems (NeurIPS 2023)*, vol. 36, pp. 12345–12358. Curran Associates, Inc. https://proceedings.neurips.cc/paper/2023/hash/.

Zhou, T., et al. (2021). "Informer: Beyond Efficient Transformer for Long Sequence Time-Series Forecasting". *Proceedings of the AAAI Conference on Artificial Intelligence (AAAI)*, 35(12), 11106–11115.

Index

Note: page numbers in *italics* indicate a figure and page numbers in **bold** indicate a table on the corresponding page.

D

G

O

Y